MATHEMATICAL SURVEYS

Number II

THE THEORY OF RINGS

BY

NATHAN JACOBSON

AMERICAN MATHEMATICAL SOCIETY

Providence, Rhode Island

1943

3rd printing, 1966

Printed in the United States of America

TABLE OF CONTENTS

PREFACE

The theory that forms the subject of this book had its beginning with Artin's extension in 1927 of Wedderburn's structure theory of algebras to rings satisfying the chain conditions. Since then the theory has been considerably extended and simplified. The only exposition of the subject in book form that has appeared to date is Deuring's *Algebren* published in the Ergebnisse series in 1935. Much progress has been made since then and this perhaps justifies a new exposition of the subject.

The present account is almost completely self-contained. That this has been possible in a book dealing with results of the significance of Wedderburn's theorems, the Albert-Brauer-Noether theory of simple algebras and the arithmetic ideal theory is another demonstration of one of the most remarkable characteristics of modern algebra, namely, the simplicity of its logical structure.

Roughly speaking our subject falls into three parts: structure theory, representation theory and arithmetic ideal theory. The first of these is an outgrowth of the structure theory of algebras. It was motivated originally by the desire to discover and to classify "hypercomplex" extensions of the field of real numbers. The most important names connected with this phase of the development of the theory are those of Molien, Dedekind, Frobenius and Cartan. The structure theory for algebras over a general field dates from the publication of Wedderburn's thesis in 1907; the extension to rings, from Artin's paper in 1927. The theory of representations was originally concerned with the problem of representing a group by matrices. This was extended to rings and was formulated as a theory of modules by Emmy Noether. The study of modules also forms an important part of the arithmetic ideal theory. This part of the theory of rings had its origin in Dedekind's ideal theory of algebraic number fields and more immediately in Emmy Noether's axiomatic foundation of this theory.

Throughout this book we have placed particular emphasis on the study of rings of endomorphisms. By using the regular representations the theory of abstract rings is obtained as a special case of the more concrete theory of endomorphisms. Moreover, the theory of modules, and hence representation theory, may be regarded as the study of a set of rings of endomorphisms all of which are homomorphic images of a fixed ring $\mathfrak{o}$. Chapter 1 lays the foundations of the theory of endomorphisms of a group. The concepts and results developed here are fundamental in all the subsequent work. Chapter 2 deals with vector spaces and contains some material that, at any rate in the commutative case, might have been assumed as known. For the sake of completeness this has been included. Chapter 3 is concerned with the arithmetic of non-commutative principal ideal domains. Much of this chapter can be regarded as a special case of the general arithmetic ideal theory developed in Chapter 6. The methods of Chapter 3 are, however, of a much more elementary character and this fact may be of interest to the student of geometry, since the results of this chapter have many applications in that field. A reader who is primarily interested in structure theory or in representation theory may omit Chapter 3 with the exception of **3**. Chapter 4 is devoted to the development of these theories and to some applications to the problem of representation of groups by projective transformations and to the Galois theory of division rings. In Chapter 5 we take up the study of algebras. In the first part of this chapter we consider the theory of simple algebras over a general field. The second part is concerned with the theory of characteristic and minimum polynomials of an algebra and the trace criterion for separability of an algebra.

In recent years there has been a considerable interest in the study of rings that do not satisfy the chain conditions but instead are restricted by topological or metric conditions. We mention von Neumann and Murray's investigation of rings of transformations in Hilbert space, von Neumann's theory of regular rings and Gelfand's theory of normed rings. There are many important applications of these theories to analysis. Because of the conditions that we have imposed on the rings considered in this work, our discussion is not directly applicable to these problems in topological algebra. It may be hoped, however, that the methods and results of the purely algebraic theory will point the way for further development of the topological algebraic theory.

This book was begun during the academic year 1940–1941 when I was a visiting lecturer at Johns Hopkins University. It served as a basis of a course given there and it gained materially from the careful reading and criticism of Dr. Irvin Cohen who at that time was one of the auditors of my lectures. My thanks are due to him and also to Professors Albert, Schilling and Hurewicz for their encouragement and for many helpful suggestions.

N. Jacobson.

Chapel Hill, N. C.,
March 7, 1943.

CHAPTER 1

GROUPS AND ENDOMORPHISMS

1. Rings of endomorphisms. With any commutative group $\mathfrak{M}$ we may associate a ring $\mathfrak{E}(\mathfrak{M})$, the ring of endomorphisms (homomorphisms of $\mathfrak{M}$ into itself) of $\mathfrak{M}$. On the other hand, as we shall see, any ring with an identity may be obtained as a subring of the ring of endomorphisms of its additive group. Because of this fact, we may use the theory of rings of endomorphisms to obtain the theory of abstract rings. This method of studying rings is one of the most important ones that we shall use in this book. It will therefore be well to begin our discussion with a brief survey of that part of the theory of groups and endomorphisms that will be required later.

Our primary concern in the sequel is with commutative groups. However, since most of the results of this chapter are valid for an arbitrary group $\mathfrak{M}$, we shall not assume at the outset that $\mathfrak{M}$ is commutative. Nevertheless, we shall find it convenient to use the additive notation in $\mathfrak{M}$: The group operation will be denoted as $+$, the identity element as 0, the inverse of a as $-a$, etc.

Consider the collection $\mathfrak{T}(\mathfrak{M})$ of single-valued transformations of $\mathfrak{M}$ into itself, i.e. onto a subset of $\mathfrak{M}$. As always for transformations, we regard $A = B$ if the images xA and xB are identical for all x in $\mathfrak{M}$. Now we shall turn $\mathfrak{T}$ into an algebraic system by introducing into it two fundamental operations. First, if A and B are in $\mathfrak{T}$, the *sum* $A + B$ is defined as the transformation whose effect on any x in $\mathfrak{M}$ is obtained by adding the images xA and xB. In other terms

$$x(A + B) = xA + xB.$$

The *product* AB is the resultant of A and B:

$$x(AB) = (xA)B.[1]$$

The following facts concerning the algebraic system $\mathfrak{T}$ are readily verified:

1) $\mathfrak{T}$ is a group relative to $+$. The identity element of this group is the transformation 0 that is defined by the equation $x0 = 0$. The negative of A, $-A$, is given by the defining equation $x(-A) = -xA$.

2) $\mathfrak{T}$ is a semi-group with an identity relative to multiplication, i.e. $(AB)C = A(BC)$ and the identity element of $\mathfrak{T}$ is the identity transformation 1 ($x1 = x$).

3) The distributive law

$$A(B + C) = AB + AC$$

holds.

The system $\mathfrak{T}$ is therefore very nearly a ring. It fails to be one since the

[1] This equation justifies our notation xA. For by using it, the order of writing corresponds to the order of performance of the transformations.

relations $A + B = B + A$ and $(B + C)A = BA + CA$ are not universally valid. We may satisfy the first of these conditions if we suppose that $\mathfrak{M}$ is commutative, but even in this case, the second condition fails.

Example. Let $\mathfrak{M}$ be the cyclic group of order 2 with elements 0, 1 where $1 + 1 = 0$. $\mathfrak{T}$ contains four elements

$$0 = \begin{pmatrix} 0 & 1 \\ 0 & 0 \end{pmatrix}, \quad 1 = \begin{pmatrix} 0 & 1 \\ 0 & 1 \end{pmatrix}, \quad A = \begin{pmatrix} 0 & 1 \\ 1 & 0 \end{pmatrix}, \quad B = \begin{pmatrix} 0 & 1 \\ 1 & 1 \end{pmatrix}$$

where, in general, $\begin{pmatrix} 0 & 1 \\ a & b \end{pmatrix}$ denotes the transformation $0 \to a$, $1 \to b$. The addition and multiplication tables in $\mathfrak{T}$ are, respectively,

	0	1	A	B
0	0	1	A	B
1	1	0	B	A
A	A	B	0	1
B	B	A	1	0

	0	1	A	B
0	0	0	B	B
1	0	1	A	B
A	0	A	1	B
B	0	B	0	B

Since $0A \neq 0$, it is clear that the second distributive law does not hold.

We consider next the subset $\mathfrak{E}(\mathfrak{M})$ of $\mathfrak{T}$ consisting of the endomorphisms of $\mathfrak{M}$ (an arbitrary group). We recall the definition: A transformation A of a group is an *endomorphism* if it is a homomorphism of the group into itself, that is,

$$(x + y)A = xA + yA.$$

It is clear that $\mathfrak{E}$ is closed relative to the multiplication defined in $\mathfrak{T}$. Moreover, if B and C are arbitrary elements of $\mathfrak{T}$ and A is in $\mathfrak{E}$, then

$$(B + C)A = BA + CA.$$

From our point of view the system $\mathfrak{E}$ is not particularly interesting when $\mathfrak{M}$ is an arbitrary group, for then $\mathfrak{E}$ need not be closed relative to the addition that we defined in $\mathfrak{T}$, However, the situation is quite different when $\mathfrak{M}$ is commutative. In this case it is readily seen that if A and B are in $\mathfrak{E}$, then $A + B = B + A$, 0 and $-A$ all belong to $\mathfrak{E}$. Since the associative and distributive laws for multiplication hold, $\mathfrak{E}$ is a ring. This is the fundamental

THEOREM 1. *If $\mathfrak{M}$ is a commutative group, then the set $\mathfrak{E}(\mathfrak{M})$ of endomorphisms of $\mathfrak{M}$ is a ring relative to the operations $A + B$ and AB that are defined by the equations $x(A + B) = xA + xB$, $x(AB) = (xA)B$.*

Examples. 1) Let $\mathfrak{M}$ be the group of rational integers under ordinary addition. Since $\mathfrak{M}$ is a cyclic group with 1 as a generator, any endomorphism A is determined by its effect on 1. For if $1A = a$ and $x = \overbrace{1 + \cdots + 1}^{x}$, then $xA = xa$ the ordinary product of the integers x and a. Since $(-x)A = -xA$, this equation holds also for negative x's and since $0A = 0 = 0a$, it holds for 0. Thus any endomorphism A of $\mathfrak{M}$ is a transformation that multiplies the element

x of $\mathfrak{M}$ by a fixed element a. The element a is uniquely determined by A, and it is clear that every integer a arises from some endomorphism in this way. Hence $\mathfrak{E}$ is in $(1 - 1)$ correspondence with $\mathfrak{M}$. If $A \rightarrow a$ and $B \rightarrow b$ in our correspondence, then $x(A + B) = xA + xB = xa + xb = x(a + b)$ and similarly $x(AB) = x(ab)$. Hence $A + B \rightarrow a + b$ and $AB \rightarrow ab$, i.e. $\mathfrak{E}$ is isomorphic to the *ring* of rational integers $\mathfrak{M}$.

2) As a generalization of 1) we let $\mathfrak{M}$ be a direct sum of n infinite cyclic groups. If $e_1, \cdots, e_n$ are generators of $\mathfrak{M}$, any endomorphism A is completely determined by the images $e_iA = f_i$. On the other hand, we may choose elements f_i arbitrarily in $\mathfrak{M}$ and define $(\Sigma e_i x_i)A = \Sigma f_i x_i$, x_i integers. Then A is an endomorphism. If

$$e_iA = e_1a_{1i} + \cdots + e_na_{ni}, \qquad (i = 1, \cdots, n),$$

a_{ij} rational integers, then the correspondence $A \rightarrow (a_{ij})$ is $(1 - 1)$ between $\mathfrak{E}$ and the ring of $n \times n$ matrices with rational integral elements. If $B \rightarrow (b_{ij})$, we may verify that $A + B \rightarrow (a_{ij}) + (b_{ij})$ and $AB \rightarrow (b_{ij})(a_{ij})$. Hence the correspondence is an anti-isomorphism between $\mathfrak{E}$ and the ring of rational integral matrices.[2] It may be remarked that the associative and distributive laws for these matrices may be deduced by means of our correspondence from the associative and distributive laws for endomorphisms.

3) If $\mathfrak{M}$ is a direct sum of cyclic groups of order m, a similar discussion shows that the ring of endomorphisms of $\mathfrak{M}$ is anti-isomorphic to the ring of matrices with elements in the ring of rational integers reduced modulo m.

We return to the consideration of an arbitrary group $\mathfrak{M}$. Let $\mathfrak{G}(\mathfrak{M})$ be the set of $(1 - 1)$ transformations of $\mathfrak{M}$ onto itself. It is clear that if A is in $\mathfrak{G}(\mathfrak{M})$, then the inverse transformation A^{-1} is defined. It follows that $\mathfrak{G}(\mathfrak{M})$ is a group under multiplication.

Now if A is an endomorphism, A^{-1} is also an endomorphism. Hence the intersection $\mathfrak{A}(\mathfrak{M}) = \mathfrak{E}(\mathfrak{M}) \wedge \mathfrak{G}(\mathfrak{M})$ is also a group under multiplication. The elements of this group, the $(1 - 1)$ endomorphisms of $\mathfrak{M}$ onto itself, are the *automorphisms* of $\mathfrak{M}$. Of particular interest among these transformations are the *inner automorphisms.* If $s \epsilon \mathfrak{M}$, then the inner automorphism corresponding to s is the transformation S defined by the equation $xS = -s + x + s$. If A is an arbitrary automorphism, then $x(A^{-1}SA) = -sA + x + sA$, i.e. $A^{-1}SA$ is the inner automorphism associated with the element sA. This shows that the totality of inner automorphisms constitutes an invariant subgroup of the complete group of automorphisms.

We recall that in any ring with an identity, an element u is a *unit* if it has both a left and a right inverse relative to the identity. It follows immediately that these two inverses are equal and that no other element in the ring can satisfy either of the equations $ux = 1$ or $xu = 1$. As usual we denote the inverse of

[2] If we use the correspondence $A \rightarrow (a_{ij})^*$, the transposed matrix of (a_{ij}), we obtain an isomorphism. However, in a similar situation that will be encountered later, it is impossible to effect this passage from an anti-isomorphism to an isomorphism. For this reason we prefer to emphasize the correspondence $A \rightarrow (a_{ij})$.

u by u^{-1}. It may be proved directly that the set of units of any ring is a group relative to the multiplication defined in the ring. Now consider any commutative group $\mathfrak{M}$, its ring of endomorphisms $\mathfrak{E}$ and its group of automorphisms $\mathfrak{A}$. Since the $(1 - 1)$ transformations of a set are the only ones that possess two-sided inverses, it is evident that $\mathfrak{A}$ is the group of units of $\mathfrak{E}$. As an application of this fact, we see that the group of automorphisms of the direct sum $\mathfrak{M}$ of n infinite cyclic groups is isomorphic to the extended unimodular group of $n \times n$ rational integral matrices having determinants $+1$ or -1. For we have seen that the ring of endomorphisms of $\mathfrak{M}$ is isomorphic to the ring of $n \times n$ rational integral matrices, and by using the multiplicative property of determinants, we see that the units of the latter ring are the matrices of determinants ± 1.

2. Groups relative to a set of endomorphisms. In many algebraic problems we are interested in studying a group $\mathfrak{M}$ relative to a fixed set of endomorphisms Ω acting in $\mathfrak{M}$. We fix our attention on the subgroups, called *Ω-subgroups* (allowable), which are transformed into themselves by every endomorphism belonging to Ω. Although, in our applications, $\mathfrak{M}$ will usually be an infinite group, the following examples indicate that this point of view is fruitful even in the study of finite groups.

Examples. 1) Ω is vacuous. All subgroups are allowable. 2) Ω consists of the inner automorphisms. Here the Ω-subgroups are the invariant subgroups. 3) Ω is the complete set of automorphisms. The Ω-subgroups are the characteristic subgroups of $\mathfrak{M}$.

We suppose now that $\mathfrak{M}$ and Ω are fixed. If $\mathfrak{N}_1$ and $\mathfrak{N}_2$ are Ω-subgroups, evidently the intersection $\mathfrak{N}_1 \wedge \mathfrak{N}_2$ is also an Ω-subgroup. The join $(\mathfrak{N}_1, \mathfrak{N}_2)$, defined as the smallest subgroup containing $\mathfrak{N}_1$ and $\mathfrak{N}_2$, may be characterized as the set of finite sums of elements in $\mathfrak{N}_1$ and $\mathfrak{N}_2$. It follows that $(\mathfrak{N}_1, \mathfrak{N}_2)$ is an Ω-subgroup. If $\mathfrak{N}_1$ is invariant, $(\mathfrak{N}_1, \mathfrak{N}_2) = \mathfrak{N}_1 + \mathfrak{N}_2 = \mathfrak{N}_2 + \mathfrak{N}_1$ where $\mathfrak{N}_1 + \mathfrak{N}_2$ denotes the set of elements $x_1 + x_2$, x_i in $\mathfrak{N}_i$.

If $\mathfrak{N}$ is an Ω-subgroup, the endomorphism α of Ω induces in $\mathfrak{N}$ an endomorphism which we shall also denote as α. Of course, distinct mappings α and β in $\mathfrak{M}$ may coincide when regarded as mappings in $\mathfrak{N}$. We note that if $\alpha\beta = \gamma \in \Omega$ or $\alpha + \beta = \delta \in \Omega$, then these relations hold also for the induced transformations in $\mathfrak{N}$.

Now suppose that $\mathfrak{N}$ and $\mathfrak{P}$ are Ω-subgroups and that $\mathfrak{P}$ is invariant in $\mathfrak{N}$. We consider the difference group consisting of the cosets $\mathfrak{P} + y$, y in $\mathfrak{N}$. If $\alpha \in \Omega$, α determines a transformation in $\mathfrak{N} - \mathfrak{P}$ in the following way. If $\mathfrak{P} + y$ is an arbitrary coset, then the coset $\mathfrak{P} + y\alpha$ does not depend on the choice of the representative y and so it is uniquely determined by the coset $\mathfrak{P} + y$ and by the endomorphism α. Hence the correspondence $\mathfrak{P} + y \rightarrow \mathfrak{P} + y\alpha$ is a single-valued transformation. Again, we denote this transformation in $\mathfrak{N} - \mathfrak{P}$ by α, i.e. $(\mathfrak{P} + y)\alpha = \mathfrak{P} + y\alpha$. It is clear that α is an endomorphism in $\mathfrak{N} - \mathfrak{P}$. As in the case of subgroups, $\alpha\beta = \gamma$ or $\alpha + \beta = \delta$ in $\mathfrak{N}$ implies the same relation for the induced transformations in $\mathfrak{N} - \mathfrak{P}$. We may repeat these processes, forming difference groups of difference groups, subgroups of difference groups, etc. In this way a whole hierarchy $\mathfrak{K}$ of groups is generated in which the original endomorphisms α induce uniquely defined endomorphisms. We shall call the members of $\mathfrak{K}$, *Ω-groups.*

Let $\mathfrak{N}$ and $\bar{\mathfrak{N}}$ be any two Ω-groups. A mapping A of $\mathfrak{N}$ into the whole of $\bar{\mathfrak{N}}$ is an *Ω-homomorphism* if it is an ordinary homomorphism and $\alpha A = A\alpha$ for all α in Ω. Then $\mathfrak{N}$ and $\bar{\mathfrak{N}}$ are *Ω-homomorphic.*[3] If A is $(1 - 1)$, it is an *Ω-isomorphism* and then $\mathfrak{N}$ and $\bar{\mathfrak{N}}$ are *Ω-isomorphic.* If $\bar{\mathfrak{N}} \leqq \mathfrak{N}$, we use the term *$\Omega$-endomorphism* for Ω-homomorphism and if $\bar{\mathfrak{N}} = \mathfrak{N}$, we use the term *$\Omega$-automorphism* for Ω-isomorphism.

3. The isomorphism theorems. Let $\mathfrak{N}$ and $\mathfrak{P}$ be Ω-groups, $\mathfrak{P}$ invariant in $\mathfrak{N}$. It is well known that the correspondence $x \rightarrow \mathfrak{P} + x$ is a homomorphism A between $\mathfrak{N}$ and $\mathfrak{N} - \mathfrak{P}$. Since $(\mathfrak{P} + x)\alpha = \mathfrak{P} + x\alpha$, $A\alpha = \alpha A$ and A is an Ω-homomorphism. Now suppose that $\mathfrak{N}$ and $\bar{\mathfrak{N}}$ are two Ω-groups and that $x \rightarrow \bar{x} = xA$ is an Ω-homomorphism between them. If $\mathfrak{P}$ is the set of elements of $\mathfrak{N}$ sent into 0, we know that $\mathfrak{P}$ is an invariant subgroup of $\mathfrak{N}$ and that the correspondence $\mathfrak{P} + x \rightarrow \bar{x} = xA$ is an isomorphism between $(\mathfrak{N} - \mathfrak{P})$ and $\bar{\mathfrak{N}}$. Since $(y\alpha)A = (yA)\alpha = 0\alpha = 0$ if $y \in \mathfrak{P}$, $\mathfrak{P}$ is an Ω-subgroup and since $(\mathfrak{P} + x)\alpha = (\mathfrak{P} + x\alpha) \rightarrow (x\alpha)A = (xA)\alpha$, the isomorphism is an Ω-isomorphism between $\mathfrak{N} - \mathfrak{P}$ and $\bar{\mathfrak{N}}$. This proves the fundamental theorem on Ω-homomorphisms:

THEOREM 2. *If $\mathfrak{N}$ and $\mathfrak{P}$ are Ω-groups and $\mathfrak{P}$ is invariant in $\mathfrak{N}$, then $\mathfrak{N}$ and $\mathfrak{N} - \mathfrak{P}$ are Ω-homomorphic. Conversely if $\mathfrak{N}$ is Ω-homomorphic to an Ω-group $\bar{\mathfrak{N}}$ and $\mathfrak{P}$ is the set of elements mapped into* 0 *by the homomorphism, $\mathfrak{P}$ is an invariant Ω-subgroup of $\mathfrak{N}$ and $\mathfrak{N} - \mathfrak{P}$ and $\bar{\mathfrak{N}}$ are Ω-isomorphic.*

If A is an Ω-homomorphism between $\mathfrak{N}$ and $\bar{\mathfrak{N}}$ and $\mathfrak{R}$ is an Ω-subgroup of $\mathfrak{N}$, then its image $\mathfrak{R}A$ is an Ω-subgroup of $\bar{\mathfrak{N}}$. If $\mathfrak{R}$ is invariant in $\mathfrak{N}$, $\mathfrak{R}A$ is invariant in $\mathfrak{N}A = \bar{\mathfrak{N}}$. On the other hand, if $\bar{\mathfrak{R}}$ is an Ω-subgroup of $\bar{\mathfrak{N}}$ and $\mathfrak{R}$ is the set of elements y of $\mathfrak{N}$ such that $yA \in \bar{\mathfrak{R}}$, then $\mathfrak{R}$ is an Ω-subgroup of $\mathfrak{N}$ containing $\mathfrak{P}$, the set of elements mapped into 0 by the homomorphism. Again, the invariance of $\bar{\mathfrak{R}}$ implies that of $\mathfrak{R}$. If $\mathfrak{R}$ is an Ω-subgroup containing $\mathfrak{P}$, any element of $\mathfrak{N}$ mapped into an element of $\mathfrak{R}A$ is in $\mathfrak{R}$. For if $xA = yA$ for x in $\mathfrak{N}$ and y in $\mathfrak{R}$, $(x - y)A = 0$ and $x - y \in \mathfrak{P}$. Hence $x = (x - y) + y \in \mathfrak{R}$. These results may be stated as follows:

THEOREM 3. *Let $\mathfrak{N}$ be Ω-homomorphic to $\bar{\mathfrak{N}}$ under the Ω-homomorphism A and let $\mathfrak{P}$ be the set of elements mapped into* 0 *by A. Then the correspondence $\mathfrak{R} \rightarrow \mathfrak{R}A = \bar{\mathfrak{R}}$ is* $(1 - 1)$ *between the Ω-subgroups $\mathfrak{R}$ containing $\mathfrak{P}$ and the Ω-subgroups of $\bar{\mathfrak{N}}$. The group $\mathfrak{R}$ is invariant in $\mathfrak{N}$ if and only if $\bar{\mathfrak{R}}$ is invariant in $\bar{\mathfrak{N}}$.*

[3] If $\mathfrak{M}_i$ $(i = 1, 2)$ is a group and Ω_i a fixed set of endomorphisms, we may define $\mathfrak{M}_1$ and $\mathfrak{M}_2$ to be (Ω_1, Ω_2)-homomorphic if there is a single-valued mapping $x_1 \rightarrow x_2$ of $\mathfrak{M}_1$ into the whole of $\mathfrak{M}_2$ and a single-valued mapping $\alpha_1 \rightarrow \alpha_2$ of Ω_1 into the whole of Ω_2 such that $x_1 + y_1 \rightarrow x_2 + y_2$, $x_1\alpha_1 \rightarrow x_2\alpha_2$ if $x_1 \rightarrow x_2$, $y_1 \rightarrow y_2$ and $\alpha_1 \rightarrow \alpha_2$. This differs from the definition of Ω-homomorphism, for in the latter the mapping between the transformations is completely determined by the original group $\mathfrak{M}$. The concept of Ω-homomorphism is the important one for our purposes.

Now let $\bar{\mathfrak{K}}$ be an invariant Ω-subgroup of $\bar{\mathfrak{N}}$. If we apply the Ω-homomorphism between $\bar{\mathfrak{N}}$ and $\bar{\mathfrak{N}} - \bar{\mathfrak{K}}$ after that between $\mathfrak{N}$ and $\bar{\mathfrak{N}}$, we obtain an Ω-homomorphism between $\mathfrak{N}$ and $\bar{\mathfrak{N}} - \bar{\mathfrak{K}}$. The elements mapped into 0 of $\bar{\mathfrak{N}} - \bar{\mathfrak{K}}$ are those in $\mathfrak{K}$. Hence we have the

FIRST ISOMORPHISM THEOREM. *Suppose that $\mathfrak{N}$ is Ω-homomorphic to $\bar{\mathfrak{N}}$, and* let *$\bar{\mathfrak{K}}$ be an invariant Ω-subgroup of $\bar{\mathfrak{N}}$ and $\mathfrak{K}$ the totality of elements mapped into $\bar{\mathfrak{N}}$.* Then *$\mathfrak{N} - \mathfrak{K}$ and $\bar{\mathfrak{N}} - \bar{\mathfrak{K}}$ are Ω-isomorphic.*

Evidently this implies the

COROLLARY. *If $\mathfrak{K}$ is an Ω-subgroup of $\mathfrak{N}$ containing the invariant Ω-subgroup $\mathfrak{P}$ of $\mathfrak{N}$ and $(\mathfrak{K} - \mathfrak{P})$ is invariant in $(\mathfrak{N} - \mathfrak{P})$, then $\mathfrak{K}$ is invariant in $\mathfrak{N}$ and $\mathfrak{N} - \mathfrak{K}$ is Ω-isomorphic to $(\mathfrak{N} - \mathfrak{P}) - (\mathfrak{K} - \mathfrak{P})$.*

Suppose that $\mathfrak{N}_1$, $\mathfrak{N}_2$, $\mathfrak{M}_1$ are Ω-groups; $\mathfrak{N}_i \leqq \mathfrak{M}_1$ and $\mathfrak{N}_2$ invariant in $\mathfrak{M}_1$. Then the smallest subgroup containing $\mathfrak{N}_1$ and $\mathfrak{N}_2$ is $\mathfrak{N} \equiv \mathfrak{N}_1 + \mathfrak{N}_2 = \mathfrak{N}_2 + \mathfrak{N}_1$. The group $\mathfrak{N}_2$ is invariant in $\mathfrak{N}$ and the cosets in the difference group $\mathfrak{N} - \mathfrak{N}_2$ have the form $\mathfrak{N}_2 + x_1$, x_1 in $\mathfrak{N}_1$. It follows that the correspondence $x_1 \to \mathfrak{N}_2 + x_1$ is an Ω-homomorphism between $\mathfrak{N}_1$ and $\mathfrak{N} - \mathfrak{N}_2$. Since the elements mapped into 0 are those of $\mathfrak{N}_1 \wedge \mathfrak{N}_2$, we have the

SECOND ISOMORPHISM THEOREM. *If $\mathfrak{N}_1$, $\mathfrak{N}_2$, $\mathfrak{M}_1$ are Ω-groups, $\mathfrak{N}_i \leqq \mathfrak{M}_1$ and $\mathfrak{N}_2$ is invariant in $\mathfrak{M}_1$, then* 1) $\mathfrak{N}_1 + \mathfrak{N}_2 = \mathfrak{N}_2 + \mathfrak{N}_1$, 2) $\mathfrak{N}_1 \wedge \mathfrak{N}_2$ *is invariant in $\mathfrak{N}_1$ and* 3) $(\mathfrak{N}_1 + \mathfrak{N}_2) - \mathfrak{N}_2$ *is Ω-isomorphic to* $\mathfrak{N}_1 - (\mathfrak{N}_1 \wedge \mathfrak{N}_2)$.

4. The Jordan-Hölder-Schreier theorem. A chain of Ω-groups $\mathfrak{M}_1 \geqq \mathfrak{M}_2 \geqq \cdots \geqq \mathfrak{M}_{s+1} = 0$ is a *normal series* for $\mathfrak{M}_1$ if each $\mathfrak{M}_i$ is invariant in $\mathfrak{M}_{i-1}$. The difference groups $\mathfrak{M}_{i-1} - \mathfrak{M}_i$ are called the *factors* of the series while a second chain is a *refinement* of the first if it contains all of the $\mathfrak{M}_i$. We shall call two normal series *equivalent* if there is a $(1 - 1)$ correspondence between their factors such that the paired factors are Ω-isomorphic.

THEOREM 4 (Schreier). *Any two normal series for $\mathfrak{M}_1$ have equivalent refinements.*

Let $\mathfrak{M}_1 \geqq \cdots \geqq \mathfrak{M}_{s+1} = 0$ and $\mathfrak{M}_1 = \mathfrak{N}_1 \geqq \cdots \geqq \mathfrak{N}_{t+1} = 0$ be the two normal series. Define $\mathfrak{M}_{ij} = \mathfrak{M}_{i+1} + (\mathfrak{M}_i \wedge \mathfrak{N}_j)$ for $j = 1, \cdots, t+1$ and $i = 1, \cdots, s$, $\mathfrak{M}_{s+1,1} = 0$. Then $\mathfrak{M}_{i,t+1} = \mathfrak{M}_{i+1,1}$ and $(\mathfrak{M}_1 =)\ \mathfrak{M}_{11} \geqq \cdots \geqq \mathfrak{M}_{1t} \geqq \mathfrak{M}_{21} \geqq \cdots \geqq \mathfrak{M}_{2t} \geqq \cdots \geqq \mathfrak{M}_{s,t} \geqq 0$. Similarly, set $\mathfrak{N}_{ji} = \mathfrak{N}_{j+1} + (\mathfrak{N}_j \wedge \mathfrak{M}_i)$ for $i = 1, \cdots, s+1$ and $j = 1, \cdots, t$, $\mathfrak{N}_{t+1,1} = 0$ and obtain $\mathfrak{N}_{j,s+1} = \mathfrak{N}_{j+1,1}$ and $(\mathfrak{N}_1 =)\ \mathfrak{N}_{11} \geqq \cdots \geqq \mathfrak{N}_{1s} \geqq \mathfrak{N}_{21} \geqq \cdots \geqq \mathfrak{N}_{2s} \geqq \cdots \geqq \mathfrak{N}_{t,s} \geqq 0$. Thus in each chain we have $st + 1$ terms. We may pair $\mathfrak{M}_{ij} - \mathfrak{M}_{i,j+1}$ with $\mathfrak{N}_{ji} - \mathfrak{N}_{j,i+1}$ to obtain the theorem as a consequence of the following

LEMMA (Zassenhaus). *Let $\mathfrak{N}_1$, $\mathfrak{N}_1'$, $\mathfrak{N}_2$, $\mathfrak{N}_2'$, $\mathfrak{M}_1$ be Ω-groups where $\mathfrak{N}_i \leqq \mathfrak{M}_1$, $\mathfrak{N}_i' \leqq \mathfrak{N}_i$ and $\mathfrak{N}_i'$ is invariant in $\mathfrak{N}_i$. Then $\mathfrak{N}_1' + (\mathfrak{N}_1 \wedge \mathfrak{N}_2')$ is invariant in $\mathfrak{N}_1' + (\mathfrak{N}_1 \wedge \mathfrak{N}_2)$, $\mathfrak{N}_2' + (\mathfrak{N}_2 \wedge \mathfrak{N}_1')$ is invariant in $\mathfrak{N}_2' + (\mathfrak{N}_2 \wedge \mathfrak{N}_1)$ and the corresponding difference groups are Ω-isomorphic.*

By the Second Isomorphism Theorem, $\mathfrak{N}_2 \wedge \mathfrak{N}_1' = (\mathfrak{N}_2 \wedge \mathfrak{N}_1) \wedge \mathfrak{N}_1'$ is invariant in $\mathfrak{N}_1 \wedge \mathfrak{N}_2$ and $(\mathfrak{N}_1 \wedge \mathfrak{N}_2) - (\mathfrak{N}_1' \wedge \mathfrak{N}_2)$ and $(\mathfrak{N}_1' + (\mathfrak{N}_1 \wedge \mathfrak{N}_2)) - \mathfrak{N}_1'$ are Ω-isomorphic. Similarly, $\mathfrak{N}_1 \wedge \mathfrak{N}_2'$ is invariant in $\mathfrak{N}_1 \wedge \mathfrak{N}_2$ and hence $(\mathfrak{N}_1' \wedge \mathfrak{N}_2) + (\mathfrak{N}_1 \wedge \mathfrak{N}_2')$ is invariant. In the homomorphism between $\mathfrak{N}_1 \wedge \mathfrak{N}_2$ and $(\mathfrak{N}_1' + (\mathfrak{N}_1 \wedge \mathfrak{N}_2)) - \mathfrak{N}_1'$, the group $((\mathfrak{N}_1' \wedge \mathfrak{N}_2) + (\mathfrak{N}_1 \wedge \mathfrak{N}_2'))$ is mapped into $((\mathfrak{N}_1' \wedge \mathfrak{N}_2) + (\mathfrak{N}_1 \wedge \mathfrak{N}_2') + \mathfrak{N}_1') - \mathfrak{N}_1' = ((\mathfrak{N}_1 \wedge \mathfrak{N}_2') + \mathfrak{N}_1') - \mathfrak{N}_1'$. Hence by the above corollary $(\mathfrak{N}_1 \wedge \mathfrak{N}_2') + \mathfrak{N}_1'$ is invariant in $(\mathfrak{N}_1 \wedge \mathfrak{N}_2) + \mathfrak{N}_1'$ and $(\mathfrak{N}_1' + (\mathfrak{N}_1 \wedge \mathfrak{N}_2)) - (\mathfrak{N}_1' + (\mathfrak{N}_1 \wedge \mathfrak{N}_2'))$ and $(\mathfrak{N}_1 \wedge \mathfrak{N}_2) - ((\mathfrak{N}_1 \wedge \mathfrak{N}_2') + (\mathfrak{N}_1' \wedge \mathfrak{N}_2))$ are Ω-isomorphic. By symmetry $(\mathfrak{N}_1 \wedge \mathfrak{N}_2) - ((\mathfrak{N}_1 \wedge \mathfrak{N}_2') + (\mathfrak{N}_1' \wedge \mathfrak{N}_2))$ and $(\mathfrak{N}_2' + (\mathfrak{N}_1 \wedge \mathfrak{N}_2)) - (\mathfrak{N}_2' + (\mathfrak{N}_2 \wedge \mathfrak{N}_1'))$ are Ω-isomorphic. Comparing the second members of these isomorphic pairs, we obtain the lemma.

5. Chain conditions. If $\mathfrak{N}$ is an Ω-group, we shall at various times assume one or both of the following finiteness conditions:

Descending chain condition. If $\mathfrak{N} = \mathfrak{N}_1 > \mathfrak{N}_2 > \cdots$ where $\mathfrak{N}_i$ is an invariant Ω-subgroup of $\mathfrak{N}_{i-1}$, then the sequence has only a finite number of terms.

Ascending chain condition. If $\mathfrak{N} = \mathfrak{N}_1 > \cdots > \mathfrak{N}_k = \mathfrak{P} > 0$ is a normal series for $\mathfrak{N}$, then any chain of Ω-subgroups $0 < \mathfrak{P}_1 < \mathfrak{P}_2 < \cdots$ all of which are invariant in $\mathfrak{P}$ is finite.

Of course both chain conditions hold if $\mathfrak{N}$ is of finite order. On the other hand, we shall see that these conditions may be used in place of the assumption of finiteness of order to obtain extensions of some of the classical theorems on finite groups to infinite Ω-groups. The following examples prove the independence of the two chain conditions.

Examples. 1) *The additive group of integers.* This group satisfies the ascending chain condition but not the descending chain condition. This is also true for the direct sum of a finite number of infinite cyclic groups (Cf. Chapter 3, **3**).

2) *The direct sum $\mathfrak{M}$ of an infinite number of cyclic groups of order a prime p.*[4] Let $x_1, x_2, \cdots$ be a basis for $\mathfrak{M}$ and let A be the endomorphism determined by the equations $x_1A = 0$, $x_iA = x_{i-1}$. Then $\mathfrak{M}$ satisfies the descending chain condition relative to $\Omega = \{A\}$ but not the ascending chain condition. Another example of this type is furnished by the commutative group with generators $x_1, x_2, \cdots$ satisfying the relations $px_1 = 0$, $px_i = x_{i-1}$. Here we take Ω to be vacuous.

It should be noted that if $\mathfrak{N}$ is commutative, the ascending chain condition assumes the simpler form that any chain $0 < \mathfrak{P}_1 < \mathfrak{P}_2 < \cdots$ of Ω-subgroups of $\mathfrak{N}$ is finite in length. If either chain condition holds for an (arbitrary) $\mathfrak{N}$, then it holds also for any invariant Ω-subgroup $\mathfrak{P}$ and for any difference group $\mathfrak{N} - \mathfrak{P}$. If both chain conditions hold, $\mathfrak{N}$ has a *composition series,* i.e. a normal series $\mathfrak{N} = \mathfrak{N}_1 > \cdots > \mathfrak{N}_h > 0$ that has no proper refinements. Thus a normal series is a composition series if $\mathfrak{N}_{i-1} > \mathfrak{N}_i$ and $\mathfrak{N}_{i-1} - \mathfrak{N}_i$ is *Ω-irreducible* in the sense that it has no proper invariant Ω-subgroups. To prove our assertion let $\mathfrak{N}'$ be a proper invariant Ω-subgroup. If $\mathfrak{N} - \mathfrak{N}'$ is reducible, there is an $\mathfrak{N}''$ invariant in $\mathfrak{N}$ such that $\mathfrak{N} > \mathfrak{N}'' > \mathfrak{N}' > 0$. Continuing in this way we

[4] Note that this group relative to the vacuous set of endomorphisms satisfies neither chain condition.

obtain, after a finite number of steps, an invariant Ω-subgroup $\mathfrak{N}_2$ of $\mathfrak{N} = \mathfrak{N}_1$ such that $\mathfrak{N}_1 - \mathfrak{N}_2$ is Ω-irreducible. If we repeat this process for $\mathfrak{N}_2$, we obtain an $\mathfrak{N}_3$, etc. Then we have a normal series $\mathfrak{N}_1 > \mathfrak{N}_2 > \cdots$, and by the descending chain condition this breaks off after a finite number of steps, yielding a composition series for $\mathfrak{N}$.

If Ω is the set of inner automorphisms, a composition series for $\mathfrak{N}$ is called a *principal series* and if Ω is the complete set of automorphisms, we have a *characteristic series.* The following extension of the Jordan-Hölder theorem implies, in particular, the uniqueness (in the sense of isomorphism) of the factors of these series as well as of ordinary composition series (Ω vacuous).

THEOREM 5. *Any two composition series for an Ω-group $\mathfrak{N}$ are equivalent.*

This is an immediate consequence of Schreier's theorem.

THEOREM 6. *A necessary and sufficient condition that an Ω-group have a composition series is that it satisfy both chain conditions.*

The sufficiency of this condition has already been proved. Now suppose that $\mathfrak{N}$ has a composition series of h terms. If $\mathfrak{N} = \mathfrak{N}_1 > \mathfrak{N}_2 > \cdots$ is a descending chain of Ω-subgroups, then there are at most h terms in this chain since $\mathfrak{N}_1 > \mathfrak{N}_2 > \cdots > \mathfrak{N}_k > 0$ is a normal chain and may be refined into a composition series having h terms. A similar argument applies to ascending chains.

If $\mathfrak{N}_1 > \cdots > \mathfrak{N}_h > 0$ is a composition series for $\mathfrak{N}_1$, then h is the *length* of the group $\mathfrak{N}_1$. Hence a group is Ω-irreducible if and only if it has length one. If $\mathfrak{N}'$ is an invariant Ω-subgroup of $\mathfrak{N}_1$, we may suppose that $\mathfrak{N}'$ is the term $\mathfrak{N}_{k+1}$ in a composition series. Then $\mathfrak{N}_{k+1}$ has length $h - k$. By the First Isomorphism Theorem, $(\mathfrak{N}_1 - \mathfrak{N}_{k+1}) > \cdots > (\mathfrak{N}_k - \mathfrak{N}_{k+1}) > 0$ is a composition series for $\mathfrak{N}_1 - \mathfrak{N}_{k+1}$, and so the difference group has length k.

An Ω-endomorphism A of $\mathfrak{N}$ is *normal* if it commutes with all the inner automorphisms of $\mathfrak{N}$. Then for any a and x, $-aA + xA + aA = -a + xA + a$. Thus $aA = a + c(a)$ where $c(a)$ is an element that commutes with every element of $\mathfrak{N}A$. If $\mathfrak{P}$ is an invariant Ω-subgroup, then $\mathfrak{P}A$ is invariant in $\mathfrak{N}$ for any normal A. We note also that the product of normal endomorphisms is normal.

If A is any Ω-endomorphism, the set $\mathfrak{Z}_A$ of elements z such that $zA = 0$ is an Ω-subgroup. Evidently $0 \leqq \mathfrak{Z}_A \leqq \mathfrak{Z}_{A^2} \leqq \cdots$. If $\mathfrak{Z}_{A^k} = \mathfrak{Z}_{A^{k+1}}$, we have $\mathfrak{Z}_{A^{k+1}} = \mathfrak{Z}_{A^{k+2}} = \cdots$. Thus in the chain $0 \leqq \mathfrak{Z}_A \leqq \mathfrak{Z}_{A^2} \leqq \cdots$ we have either the sign $<$ throughout or we have this sign for k $(\geqq 0)$ terms and thereafter equality. Now suppose that $\mathfrak{N}A = \mathfrak{N}$ and $\mathfrak{Z}_A \neq 0$. Then $\mathfrak{Z}_{A^2} > \mathfrak{Z}_A$. For, each z in $\mathfrak{Z}_A$ has the form xA for a suitable x and so $zA = xA^2 = 0$. Hence if $\mathfrak{Z}_{A^2} = \mathfrak{Z}_A$, $xA = 0$, i.e. every $z = 0$. Similarly we see that $0 < \mathfrak{Z}_A < \mathfrak{Z}_{A^2} < \cdots$. Hence

THEOREM 7. *If $\mathfrak{N}$ satisfies the ascending chain condition and if A is an endomorphism such that $\mathfrak{N}A = \mathfrak{N}$, then $\mathfrak{Z}_A = 0$.*

If A is a normal endomorphism, the chain $\mathfrak{N} \geqq \mathfrak{N}A \geqq \mathfrak{N}A^2 \geqq \cdots$ is a normal chain. We have either $\mathfrak{N} > \mathfrak{N}A > \cdots$ or $\mathfrak{N} > \mathfrak{N}A > \cdots > \mathfrak{N}A^k = \mathfrak{N}A^{k+1} = \cdots$. The first of these alternatives certainly holds if $\mathfrak{Z}_A = 0$ and

$\mathfrak{R} > \mathfrak{R}A$. For if $\mathfrak{R}A^k = \mathfrak{R}A^{k+1}$, $xA^{k+1} = yA^k$ for any x and a suitable y. Hence $(xA^k - yA^{k-1})A = 0$ and $xA^k = yA^{k-1}$, i.e. $\mathfrak{R}A^{k-1} = \mathfrak{R}A^k$. Thus we have

THEOREM 8. *If $\mathfrak{R}$ satisfies the descending chain condition and if A is a normal Ω-endomorphism such that $\mathfrak{Z}_A = 0$, then $\mathfrak{R} = \mathfrak{R}A$.*

If we combine the two preceding theorems, we obtain

THEOREM 9. *If $\mathfrak{R}$ satisfies both chain conditions and if A is a normal Ω-endomorphism, then either A is an automorphism or $\mathfrak{R}A < \mathfrak{R}$ and $\mathfrak{Z}_A \neq 0$.*

Assume again the ascending chain condition. Then $0 < \mathfrak{Z}_A < \cdots < \mathfrak{Z}_{A^k} = \mathfrak{Z}_{A^{k+1}} = \cdots$ for a finite k. It follows that $\mathfrak{Z}_{A^k} \wedge \mathfrak{R}A^k = 0$. For if w is in this intersection, $w = xA^k$ and $wA^k = 0$. Hence $xA^{2k} = 0$ and since $\mathfrak{Z}_{A^k} = \mathfrak{Z}_{A^{2k}}$, $xA^k = w = 0$. Since $\mathfrak{R}A^{k+1} \leqq \mathfrak{R}A^k$, A induces an Ω-endomorphism in $\mathfrak{P} = \mathfrak{R}A^k$ and since there are no elements z in $\mathfrak{P}$ other than 0 such that $zA = 0$, A is an isomorphism between $\mathfrak{P}$ and $\mathfrak{P}A$. Hence if D is any transformation in $\mathfrak{P}$ such that $DA = 0$, then $D = 0$. Evidently A induces a nilpotent endomorphism $(A^k = 0)$ in $\mathfrak{Z}_{A^k}$.

If A is normal and $\mathfrak{R}$ satisfies the descending chain condition, we have $\mathfrak{R} > \cdots > \mathfrak{R}A^l = \mathfrak{R}A^{l+1} = \cdots$. If x is any element of $\mathfrak{R}$, $xA^l = yA^{2l}$ for a suitable y and so $x = yA^l + (-yA^l + x) = (x - yA^l) + yA^l \in \mathfrak{R}A^l + \mathfrak{Z}_{A^l} = \mathfrak{Z}_{A^l} + \mathfrak{R}A^l$. The transformation induced by A in $\mathfrak{Z}_{A^l}$ is nilpotent. If D is any transformation in $\mathfrak{R}A^l$ such that $AD = 0$, where A is the induced endomorphism in $\mathfrak{R}A^l$, then $D = 0$.

If both chain conditions hold, the integers k and l of the last two paragraphs are equal. For $\mathfrak{R}A^k \wedge \mathfrak{Z}_{A^k} = 0$ and hence the only element of $\mathfrak{R}A^k$ mapped into 0 by A is 0. It follows that $\mathfrak{R}A^k = \mathfrak{R}A^{k+1}$ so that $l \leqq k$. On the other hand, $\mathfrak{R}A^l = (\mathfrak{R}A^l)A$ implies that $\mathfrak{R}A^l \wedge \mathfrak{Z}_A = 0$. Thus if $yA^{l+1} = (yA^l)A = 0$, $yA^l = 0$; hence $\mathfrak{Z}_{A^{l+1}} = \mathfrak{Z}_{A^l}$ and $k \leqq l$. Hence we have proved the important

LEMMA (Fitting). *Suppose that both chain conditions hold for $\mathfrak{R}$ and that A is a normal Ω-endomorphism. Then for a suitable k we have $\mathfrak{R} = \mathfrak{R}A^k + \mathfrak{Z}_{A^k}$, $\mathfrak{R}A^k \wedge \mathfrak{Z}_{A^k} = 0$ where A is nilpotent in $\mathfrak{Z}_{A^k}$ and an automorphism in $\mathfrak{R}A^k$.*

Remark. We need not suppose that A is an Ω-endomorphism in the above discussion. Instead let Ω contain the inner automorphisms and let A satisfy the condition that $A\Omega = \Omega A$, i.e. for each α in Ω there is an α' and an α'' in Ω such that $A\alpha = \alpha' A$, $\alpha A = A\alpha''$. Since Ω contains the inner automorphisms, Ω-subgroups are invariant. The groups $\mathfrak{R}A$ and $\mathfrak{Z}_A$ are Ω-subgroups and one may carry over the above arguments without change. However, we shall sketch a more direct proof of the final result. Consider the chains $\mathfrak{R} \geq \mathfrak{R}A \geqq \cdots$ and $0 \leqq \mathfrak{Z}_A \leqq \cdots$. The terms of these chains are Ω-subgroups and so by the chain conditions there is an integer m such that $\mathfrak{R}A^m = \mathfrak{R}A^{m+1} = \cdots$ and $\mathfrak{Z}_{A^m} = \mathfrak{Z}_{A^{m+1}} = \cdots$. Set $A^m = B$. Then $\mathfrak{R}B = \mathfrak{R}B^2$, $\mathfrak{Z}_B = \mathfrak{Z}_{B^2}$ and hence by the chain conditions $\mathfrak{R}B \wedge \mathfrak{Z}_B = 0$. If x is any element of $\mathfrak{R}$, $xB^2 = yB$ for a suitable y and so $x = yB + (-yB + x) \in \mathfrak{R}B + \mathfrak{Z}_B$.

6. Direct sums. In the remainder of this chapter we consider an Ω-group $\mathfrak{R}$ for which the endomorphisms in Ω induce all of the inner automorphisms of $\mathfrak{R}$.

We shall also suppose that $\mathfrak{N}$ satisfies both chain conditions. As we have seen, the first assumption implies that every Ω-subgroup is invariant and that Ω-endomorphisms are normal. The ascending chain condition may be stated in the simpler form: Every ascending chain $0 < \mathfrak{N}_1 < \mathfrak{N}_2 \cdots$ terminates after a finite number of terms.

We say that $\mathfrak{N}$ is a *direct sum* of the Ω-subgroups $\mathfrak{N}_i$, $i = 1, \cdots, h$ if

$$\mathfrak{N} = \mathfrak{N}_1 + \cdots + \mathfrak{N}_h$$

and

$$\mathfrak{N}_i \wedge (\mathfrak{N}_1 + \cdots + \mathfrak{N}_{i-1} + \mathfrak{N}_{i+1} + \cdots + \mathfrak{N}_h) = 0$$

for all i. The decomposition is *proper* if all $\mathfrak{N}_i \neq 0$. If no proper decomposition exists other than $\mathfrak{N} = \mathfrak{N}$, $\mathfrak{N}$ is *indecomposable*. We use the notation $\mathfrak{N} = \mathfrak{N}_1 \oplus \cdots \oplus \mathfrak{N}_h$ for a direct sum. Since the $\mathfrak{N}_i$ are invariant, $\mathfrak{N}_i + \mathfrak{N}_j = \mathfrak{N}_j + \mathfrak{N}_i$ and we may equally well write $\mathfrak{N} = \mathfrak{N}_{1'} \oplus \cdots \oplus \mathfrak{N}_{h'}$ for any permutation $1', \cdots, h'$ of $1, \cdots, h$. If $a \in \mathfrak{N}_i$ and $b \in \mathfrak{N}_j$, $j \neq i$, then the commutator $-a - b + a + b \in \mathfrak{N}_i \wedge \mathfrak{N}_j = 0$. Hence $a + b = b + a$ and any element of $\mathfrak{N}_i$ commutes with any in $\mathfrak{N}_j$.

A necessary and sufficient condition that $\mathfrak{N} = \mathfrak{N}_1 \oplus \cdots \oplus \mathfrak{N}_h$, where the $\mathfrak{N}_i$ are Ω-subgroups, is that every x in $\mathfrak{N}$ be expressible in one and only one way in the form $x_1 + \cdots + x_h$, x_i in $\mathfrak{N}_i$. This implies directly that if $\mathfrak{N} = \mathfrak{N}_1 \oplus \cdots \oplus \mathfrak{N}_h$, then $\mathfrak{N}_1' = \mathfrak{N}_1 + \cdots + \mathfrak{N}_{k_1} = \mathfrak{N}_1 \oplus \cdots \oplus \mathfrak{N}_{k_1}$ and if $\mathfrak{N}_2' = \mathfrak{N}_{k_1+1} + \cdots + \mathfrak{N}_{k_1+k_2}, \cdots, \mathfrak{N}_l' = \mathfrak{N}_{k_1+\cdots+k_{l-1}+1} + \cdots + \mathfrak{N}_{k_1+\cdots+k_l}$, then $\mathfrak{N} = \mathfrak{N}_1' \oplus \cdots \oplus \mathfrak{N}_l'$. Conversely, if $\mathfrak{N} = \mathfrak{N}_1' \oplus \cdots \oplus \mathfrak{N}_l'$ and $\mathfrak{N}_1' = \mathfrak{N}_1 \oplus \cdots \oplus \mathfrak{N}_{k_1}, \cdots, \mathfrak{N}_l' = \mathfrak{N}_{k_1+\cdots+k_{l-1}+1} \oplus \cdots \oplus \mathfrak{N}_{k_1+\cdots+k_l}$, then $\mathfrak{N} = \mathfrak{N}_1 \oplus \cdots \oplus \mathfrak{N}_h$, $h = k_1 + \cdots + k_l$.

If $\mathfrak{N} = \mathfrak{N}_1 \oplus \mathfrak{N}_2$, the Second Isomorphism Theorem implies that $\mathfrak{N}_2$ is Ω-isomorphic to $\mathfrak{N} - \mathfrak{N}_1$. Evidently the length of $\mathfrak{N}$ = length $\mathfrak{N}_1$ + length $\mathfrak{N}_2$. If $\mathfrak{N}_1$ and $\mathfrak{N}_2$ are Ω-subgroups of $\mathfrak{N}$ such that $\mathfrak{N} = \mathfrak{N}_1 + \mathfrak{N}_2$, and $\mathfrak{N}_3 = \mathfrak{N}_1 \wedge \mathfrak{N}_2$, then $\mathfrak{N} - \mathfrak{N}_3 = (\mathfrak{N}_1 - \mathfrak{N}_3) \oplus (\mathfrak{N}_2 - \mathfrak{N}_3)$. It follows that

$$\text{length } \mathfrak{N} + \text{length } (\mathfrak{N}_1 \wedge \mathfrak{N}_2) = \text{length } \mathfrak{N}_1 + \text{length } \mathfrak{N}_2 .$$

We may, of course, replace $\mathfrak{N}$ by $\mathfrak{N}_1 + \mathfrak{N}_2$ and obtain this relation for arbitrary Ω-subgroups of $\mathfrak{N}$.

If $\mathfrak{N} = \mathfrak{N}_1 \oplus \cdots \oplus \mathfrak{N}_h$ so that we have, for every x, $x = x_1 + \cdots + x_h$, x_i in $\mathfrak{N}_i$, then we define the mapping E_i by $xE_i = x_i$. Since the expression for x is unique, E_i is single valued. If $y = y_1 + \cdots + y_h$, $x + y = (x_1 + y_1) + \cdots + (x_h + y_h)$. Hence $(x + y)E_i = xE_i + yE_i$. If $\alpha \in \Omega$, $x\alpha = x_1\alpha + \cdots + x_n\alpha$ so that $\alpha E_i = E_i\alpha$. The E_i are therefore Ω-endomorphisms. Evidently the following relations hold:

$$(1) \qquad E_i^2 = E_i, \qquad E_iE_j = 0 \quad \text{if} \quad i \neq j, \qquad E_1 + \cdots + E_h = 1.$$

We note also that $E_i + E_j = E_j + E_i$ and that any partial sum $E_{i_1} + \cdots + E_{i_m}$, i_k distinct, is an endomorphism.

An Ω-endomorphism E that is idempotent ($E^2 = E$) will be called a *projection*.

The E_i determined by the direct decomposition are of this type. Now suppose, conversely, that the E_i are arbitrary projections that satisfy (1). Then $\mathfrak{N}E_i \equiv \mathfrak{N}_i$ are Ω-subgroups such that $\mathfrak{N} = \mathfrak{N}_1 \oplus \cdots \oplus \mathfrak{N}_h$ and the E_i are the projections determined by this decomposition. Furthermore if E is any projection and $\mathfrak{Z}_E$ is the set of elements z such that $zE = 0$, then by Fitting's lemma, or directly, we have $\mathfrak{N} = \mathfrak{N}E \oplus \mathfrak{Z}_E$. Hence there is a projection E' such that $E + E' = E' + E = 1$, $EE' = E'E = 0$. We shall call an idempotent element E of any ring *primitive* if it is impossible to write $E = E' + E''$ where E' and E'' are idempotent elements $\neq 0$ of the ring and $E'E'' = E''E' = 0$. Thus $\mathfrak{N}$ is indecomposable if and only if 1 is a primitive projection.

By Fitting's lemma we have

THEOREM 10. *Let $\mathfrak{N}$ be an Ω-group for which Ω contains all the inner automorphisms of $\mathfrak{N}$ and both chain conditions hold. If $\mathfrak{N}$ is indecomposable, then any Ω-endomorphism is either nilpotent or an automorphism.*

7. The Krull-Schmidt theorems. Suppose that $\mathfrak{N}$ is decomposable so that $\mathfrak{N} = \mathfrak{N}_1 \oplus \mathfrak{N}_2$, $\mathfrak{N}_i \neq 0$. If $\mathfrak{N}_1$ is decomposable, $\mathfrak{N}_1 = \mathfrak{N}_{11} \oplus \mathfrak{N}_{12}$ and $\mathfrak{N} = \mathfrak{N}_{11} \oplus \mathfrak{N}_{12} \oplus \mathfrak{N}_2$. Thus $\mathfrak{N} > \mathfrak{N}_1 > \mathfrak{N}_{11} \neq 0$ and continuing in this way, we obtain an indecomposable $\mathfrak{N}_{1\cdots 1}$ such that $\mathfrak{N} = \mathfrak{N}_{1\cdots 1} \oplus \mathfrak{N}_1'$. We simplify the notation and write $\mathfrak{N} = \mathfrak{N}_1 \oplus \mathfrak{N}_1'$ where $\mathfrak{N}_1$ is indecomposable and $\neq 0$. If $\mathfrak{N}_1'$ is decomposable, we have $\mathfrak{N}_1' = \mathfrak{N}_2 \oplus \mathfrak{N}_2'$ where $\mathfrak{N}_2$ is indecomposable and $\neq 0$. Then $\mathfrak{N} = \mathfrak{N}_1 \oplus \mathfrak{N}_2 \oplus \mathfrak{N}_2'$. This process yields a descending chain $\mathfrak{N}_1' > \mathfrak{N}_2' > \cdots$. Hence it breaks off and we obtain $\mathfrak{N} = \mathfrak{N}_1 \oplus \cdots \oplus \mathfrak{N}_h$ where the $\mathfrak{N}_i$ are indecomposable and $\neq 0$.

Now suppose that $\mathfrak{N} = \mathfrak{P}_1 \oplus \cdots \oplus \mathfrak{P}_k$ is a second decomposition where the Ω-subgroups $\mathfrak{P}_j$ are indecomposable and $\neq 0$. Let E_i and F_j be the projections determined by the two decompositions. Since any sum $E_{i_1} + \cdots + E_{i_n}$, i_m distinct, is an endomorphism, this is true also for $AE_{i_1} + \cdots + AE_{i_n} = A(E_{i_1} + \cdots + E_{i_n})$ and $E_{i_1}A + \cdots + E_{i_n}A = (E_{i_1} + \cdots + E_{i_n})A$ for any endomorphism A. If we apply the endomorphism F_jE_1 to $\mathfrak{N}_1$, we obtain an endomorphism in this group and we have $F_1E_1 + \cdots + F_kE_1 = E_1$ as the identity in $\mathfrak{N}_1$. We wish to show that at least one of the F_jE_1 is an automorphism in $\mathfrak{N}_1$. This will follow from the following lemma.

LEMMA. *Let $\mathfrak{N}$ be an Ω-group for which Ω contains all the inner automorphisms of $\mathfrak{N}$ and both chain conditions hold. If $\mathfrak{N}$ is indecomposable and A and B are Ω-endomorphisms such that $A + B = 1$, then either A or B is an automorphism.*

Since $A + B = 1$ and A and B are endomorphisms, $A^2 + AB = A^2 + BA$ and hence $AB = BA$. If neither A nor B is an automorphism, both are nilpotent. Then $1 = (A + B)^m$ is a sum of terms of the type A^rB^s where $r + s = m$. If m is sufficiently large, we have either $A^r = 0$ or $B^s = 0$, and so we obtain the contradiction $1 = 0$.

We apply this to $F_1E_1 = A$ and $F_2E_1 + \cdots + F_kE_1 = B$ acting in $\mathfrak{N}_1$. If F_1E_1 is not an automorphism, then B is and hence B^{-1} exists. It follows that $F_2E_1B^{-1} + \cdots + F_kE_1B^{-1} = 1$. Either $F_2E_1B^{-1}$ is an automorphism or

$F_3E_1B^{-1} + \cdots + F_kE_1B^{-1}$ is. If we continue in this way, we obtain the result that for some j, $F_jE_1B^{-1}C^{-1} \cdots G^{-1}$ is an automorphism where B^{-1}, C^{-1}, $\cdots$ are automorphisms. It follows that F_jE_1 is an automorphism in $\mathfrak{N}_1$. For simplicity we write $j = 1$.

Consider the Ω-homomorphism F_1 between $\mathfrak{N}_1$ and $\mathfrak{N}_1F_1 \leqq \mathfrak{P}_1$. Since F_1E_1 is an automorphism, F_1 is an isomorphism. Now $\mathfrak{N}_1F_1$ is an Ω-subgroup of $\mathfrak{P}_1$, as is also $\bar{\mathfrak{P}}_1$, the subset of $\mathfrak{P}_1$ of elements z such that $zE_1 = 0$. If y is any element of $\mathfrak{P}_1$, $yE_1 = wF_1E_1$ for some w in $\mathfrak{N}_1$. Hence $y = (y - wF_1) + wF_1$ where $y - wF_1$ is in $\bar{\mathfrak{P}}_1$. Since $\bar{\mathfrak{P}}_1 \wedge \mathfrak{N}_1F_1 = 0$, this contradicts the indecomposability of $\mathfrak{P}_1$ unless $\bar{\mathfrak{P}}_1 = 0$ and $\mathfrak{N}_1F_1 = \mathfrak{P}_1$. Thus $\mathfrak{N}_1F_1 = \mathfrak{P}_1$ and hence F_1 is an isomorphism between $\mathfrak{N}_1$ and $\mathfrak{P}_1$, and E_1 is an isomorphism between $\mathfrak{P}_1$ and $\mathfrak{N}_1$. We assert that $H_1 = E_1F_1 + E_2 + \cdots + E_h$ is an Ω-endomorphism. This is a consequence of the following general remark: Suppose that $\mathfrak{N} = \mathfrak{N}_1 \oplus \cdots \oplus \mathfrak{N}_h$ and that $\mathfrak{N}' = \mathfrak{N}_1' \oplus \cdots \oplus \mathfrak{N}_h'$ is an Ω-subgroup of $\mathfrak{N}$. If A_i is an Ω-homomorphism between $\mathfrak{N}_i$ and $\mathfrak{N}_i'$, then $E_1A_1 + \cdots + E_hA_h$ is an Ω-endomorphism in $\mathfrak{N}$. Our result follows by noting that $\mathfrak{P}_1 \wedge (\mathfrak{N}_2 + \cdots + \mathfrak{N}_h) = 0$ so that $\mathfrak{N}' \equiv \mathfrak{P}_1 + \mathfrak{N}_2 + \cdots + \mathfrak{N}_h = \mathfrak{P}_1 \oplus \mathfrak{N}_2 \oplus \cdots \oplus \mathfrak{N}_h$. Since $zH_1 = 0$ implies that $z = 0$, H_1 is an automorphism, i.e. $\mathfrak{N}' = \mathfrak{N}$.

Now suppose that we have already obtained a pairing between $\mathfrak{P}_i$ and $\mathfrak{N}_i$ for $i = 1, \cdots, r$ such that E_i is an Ω-isomorphism between $\mathfrak{P}_i$ and $\mathfrak{N}_i$ and F_i is one between $\mathfrak{N}_i$ and $\mathfrak{P}_i$. Suppose also that $\mathfrak{N} = \mathfrak{P}_1 \oplus \cdots \oplus \mathfrak{P}_r \oplus \mathfrak{N}_{r+1} \oplus \cdots \oplus \mathfrak{N}_h$, and $H_r = E_1F_1 + \cdots + E_rF_r + E_{r+1} + \cdots + E_h$ is an automorphism. Since the inner automorphisms of a difference group are induced by inner automorphism of the group, $\bar{\mathfrak{N}} = \mathfrak{N} - (\mathfrak{P}_1 + \cdots + \mathfrak{P}_r)$ satisfies our conditions. We have

$$\bar{\mathfrak{N}} = \bar{\mathfrak{N}}_{r+1} \oplus \cdots \oplus \bar{\mathfrak{N}}_h = \bar{\mathfrak{P}}_{r+1} \oplus \cdots \oplus \bar{\mathfrak{P}}_k$$

where $\bar{\mathfrak{N}}_l = (\mathfrak{P}_1 + \cdots + \mathfrak{P}_r + \mathfrak{N}_l) - (\mathfrak{P}_1 + \cdots + \mathfrak{P}_l)$, $\bar{\mathfrak{P}}_j = (\mathfrak{P}_1 + \cdots + \mathfrak{P}_r + \mathfrak{P}_j) - (\mathfrak{P}_1 + \cdots + \mathfrak{P}_r)$ are Ω-isomorphic to $\mathfrak{N}_l$ and $\mathfrak{P}_j$ respectively. By the above discussion we may pair $\bar{\mathfrak{P}}_{r+1}$ with, say, $\bar{\mathfrak{N}}_{r+1}$ so that the corresponding projections $\bar{E}_{r+1}$, $\bar{F}_{r+1}$ are isomorphisms between $\bar{\mathfrak{P}}_{r+1}$ and $\bar{\mathfrak{N}}_{r+1}$. We also have the equation $\bar{\mathfrak{N}} = \bar{\mathfrak{P}}_{r+1} \oplus \bar{\mathfrak{N}}_{r+2} \oplus \cdots \oplus \bar{\mathfrak{N}}_h$. If $x \in (\mathfrak{P}_1 + \cdots + \mathfrak{P}_{r+1}) \wedge (\mathfrak{N}_{r+2} + \cdots + \mathfrak{N}_h)$, the coset $\bar{x} = x + (\mathfrak{P}_1 + \cdots + \mathfrak{P}_r) \in \bar{\mathfrak{P}}_{r+1} \wedge (\bar{\mathfrak{N}}_{r+2} + \cdots + \bar{\mathfrak{N}}_h)$. Hence $\bar{x} = 0$ and $x \in \mathfrak{P}_1 + \cdots + \mathfrak{P}_r$. Since $(\mathfrak{P}_1 + \cdots + \mathfrak{P}_r) \wedge (\mathfrak{N}_{r+1} + \cdots + \mathfrak{N}_h) = 0$, $x = 0$. Thus

$$\mathfrak{P}_1 + \cdots + \mathfrak{P}_{r+1} + \mathfrak{N}_{r+1} + \cdots + \mathfrak{N}_h = \mathfrak{P}_1 \oplus \cdots \oplus \mathfrak{P}_{r+1} \oplus \mathfrak{N}_{r+2} \oplus \cdots \oplus \mathfrak{N}_h.$$

Hence $H_{r+1} = E_1F_1 + \cdots + E_{r+1}F_{r+1} + E_{r+2} + \cdots + E_h$ is an endomorphism. Since $\bar{F}_{r+1}$ is an isomorphism between $\bar{\mathfrak{N}}_{r+1}$ and $\bar{\mathfrak{P}}_{r+1}$, $z_{r+1}F_{r+1} \neq 0$ if $z_{r+1} \neq 0$ is in $\mathfrak{N}_{r+1}$. Hence $zH_{r+1} = 0$ only if $z = 0$; H_{r+1} is an automorphism and $\mathfrak{N} = \mathfrak{P}_1 \oplus \cdots \oplus \mathfrak{P}_{r+1} \oplus \mathfrak{N}_{r+2} \oplus \cdots \oplus \mathfrak{N}_h$. This proves the following theorems.

THEOREM 11 (Krull-Schmidt). *Let $\mathfrak{N}$ be an Ω-group such that Ω contains all the inner automorphisms and both chain conditions hold. Suppose that $\mathfrak{N} = \mathfrak{N}_1 \oplus \cdots \oplus \mathfrak{N}_h = \mathfrak{P}_1 \oplus \cdots \oplus \mathfrak{P}_k$ are two decompositions of $\mathfrak{N}$ as direct sums of in-*

decomposable groups $\neq 0$. *Then* $h = k$ *and there is an* Ω*-automorphism* H *and a suitable ordering of the* $\mathfrak{P}$*'s such that* $\mathfrak{N}_i H = \mathfrak{P}_i$ *and* $\mathfrak{N} = \mathfrak{P}_1 \oplus \cdots \oplus \mathfrak{P}_r \oplus \mathfrak{N}_{r+1} \oplus \cdots \oplus \mathfrak{N}_h$.

THEOREM 11′ (Krull-Schmidt, second formulation). *Under the above assumptions let* E_i, F_j *be sets of primitive projections* $\neq 0$ *such that*

$$E_1 + \cdots + E_h = 1, \qquad E_i E_{i'} = 0 \quad \text{if} \quad i \neq i'$$

$$F_1 + \cdots + F_k = 1, \qquad F_j F_{j'} = 0 \quad \text{if} \quad j \neq j'.$$

Then $h = k$, *and we may order the* F*'s so that there exists an* Ω*-automorphism* H *satisfying* $F_i = H^{-1} E_i H$ *and so that* $H_r = E_1 F_1 + \cdots + E_r F_r + E_{r+1} + \cdots + E_h$ *is an* Ω*-automorphism.*

In both theorems we take $H = E_1 F_1 + \cdots + E_h F_h$.

If $\mathfrak{N} = \mathfrak{N}' \oplus \mathfrak{N}''$ is any direct decomposition, there is a refinement of this decomposition to a direct sum of indecomposable groups. It follows that if the $\mathfrak{P}_i$ above are suitably ordered, then there is an Ω-automorphism H such that $\mathfrak{N}'H = \mathfrak{P}_1 \oplus \cdots \oplus \mathfrak{P}_t$ and $\mathfrak{N}''H = \mathfrak{P}_{t+1} \oplus \cdots \oplus \mathfrak{P}_h$.

8. Complete reducibility. If, as in the present situation, the Ω-subgroups of $\mathfrak{N}$ are invariant, they constitute a modular lattice (Dedekind structure) $\mathfrak{L}$ relative to the operations $\wedge$ and $+$. For Dedekind's distributive law:

$$\mathfrak{N}_1 \wedge (\mathfrak{N}_2 + \mathfrak{N}_3) = \mathfrak{N}_2 + \mathfrak{N}_1 \wedge \mathfrak{N}_3 \quad \text{if} \quad \mathfrak{N}_1 \geqq \mathfrak{N}_2$$

is valid. The concepts of reducibility and decomposability are lattice concepts. Similarly, we say that $\mathfrak{N}$ is *completely reducible* if the lattice $\mathfrak{L}$ is completely reducible, that is, if for every $\mathfrak{N}'$ in $\mathfrak{N}$ there is an $\mathfrak{N}''$ such that $\mathfrak{N} = \mathfrak{N}' \oplus \mathfrak{N}''$. The element $\mathfrak{N}''$ is a *complement* of $\mathfrak{N}'$ relative to $\mathfrak{N}$.

If $\mathfrak{P}, \mathfrak{P}' \in \mathfrak{L}$ and $\mathfrak{P}' \leqq \mathfrak{P}$, let $\mathfrak{P}''$ be a complement of $\mathfrak{P}'$ relative to $\mathfrak{N}$. From $\mathfrak{N} = \mathfrak{P}' + \mathfrak{P}''$ and Dedekind's law we have $\mathfrak{P} = \mathfrak{P}' + (\mathfrak{P}'' \wedge \mathfrak{P})$. Since $\mathfrak{P}' \wedge \mathfrak{P}'' = 0$, $(\mathfrak{P} \wedge \mathfrak{P}'')$ is a complement of $\mathfrak{P}'$ relative to $\mathfrak{P}$. Thus any Ω-subgroup $\mathfrak{P}$ of a completely reducible group $\mathfrak{N}$ is completely reducible. If $\mathfrak{N} = \mathfrak{N}_1 > \mathfrak{N}_2 > \cdots$ is an infinite descending chain of elements in $\mathfrak{L}$, there exist elements $\mathfrak{N}_i' \neq 0$ for $i = 2, 3, \cdots$ such that $\mathfrak{N}_{i-1} = \mathfrak{N}_i \oplus \mathfrak{N}_i'$. Then

$$\mathfrak{N}_1 = \mathfrak{N}_2' \oplus \mathfrak{N}_2 = \cdots = \mathfrak{N}_2' \oplus \cdots \oplus \mathfrak{N}_i' \oplus \mathfrak{N}_i$$

and $\mathfrak{N}_2' < \mathfrak{N}_2' \oplus \mathfrak{N}_3' < \mathfrak{N}_2' \oplus \mathfrak{N}_3' \oplus \mathfrak{N}_4' < \cdots$ is an infinite ascending chain. Hence, if the ascending chain condition holds for a completely reducible group then the descending chain condition holds. Now suppose that $0 < \mathfrak{N}_1 < \mathfrak{N}_2 < \cdots$ is an infinite ascending chain. Determine $\mathfrak{N}_1'$ so that $\mathfrak{N} = \mathfrak{N}_1 \oplus \mathfrak{N}_1'$ and $\mathfrak{N}_i'$ for $i > 1$ so that $\mathfrak{N}_{i-1}' = (\mathfrak{N}_{i-1}' \wedge \mathfrak{N}_i) \oplus \mathfrak{N}_i'$. Then $\mathfrak{N}_i + \mathfrak{N}_i' \geqq \mathfrak{N}_{i-1}'$ and $\geqq \mathfrak{N}_{i-1}$. Hence $\mathfrak{N} = \mathfrak{N}_i + \mathfrak{N}_i'$. Since $\mathfrak{N}_i' \wedge (\mathfrak{N}_{i-1}' \wedge \mathfrak{N}_i) = 0$, $(\mathfrak{N}_i' \wedge \mathfrak{N}_i) \wedge \mathfrak{N}_{i-1}' = 0$ and since $(\mathfrak{N}_i' \wedge \mathfrak{N}_i) \leqq \mathfrak{N}_{i-1}'$, it follows that $\mathfrak{N}_i' \wedge \mathfrak{N}_i = 0$ and $\mathfrak{N} = \mathfrak{N}_i \oplus \mathfrak{N}_i'$. This, together with the relation $\mathfrak{N}_{i-1}' > \mathfrak{N}_i'$, implies that

$\mathfrak{N}_1' > \mathfrak{N}_2' > \cdots$ is an infinite descending chain. The descending chain condition therefore implies the ascending chain condition.

THEOREM 12. *If $\mathfrak{N}$ satisfies either chain condition and is completely reducible, then $\mathfrak{N} = \mathfrak{N}_1 \oplus \cdots \oplus \mathfrak{N}_h$ where the $\mathfrak{N}_i$ are irreducible. Conversely, if $\mathfrak{N} = \mathfrak{N}_1 + \cdots + \mathfrak{N}_h$ where the $\mathfrak{N}_i$ are irreducible, then $\mathfrak{N}$ is completely reducible and satisfies both chain conditions.*

First suppose that $\mathfrak{N}$ is completely reducible and that $\mathfrak{N} = \mathfrak{N}_1 \oplus \cdots \oplus \mathfrak{N}_h$ where the $\mathfrak{N}_i$ are indecomposable. If $\mathfrak{N}_i$ is reducible, it contains an $\mathfrak{N}_i' \neq 0$ and $\neq \mathfrak{N}_i$. Since $\mathfrak{N}_i \leqq \mathfrak{N}$, it is completely reducible also. Hence $\mathfrak{N}_i = \mathfrak{N}_i' \oplus \mathfrak{N}_i''$ for a suitable $\mathfrak{N}_i'' \neq 0$ contrary to the indecomposability of $\mathfrak{N}_i$. Now let $\mathfrak{N} = \mathfrak{N}_1 + \cdots + \mathfrak{N}_h$, $\mathfrak{N}_i$ irreducible, and let $\mathfrak{P}_1$ be an Ω-subgroup of $\mathfrak{N}$. Then $\mathfrak{P}_1 \wedge \mathfrak{N}_i \in \mathfrak{L}$ and by the irreducibility of $\mathfrak{N}_i$ either $\mathfrak{P}_1 \wedge \mathfrak{N}_i = 0$ or $\mathfrak{P}_1 \wedge \mathfrak{N}_i = \mathfrak{N}_i$ so that $\mathfrak{P}_1 \geqq \mathfrak{N}_i$. If the second condition holds for all i, $\mathfrak{P}_1 = \mathfrak{N}$. Otherwise let i_1 be an index such that $\mathfrak{P}_1 \wedge \mathfrak{N}_{i_1} = 0$. Then $\mathfrak{P}_2 \equiv \mathfrak{P}_1 + \mathfrak{N}_{i_1} = \mathfrak{P}_1 \oplus \mathfrak{N}_{i_1}$. If we use $\mathfrak{P}_2$ in place of $\mathfrak{P}_1$, we obtain either $\mathfrak{P}_2 = \mathfrak{N}$, or there is an $\mathfrak{N}_{i_2}$, $i_2 \neq i_1$, such that $\mathfrak{P}_2 \wedge \mathfrak{N}_{i_2} \neq 0$. Then $\mathfrak{P}_3 \equiv \mathfrak{P}_2 + \mathfrak{N}_{i_2} = \mathfrak{P}_2 \oplus \mathfrak{N}_{i_2}$. Continuing in this way we finally reach a k such that $\mathfrak{N} = \mathfrak{P}_{k+1} = \mathfrak{P}_k \oplus \mathfrak{N}_{i_k} = \mathfrak{P}_1 \oplus \mathfrak{N}_{i_1} \oplus \cdots \oplus \mathfrak{N}_{i_k}$. Thus $\mathfrak{N}_{i_1} \oplus \cdots \oplus \mathfrak{N}_{i_k}$ is a complement of $\mathfrak{P}_1$ relative to $\mathfrak{N}$. Now if we begin with $\mathfrak{P}_1 = 0$, we obtain the decomposition $\mathfrak{N} = \mathfrak{N}_{i_1} \oplus \cdots \oplus \mathfrak{N}_{i_k}$. Hence

$$\mathfrak{N} = (\mathfrak{N}_{i_1} \oplus \cdots \oplus \mathfrak{N}_{i_k}) > (\mathfrak{N}_{i_2} \oplus \cdots \oplus \mathfrak{N}_{i_k}) > \cdots > \mathfrak{N}_{i_k} > 0$$

is a composition series for $\mathfrak{N}$ and by Theorem 6, both chain conditions hold.

9. $\mathfrak{o}$-modules. We shall now introduce the concept of a module, which is of particular importance in the theory of representations. We define a *representation* of an abstract ring $\mathfrak{o}$ as a homomorphism between $\mathfrak{o}$ and a subring of the ring of endomorphisms of a commutative group $\mathfrak{M}$. We denote the endomorphism corresponding to a in $\mathfrak{o}$ by A. However, we shall find it more convenient to denote the effect xA simply as xa and to regard this element as the "product" of x in $\mathfrak{M}$ and a in $\mathfrak{o}$. The following conditions hold:

$$\begin{aligned} (x + y)a &= xa + ya \\ x(a + b) &= xa + xb \qquad (2) \\ x(ab) &= (xa)b \end{aligned}$$

for all x, y in $\mathfrak{M}$ and all a, b in $\mathfrak{o}$. Now we shall call a commutative group $\mathfrak{M}$ an $\mathfrak{o}$-*module* if a product xa in $\mathfrak{M}$ is defined for each x in $\mathfrak{M}$ and each a in $\mathfrak{o}$ such that (2) holds. Thus we have shown that the group $\mathfrak{M}$ in which a representation of $\mathfrak{o}$ is defined may be regarded as an $\mathfrak{o}$-module. On the other hand, any $\mathfrak{o}$-module defines a representation. For by the first of equations (2), the mapping $x \to xa$ is an endomorphism in $\mathfrak{M}$. By the second and third equations, the set $\mathfrak{O}$ of these endomorphisms is closed under addition and multiplication. Moreover, we may deduce easily that $x0 = 0$ and that $x(-a) = -xa$ so that $\mathfrak{O}$ contains the 0 endomorphism and the negative of any endomorphism in this set. Thus $\mathfrak{O}$ is a ring. Now by the second and third equations, the correspondence between a and the endomorphism $x \rightarrow xa$ is a homomorphism between $\mathfrak{o}$ and $\mathfrak{O}$.

Since the ring $\mathfrak{o}$ has an existence independent of $\mathfrak{M}$, there is a natural way of comparing different $\mathfrak{o}$-modules. We define an $\mathfrak{o}$-*homomorphism* H between the $\mathfrak{o}$-module $\mathfrak{M}$ and the $\mathfrak{o}$-module $\mathfrak{N}$ as a homomorphism between $\mathfrak{M}$ and $\mathfrak{N}$ such that $(xa)H = (xH)a$ for all x in $\mathfrak{M}$ and all a in $\mathfrak{o}$. If H is $(1 - 1)$, we have an $\mathfrak{o}$-*isomorphism*. In a similar fashion we define an $\mathfrak{o}$-*endomorphism* and an $\mathfrak{o}$-*automorphism*. If $\mathfrak{N}$ is a subgroup of an $\mathfrak{o}$-module, having the property that $ya \in \mathfrak{N}$ for all y in $\mathfrak{N}$ and all a in $\mathfrak{o}$, $\mathfrak{N}$ is a module relative to the product ya. Then $\mathfrak{N}$ is called a *submodule* of $\mathfrak{M}$. We may set $(x + \mathfrak{N})a = xa + \mathfrak{N}$ and observe that this function is single valued for the pairs $x + \mathfrak{N}$ in $\mathfrak{M} - \mathfrak{N}$ and a in $\mathfrak{o}$. The rules (2) hold and so $\mathfrak{M} - \mathfrak{N}$ is a module, the *difference module* of $\mathfrak{M}$ with respect to $\mathfrak{N}$. The module $\mathfrak{M}$ is *reducible* if it contains a proper $\mathfrak{o}$-submodule. *Decomposability* and *complete reducibility* are defined in a similar fashion.

We shall see later that the following representation is fundamental in the structure theory of rings. We consider $\mathfrak{o}$ as a commutative group relative to the addition defined in the ring $\mathfrak{o}$. Now we may turn this group into an $\mathfrak{o}$-module by taking xa to be the ring product of x and a. Then the equations (2) follow from the distributive and associative laws. Hence the group $\mathfrak{o}$ is an $\mathfrak{o}$-module. It follows that the correspondence between the ring element a and the endomorphism $x \rightarrow xa$ is a representation. We shall denote the endomorphism $x \rightarrow xa$ by a_r and we shall call this mapping the *right multiplication* determined by a. The representation $a \rightarrow a_r$ is the (right) *regular representation* of $\mathfrak{o}$. If $\mathfrak{o}$ has an identity 1, $1a_r = 1b_r$ implies that $a = b$; hence the regular representation is $(1 - 1)$. The $\mathfrak{o}$-submodules relative to the regular representation are the right ideals of $\mathfrak{o}$.

The theorems that we have proved for Ω-groups are all valid for $\mathfrak{o}$-modules. The modification in statement and proof is obvious. For example, if $\mathfrak{M}$ is an $\mathfrak{o}$-module homomorphic to the $\mathfrak{o}$-module $\mathfrak{N}$, then the set $\mathfrak{P}$ of elements mapped into 0 by the homomorphism is a submodule of $\mathfrak{M}$ and $\mathfrak{M} - \mathfrak{P}$ and $\mathfrak{N}$ are $\mathfrak{o}$-isomorphic. If $\mathfrak{o}$ contains an identity, the following device enables us to reduce the theory of $\mathfrak{o}$-modules to that of Ω-groups. If $\mathfrak{M}$ and $\mathfrak{N}$ are $\mathfrak{o}$-modules, we form the direct sum $\mathfrak{S} = \mathfrak{M} \oplus \mathfrak{N} \oplus \mathfrak{o}$ and we define $(x + y + b)a = xa + ya + ba$ for x in $\mathfrak{M}$, y in $\mathfrak{N}$ and a, b in $\mathfrak{o}$. In this way we obtain a ring of endomorphisms $\mathfrak{O}$ in $\mathfrak{S}$ isomorphic to $\mathfrak{o}$. Now $\mathfrak{M}$ and $\mathfrak{N}$ are $\mathfrak{O}$-subgroups and are $\mathfrak{O}$-homomorphic ($\mathfrak{O}$-isomorphic) if and only if they are $\mathfrak{o}$-homomorphic ($\mathfrak{o}$-isomorphic).

Finally we may remark that certain problems regarding Ω-groups may be reduced to questions on representations. This is done by replacing Ω by its enveloping ring $\mathfrak{O}$ defined as the smallest subring of endomorphisms containing all the transformations in Ω. Then $\mathfrak{O}$ defines a representation of itself, or, more precisely, $\mathfrak{O}$ defines a representation of the abstract ring $\mathfrak{o}$ isomorphic to $\mathfrak{O}$. The group $\mathfrak{M}$ in which the endomorphisms act is therefore an $\mathfrak{o}$-module.

10. Left-modules. Since we are primarily interested in non-commutative rings, the concept of an anti-homomorphism is almost as important as that of a homomorphism. We recall the definition: If $\mathfrak{o}$ and $\mathfrak{o}'$ are rings, an *anti-homomorphism* is a mapping $x \rightarrow x'$ of $\mathfrak{o}$ into the whole of $\mathfrak{o}'$ such that

$$(x + y)' = x' + y' \qquad (xy)' = y'x'.$$

If, in addition, the mapping is $(1 - 1)$, then it is an *anti-isomorphism* and $\mathfrak{o}$ and $\mathfrak{o}'$ are *anti-isomorphic*. If we add the further condition that $\mathfrak{o} = \mathfrak{o}'$, we have an *anti-automorphism*. For example, the correspondence between a matrix and its transpose is an anti-isomorphism in the ring of matrices with rational integral elements.

Now if $\mathfrak{o}$ is any ring, we may form a set $\mathfrak{o}'$ whose elements x' are in $(1 - 1)$ correspondence with those of $\mathfrak{o}$ $(x \leftrightarrow x')$ and then define $x' + y'$ as $(x + y)'$ and $x'y'$ as $(yx)'$. The resulting system is a ring anti-isomorphic to $\mathfrak{o}$, the mapping $x \rightarrow x'$ being an anti-isomorphism.

We may now formulate the duals of the concepts of representation and $\mathfrak{o}$-module. We define an *anti-representation* (inverse representation) of $\mathfrak{o}$ as an anti-homomorphism $a \rightarrow A'$ between $\mathfrak{o}$ and a subring of the ring of endomorphisms of a commutative group $\mathfrak{M}$. In this case it is convenient to denote the value xA' of the function of x and a by ax. Then

$$(3) \qquad \begin{aligned} a(x + y) &= ax + ay \\ (a + b)x &= ax + bx \\ (ab)x &= a(bx). \end{aligned}$$

We are therefore led to define a *left* $\mathfrak{o}$-*module* $\mathfrak{M}$ as a commutative group for which there is defined a product of x in $\mathfrak{M}$ and a in $\mathfrak{o}$ whose values ax, in $\mathfrak{M}$, satisfy (3). Thus any anti-representation leads to a left $\mathfrak{o}$-module, and, conversely, if a left $\mathfrak{o}$-module is given, the correspondence $a \rightarrow A'$ where $xA' = ax$ is an anti-representation. The definitions of *isomorphism*, *submodule*, etc. are similar to those for ordinary modules. As is to be expected, the "anti"-theory parallels the ordinary theory. We may in fact obtain a reduction to the ordinary theory by noting that any left $\mathfrak{o}$-module may be regarded as an $\mathfrak{o}'$-module where $\mathfrak{o}'$ is anti-isomorphic to $\mathfrak{o}$. In many cases, however, we shall find it more convenient to deal directly with left modules instead of carrying out this reduction.

As before, the additive group of any ring $\mathfrak{o}$ is a left $\mathfrak{o}$-module relative to the function whose values are the products ax, x in the additive group $\mathfrak{o}$ and a in the ring $\mathfrak{o}$. We shall denote the mapping $x \rightarrow ax$ by a_l and shall call it the *left multiplication* determined by a. The anti-representation $a \rightarrow a_l$ is called the *left regular representation*. It is clear that the submodules of the left $\mathfrak{o}$-module $\mathfrak{o}$ are the left ideals of $\mathfrak{o}$.

CHAPTER 2

VECTOR SPACES

1. Definition. In this chapter we study a commutative group $\mathfrak{R}$ relative to a set Φ of endomorphisms that forms a division ring (non-commutative field). The set $\mathfrak{L}$ of Φ-endomorphisms of a group of this type is a matrix ring over a division ring Φ' anti-isomorphic to Φ. Our study is therefore equivalent to the study of matrix rings. One of the fundamental results of the structure theory of rings (obtained in Chapter 4) amounts to the statement that any simple ring satisfying certain finiteness conditions is a matrix ring over a division ring. By means of this theorem, we shall be able to elevate the present seemingly special discussion to an important place in the general theory.

The exact assumptions that we make on $\mathfrak{R}$ and on Φ are the following:

1. If $\alpha, \beta \in \Phi$, then $\alpha + \beta \in \Phi$ and $\alpha\beta \in \Phi$.
2. 0 and $1 \in \Phi$.
3. If $\alpha \in \Phi$, then $-\alpha \in \Phi$ and if $\alpha \neq 0$, then α is an automorphism and $\alpha^{-1} \in \Phi$.

Thus the set Φ is a division subring, containing the identity endomorphism, of the ring of endomorphisms of $\mathfrak{R}$. We call $\mathfrak{R}$ a *vector space* (linear space, linear set) over Φ.

As we have remarked for commutative groups relative to an arbitrary ring of endomorphisms, any vector space $\mathfrak{R}$ over Φ may be regarded as a φ-module, where φ is the abstract ring isomorphic to Φ. The module product $x1 = x$ for all x in $\mathfrak{R}$ and 1, the identity of φ. On the other hand, suppose that φ is an arbitrary division ring and that $\mathfrak{M}$ is a φ-module in which $x1 = x$ for all x. Let Φ denote the ring of endomorphisms $x \rightarrow x\alpha$ determined by the elements α in φ. Then Φ is a homomorphic image $\neq 0$ of the division ring φ. It follows that Φ is isomorphic to φ. Hence Φ is a division ring and since Φ contains the identity mapping, Φ satisfies the assumptions 1 – 3. Thus we may also define a vector space as a φ-module such that φ is a division ring and $x1 = x$ for all x in the module.

A Φ-subgroup $\mathfrak{S}$ of a vector space $\mathfrak{R}$ over Φ is called a *subspace* of $\mathfrak{R}$. We shall restrict our attention to vector spaces that are *finite* dimensional in the sense that

4. The ascending chain condition for subspaces holds.

If x is any vector (element) in $\mathfrak{R}$, the set (x) of vectors of the form $x\alpha$, α arbitrary in Φ, is a subspace. It is irreducible. For if $x \neq 0$, and $y \neq 0$ is in a subspace $\mathfrak{S}$ of (x), then $y = x\gamma$; hence $x\alpha = y\gamma^{-1}\alpha \in \mathfrak{S}$ so that $(x) = \mathfrak{S}$. Evidently the subspaces (x) are the only irreducible ones, since any subspace contains a subspace of this form.

If $\mathfrak{S}$ is a subspace $\neq \mathfrak{R}$, let y_1 be a vector not in $\mathfrak{S}$. Then $\mathfrak{S}_1 = \mathfrak{S} + (y_1) = \mathfrak{S} \oplus (y_1)$ since (y_1) is irreducible. If $\mathfrak{S}_1 \neq \mathfrak{R}$, we may find a y_2 not in $\mathfrak{S}_1$, and then $\mathfrak{S}_2 \equiv \mathfrak{S}_1 + (y_2) = \mathfrak{S}_1 \oplus (y_2) = \mathfrak{S} \oplus (y_1) \oplus (y_2)$. Continuing in this way, we obtain an increasing chain of subspaces $\mathfrak{S}_i = \mathfrak{S} \oplus (y_1) \oplus \cdots \oplus (y_i)$. Hence, by the ascending chain condition, there is an integer r such that $\mathfrak{R} = \mathfrak{S} \oplus (y_1) \oplus \cdots \oplus (y_r)$. If we set $\mathfrak{S}' = (y_1) \oplus \cdots \oplus (y_r)$, we obtain $\mathfrak{R} = \mathfrak{S} \oplus \mathfrak{S}'$. Hence $\mathfrak{R}$ is completely reducible. If we begin with $\mathfrak{S} = 0$, we obtain a set of vectors $x_1, \cdots, x_n \neq 0$ such that $\mathfrak{R} = (x_1) \oplus \cdots \oplus (x_n)$.[1] Every vector x has a unique representation as $\Sigma x_i\xi_i$, ξ_i in Φ; for $x_i\xi_i$ is determined by x and $x_i\xi_i = x_i\eta_i$ implies that $\xi_i = \eta_i$. The x's constitute a *basis* for $\mathfrak{R}$ over Φ. By either the Jordan-Hölder theorem or the Krull-Schmidt theorem their number n, the *dimensionality* of $\mathfrak{R}$ over Φ, is an invariant. In the next section we shall obtain a direct proof of this fact.

The invariance of dimensionality implies that isomorphic φ-modules have the same dimensionality. Conversely, let $\mathfrak{R}_1$ and $\mathfrak{R}_2$ be φ-modules having the same dimensionality and let $x_1^{(i)}, \cdots, x_n^{(i)}$ be a basis for $\mathfrak{R}_i$. Then $\Sigma x_j^{(1)}\alpha_j \to \Sigma x_j^{(2)}\alpha_j$, α_j in φ, is a φ-isomorphism.

If φ is any division ring, we may construct a vector space of any dimensionality n over a ring of endomorphisms Φ isomorphic to φ. For let $\mathfrak{R}$ be the set of elements $x = (\xi_1, \cdots, \xi_n)$, ξ_i in φ. We define $x = y = (\eta_1, \cdots, \eta_n)$ if $\xi_i = \eta_i$ and

$$x + y = (\xi_1 + \eta_1, \cdots, \xi_n + \eta_n), \qquad x\alpha = (\xi_1\alpha, \cdots, \xi_n\alpha)$$

for α in φ. Then $\mathfrak{R}$ is a φ-module in which $x1 = x$. Hence $\mathfrak{R}$ is a vector space over Φ, the ring of endomorphisms $x \to x\alpha$ and Φ is isomorphic to φ. If we set $x_1 = (1, 0, \cdots, 0), \cdots, x_n = (0, \cdots, 0, 1)$, we obtain $\mathfrak{R} = (x_1) \oplus \cdots \oplus (x_n)$. Hence $\mathfrak{R}$ has a composition series and therefore $\mathfrak{R}$ satisfies both chain conditions. The dimensionality of $\mathfrak{R}$ is n. The possibility of constructing a vector space for any division ring insures the applicability of the results we shall derive for division rings of endomorphisms to arbitrary division rings.

2. Change of basis. The vectors $y_1, \cdots, y_r$ of $\mathfrak{R}$ over Φ are *linearly independent* if $\mathfrak{S} \equiv (y_1) + \cdots + (y_r) = (y_1) \oplus \cdots \oplus (y_r)$ and the y_i are $\neq 0$. An equivalent condition is that $\Sigma y_i\alpha_i = 0$ only if all the $\alpha_i = 0$. Now suppose that the y's are linearly independent and that $y_i = \Sigma x_j\beta_{ji}$ are expressions for the y's in terms of the basis $x_1, \cdots, x_n$. If ρ is any element of Φ, then the vectors $y_1, y_2 + y_1\rho, y_3, \cdots, y_r$ are linearly independent. For otherwise, $y_2 + y_1\rho \in (y_1) + (y_3) + \cdots + (y_r)$ and hence $y_2 \in (y_1) + (y_3) + \cdots + (y_r)$. We note also that $\mathfrak{S} = (y_1) + (y_2 + y_1\rho) + (y_3) + \cdots + (y_r)$. Let $\beta_{n_1 1} \neq 0$ and set $\rho = -\beta_{n_1 1}^{-1}\beta_{n_1 2}$. Then the expression for the vector $y_2^{(1)} = y_2 + y_1\rho$ does

[1] If we assume the well ordering theorem, these facts may be established without using the ascending chain condition. For let $[y_\alpha]$ be the set of vectors in $\mathfrak{R}$ where α ranges over a section of ordinal numbers. If $\mathfrak{S}$ is a subspace, we define $\mathfrak{S}_\alpha$ to be the smallest subspace containing $\mathfrak{S}$ and all y_β with $\beta < \alpha$. If $y_\alpha \notin \mathfrak{S}_\alpha$, we set $y_\alpha = x_\alpha$. Then the x's are linearly independent and generate a complement of $\mathfrak{S}$. The above discussion shows that the ascending chain condition implies the descending chain condition. Now the complete reducibility of $\mathfrak{R}$ has the consequence that the ascending chain condition is implied by the descending chain condition. This may also be proved directly (Chapter 4, **12**).

not involve x_{n_1}. Similarly, we may choose vectors $y_3^{(1)}, \cdots, y_r^{(1)}$ in $(x_1) + \cdots + (x_{n_1-1}) + (x_{n_1+1}) + \cdots + (x_n)$ such that $y_1, y_2^{(1)}, \cdots, y_r^{(1)}$ are linearly independent and $\mathfrak{S} = (y_1) + y_2^{(1)} + \cdots + y_r^{(1)}$. Now suppose that $y_k^{(1)} = \Sigma x_i \beta_{ik}^{(1)}$ and that $\beta_{n_2 2}^{(1)} \neq 0$. Then we set $y_k^{(2)} = y_k^{(1)} - y_2^{(1)} \beta_{n_2 2}^{(1)-1} \beta_{n_2 k}^{(1)}$ for $k = 3, 4, \cdots$ and note that $y_k^{(2)} \neq 0$ and that the expressions for these $y_k^{(2)}$ involve neither x_{n_1} nor x_{n_2}. Moreover, $\mathfrak{S} = (y_1) \oplus (y_2^{(1)}) \oplus (y_3^{(2)}) \oplus \cdots \oplus (y_r^{(2)})$. After a number of repetitions of this process, we obtain $\mathfrak{S} = (y_1) \oplus (y_2^{(1)}) \oplus \cdots \oplus (y_r^{(r-1)})$ where $y_i^{(i-1)} = x_{n_i}\gamma_{n_i i} + \sum_{j \neq n_s} x_j \gamma_{ji} \neq 0$, $\gamma_{n_i i} \neq 0$, $s = 1, \cdots, i$ and the n_i are distinct. The correspondence between y_i and x_{n_i} clearly shows that $r \leqq n$. If $\mathfrak{S} = \mathfrak{R}$, the y's also form a basis for $\mathfrak{R}$ and so, by symmetry, $n \leqq r$. Thus the dimensionality is an invariant.

By a slight extension of the above method we obtain a basis $z_1, \cdots, z_r$ for $\mathfrak{S}$ such that $z_i = x_{n_i}\epsilon_{n_i i} + \sum_{j \neq n_t} x_j \epsilon_{ji}$, $\epsilon_{n_i i} \neq 0$, $t = 1, \cdots, r$. By multiplying these vectors by $\epsilon_{n_i i}^{-1}$ and then arranging them in proper order, we obtain finally a basis $u_1, \cdots, u_r$ for $\mathfrak{S}$ such that $u_i = x_{n_i} + \sum_{j \neq n_t} x_j \rho_{ji}$ and $n_1 < n_2 < \cdots$. If $\mathfrak{S} = \mathfrak{R}$, the invariance of dimensionality shows that $r = n$. Then the vectors u_i are merely the original vectors x_i. Suppose, conversely, that $r = n$, i.e. the y's are linearly independent and their number is the dimensionality of $\mathfrak{R}$. Then $u_i = x_i$ and so the y's as well as the x's form a basis for $\mathfrak{R}$. It should be remarked that the passage from the basis $y_1, \cdots, y_n$ to the basis $u_1 = x_1, \cdots, u_n = x_n$ has been effected by a sequence of replacements of the following types:

I. $y_i \to y_i$ for $i \neq r$ and $y_r \to y_r + y_s\rho$, $s \neq r$.

II. $y_i \to y_i$ for $i \neq r$ and $y_r \to y_r\sigma$, $\sigma \neq 0$.

III. $y_i \to y_i$ for $i \neq r, s$ and $y_r \to y_s$, $y_s \to y_r$.

The last type is required to arrange the x's in the right order.

If $x_1, \cdots, x_n$ and $y_1, \cdots, y_n$ are any two bases for $\mathfrak{R}$, we may suppose that $y_i = \Sigma x_j \beta_{ji}$ and $x_j = \Sigma y_k \alpha_{kj}$. These expressions are unique. Since

$$y_i = \sum_{k,j} y_k \alpha_{kj} \beta_{ji}, \qquad x_j = \sum_{k,i} x_k \beta_{ki} \alpha_{ij},$$

we have

$$\sum_j \alpha_{kj} \beta_{ji} = \delta_{ki}, \qquad \sum_j \beta_{kj} \alpha_{ji} = \delta_{ki}$$

(δ_{ki}, the Kronecker delta). Thus if we set $B = (\beta_{ij})$, $A = (\alpha_{ij})$, we obtain $AB = 1 = BA$ where 1 is the identity matrix in Φ_n, the ring of $n \times n$ matrices with elements in the division ring Φ.[2]

[2] If $\mathfrak{A}$ is a ring, $\mathfrak{A}_n$ denotes the set of elements $\Sigma e_{ij} a_{ij}$, a_{ij} in $\mathfrak{A}$ where $\Sigma e_{ij} a_{ij} = \Sigma e_{ij} b_{ij}$ if and only if $a_{ij} = b_{ij}$. We set $\Sigma e_{ij} a_{ij} + \Sigma e_{ij} b_{ij} = \Sigma e_{ij}(a_{ij} + b_{ij})$ and $(\Sigma e_{ij} a_{ij})(\Sigma e_{ij} b_{ij}) = \sum e_{ij}(\sum_k a_{ik} b_{kj})$. The resulting system is a ring. The subset of elements $\Sigma e_{ii} a$ is a subring

On the other hand, suppose that $B = (\beta_{ij})$ is any matrix having a right inverse A ($BA = 1$). Set $y_i = \Sigma x_j \beta_{ji}$ where $x_1, \cdots, x_n$ is a basis for $\mathfrak{R}$. Then

$$\Sigma y_k \alpha_{kj} = \Sigma x_i \beta_{ik} \alpha_{kj} = \Sigma x_i \delta_{ij} = x_j .$$

If the y's are not linearly independent, we may choose a subset $y_1, \cdots, y_r$ of them which are and such that $\mathfrak{S} \equiv (y_1) + \cdots + (y_n) = (y_1) \oplus \cdots \oplus (y_r)$, a proper decomposition. Since $x_1, \cdots, x_n \in \mathfrak{S}$, $\mathfrak{S} = \mathfrak{R}$, and we have a contradiction to the invariance of the dimensionality. Hence the y's are linearly independent and so they form a basis for $\mathfrak{R}$, and $AB = 1$.

THEOREM 1. *If Φ is a division ring, then $AB = 1$ in Φ_n if and only if $BA = 1$.*

If the y's do not form a basis, they are linearly dependent and hence $\Sigma y_i \gamma_i = 0$, γ_i not all 0. If $y_i = \Sigma x_j \beta_{ji}$, then $\Sigma \beta_{ji} \gamma_i = 0$ so that the matrix

$$C = \begin{pmatrix} \gamma_1 & 0 \cdots 0 \\ \cdot & \cdot \cdots \cdot \\ \cdot & \cdot \cdots \cdot \\ \cdot & \cdot \cdots \cdot \\ \gamma_n & 0 \cdots 0 \end{pmatrix}$$

satisfies $BC = 0$. Conversely, if $C \neq 0$ and $BC = 0$, then the vectors $\Sigma x_j \beta_{ji}$ are linearly dependent. Thus we have proved

THEOREM 2. *If Φ is a division ring, a matrix in Φ_n is a unit if and only if it is not a left zero divisor.*

Let Φ' be a division ring anti-isomorphic to Φ and $\alpha \to \alpha'$ an anti-isomorphism between Φ and Φ'. Then $(\alpha_{ij}) = A \to A^* = (\overset{*}{\alpha}_{ij})$, $\overset{*}{\alpha}_{ij} = \alpha'_{ji}$ is an anti-isomorphism between Φ_n and Φ'_n. If $AB = 1$, then $B^*A^* = 1$ in Φ'_n and if $CB = 0$, then $B^*C^* = 0$. It follows from this that we may replace the word "left" by "right" in the above theorem.

We saw that we could pass from the basis $y_1, \cdots, y_n$ to $x_1, \cdots, x_n$ by a sequence of replacements of the types I, II and III noted above. The matrices relating the new y's to the old ones are respectively

$$\begin{matrix} \\ \\ r \\ \\ s \\ \\ \end{matrix}\begin{pmatrix} 1 & & & & \\ & \ddots & & & \\ \cdots & 1 & & & \\ & \vdots & \ddots & & \\ \cdots & \rho & \cdots & 1 & \\ & & & & \ddots \\ & & & & & 1 \end{pmatrix}, \quad \begin{pmatrix} 1 & & & & \\ & \ddots & & & \\ & & 1 & & \\ & & & \sigma & \\ & & & & 1 & \\ & & & & & \ddots \\ & & & & & & 1 \end{pmatrix}, \quad \begin{matrix} \\ \\ r \\ \\ \\ s \\ \\ \\ \end{matrix}\begin{pmatrix} 1 & & & & & \\ & \ddots & & & & \\ & & 1 & & & \\ \cdots & & 0 & \cdots & 1 & \\ & & \vdots & 1 & \vdots & \\ & & \vdots & \ddots & \vdots & \\ & & \vdots & & 1 \,\vdots & \\ \cdots & & 1 & \cdots & 0 & \\ & & & & & 1 \\ & & & & & \ddots \\ & & & & & & 1 \end{pmatrix}$$

$\bar{\mathfrak{A}}$ isomorphic to $\mathfrak{A}$. We shall identify $\bar{\mathfrak{A}}$ with $\mathfrak{A}$. If $\mathfrak{A}$ has an identity 1, $\mathfrak{A}_n$ contains elements $e_{ij} = e_{ij}1$ such that

$$e_{ij} e_{kl} = \delta_{jk} e_{il}$$

and $e_{11} + \cdots + e_{nn}$ is an identity for $\mathfrak{A}_n$. Every element a in $\mathfrak{A}_n$ may be written in one and only one way as $\Sigma e_{ij} a_{ij}$ where $e_{ij} a_{ij}$ now denotes the product of e_{ij} and a_{ij}, a_{ij} in $\mathfrak{A}$. The and are called *elementary* matrices. Now if $\{x_i\}$, $\{y_i\}$ and $\{z_i\}$ are three different bases for $\mathfrak{R}$ and $y_j = \Sigma x_i\alpha_{ij}$, $z_k = \Sigma y_j\beta_{jk}$, then $z_k = \Sigma x_i\gamma_{ik}$ where $C = (\gamma_{ij}) = AB$. Hence we have

THEOREM 3. *Any unit in Φ_n is a product of elementary matrices.*

As a matter of fact, the last type is superfluous. For if $n = 2$, we have

$$\begin{pmatrix} 0 & 1 \\ 1 & 0 \end{pmatrix} = \begin{pmatrix} 1 & 1 \\ 0 & 1 \end{pmatrix}\begin{pmatrix} 1 & 0 \\ -1 & 1 \end{pmatrix}\begin{pmatrix} 1 & 1 \\ 0 & 1 \end{pmatrix}\begin{pmatrix} -1 & 0 \\ 0 & 1 \end{pmatrix},$$

and the modification for $n > 2$ is obvious.

3. Vector spaces over different division rings. A type of vector space that occurs frequently in the theory of rings and algebras is obtained as follows. Let $\mathfrak{A}$ be a ring with an identity and φ a division subring of $\mathfrak{A}$ containing the identity. We have seen that the endomorphisms $x \rightarrow x\alpha$, for x in $\mathfrak{A}$ and α in φ, form a division ring Φ isomorphic to φ. Since Φ contains the identity endomorphism, we may regard $\mathfrak{A}$ as a vector space over Φ. In a similar manner, we may use the endomorphisms $x \rightarrow \alpha x \equiv x\alpha'$ and obtain a division ring Φ' of endomorphisms anti-isomorphic to φ.

Now suppose that $\mathfrak{R}$ is any vector space over Φ and that Σ is a division subring of Φ. We denote the set of endomorphisms $\xi \rightarrow \xi\alpha$ in Φ, α in Σ, by Σ also. Suppose first that the dimensionality $(\mathfrak{R}:\Phi)$ is n and that $(\Phi:\Sigma) = m$. Then if $x_1, \cdots, x_n$ is a basis for $\mathfrak{R}$ over Φ and $\xi_1, \cdots, \xi_m$ is a basis for Φ over Σ, every vector in $\mathfrak{R}$ has one and only one representation as $\Sigma x_i\xi_j\alpha_{ij}$, α_{ij} in Σ. Hence the $x_i\xi_j$ form a basis of mn elements for $\mathfrak{R}$ over Σ. Conversely, if $\mathfrak{R}$ over Σ is finite dimensional, it is evident that $\mathfrak{R}$ over Φ is also finite dimensional since $\Phi \geqq \Sigma$. Furthermore if x is any vector $\neq 0$ in $\mathfrak{R}$, the set $x\xi$, ξ arbitrary in Φ, is a subspace of $\mathfrak{R}$ over Σ. If $x\xi_1, \cdots, x\xi_m$ is a basis for this space, then $\xi_1, \cdots, \xi_m$ is a basis for Φ over Σ and so Φ is finite dimensional over Σ. We have therefore proved

THEOREM 4. *If $\mathfrak{R}$ is a vector space of finite dimensionality over Φ and Σ is a division subring of Φ such that Φ over Σ is finite, then $(\mathfrak{R}:\Sigma) = (\mathfrak{R}:\Phi)(\Phi:\Sigma)$. Conversely, if $\mathfrak{R}$ is any vector space over Φ and $\mathfrak{R}$ is finite over Σ, a division subring of Φ, then Φ is finite over Σ.*

The same result holds for Σ', the set of endomorphisms $\xi \rightarrow \alpha\xi$. In the remainder of this chapter we consider a fixed vector space $\mathfrak{R}$ over a fixed division ring Φ.

4. The ring of linear transformations. A Φ-endomorphism A of $\mathfrak{R}$ is called a *linear transformation* of $\mathfrak{R}$ over Φ. Now, in any ring, the totality of elements that commute with the elements of a fixed subset of the ring form a subring.

elements of $\mathfrak{A}$ commute with the e_{ij}. Conversely, if $\mathfrak{B}$ is any ring containing an identity 1 and elements e_{ij} satisfying the above conditions and $\Sigma e_{ii} = 1$; and if $\mathfrak{B}$ contains a subring $\mathfrak{A}$ such that 1) $1 \in \mathfrak{A}$, 2) $ae_{ij} = e_{ij}a$ for all a in $\mathfrak{A}$ and 3) every element of $\mathfrak{B}$ may be written in one and only one way in the form $\Sigma e_{ij}a_{ij}$, then $\mathfrak{B} \cong \mathfrak{A}_n$.

Hence the set of linear transformations is a subring $\mathfrak{L}$ of the ring of endomorphisms.

If A is a linear transformation and $x_1, \cdots, x_n$ is a basis for $\mathfrak{R}$ over Φ, A is completely determined by the images x_iA of the x's. For if $x = \Sigma x_i\xi_i$, we have $xA = \Sigma(x_iA)\xi_i$. On the other hand, we may choose n elements y_i at random and verify that the mapping $\Sigma x_i\xi_i \rightarrow \Sigma y_i\xi_i$ is a linear transformation A such that $x_iA = y_i$. In particular, for each α in Φ there is a unique linear transformation α' such that $x_i\alpha' = x_i\alpha$. Of course, α' depends on the choice of the basis as well as on α. The α''s form a subring Φ' of $\mathfrak{L}$ anti-isomorphic to Φ, the correspondence $\alpha \rightarrow \alpha'$ being an anti-isomorphism.

Now we may regard $\mathfrak{R}$ as a vector space over Φ'. Since every x may be written in one and only one way in the form $\Sigma x_i\xi_i'$, $x_1, \cdots, x_n$ is a basis for $\mathfrak{R}$ over Φ' and so $\mathfrak{R}$ is n dimensional over Φ'. The endomorphisms α are linear transformations in $\mathfrak{R}$ over Φ' and since $x_i\alpha = x_i\alpha'$, α is the endomorphism associated with α' and the x's in the same way that α' is associated with α and the x's, i.e. $(\alpha')' = \alpha$.

Let E_{ij} denote the linear transformation of $\mathfrak{R}$ over Φ such that $x_rE_{ij} = \delta_{ir}x_j$. Since $x_r(E_{ij}\alpha') = x_r(\alpha'E_{ij})$, $E_{ij}\alpha' = \alpha'E_{ij}$, and E_{ij} is also linear in $\mathfrak{R}$ over Φ'. Now suppose that A is an arbitrary linear transformation in $\mathfrak{R}$ over Φ and that $x_rA = \Sigma x_j\alpha_{rj}$. Then, as is readily verified, A and $\Sigma E_{ij}\alpha'_{ij}$ have the same effect on the x's. Hence $A = \Sigma E_{ij}\alpha'_{ij}$. Conversely, if $A = \Sigma E_{ij}\alpha'_{ij}$, A is in $\mathfrak{L}$ and $x_rA = \Sigma x_j\alpha_{rj}$. It follows that every A in $\mathfrak{L}$ may be represented in one and only one way in the form $\Sigma E_{ij}\alpha'_{ij}$, α' in Φ'. Since

$$(1) \qquad E_{ij}E_{kl} = \delta_{jk}E_{il}, \qquad \Sigma E_{ii} = 1,$$

$\mathfrak{L} = \Phi'_n$.

By (1) we obtain

$$(2) \qquad \alpha'_{pq} = \sum_k E_{kp}(\Sigma E_{ij}\alpha'_{ij})E_{qk}.$$

Hence if $A = \Sigma E_{ij}\alpha'_{ij}$ commutes with all the E_{kl}, then $\alpha'_{pq} = 0$ if $p \neq q$ and $\alpha'_{pp} = A$, i.e. $A = \alpha' \in \Phi'$. Thus Φ' may be characterized as the set of elements of $\mathfrak{L}$ commutative with all the E_{ij}. We may also characterize Φ' as the complete set of endomorphisms of $\mathfrak{R}$ commutative with all the endomorphisms α and all the E_{ij}, or, more simply, with all the endomorphisms $\Sigma E_{ij}\alpha_{ij}$. For, the condition that A commute with Φ is that $A \in \mathfrak{L}$. In a similar manner we see that Φ is the complete set of $\mathfrak{L}$-endomorphisms of $\mathfrak{R}$. In particular the center[3] $\mathfrak{C}$ of $\mathfrak{L}$ is contained in Φ. Since $\mathfrak{C} \leqq \mathfrak{L}$ and the elements of $\mathfrak{C}$ commute with the E_{ij}, $\mathfrak{C} \leqq \Phi'$. It follows that $\mathfrak{C} = \Phi \wedge \Phi'$. If Φ is commutative, the transformation $\alpha' = \alpha$ and, therefore, α' is independent of the choice of the basis. The field Φ is then the center of $\mathfrak{L}$. We return to the general case where Φ is a division ring, and we shall use the specific form of $\mathfrak{L}$ to obtain a number of important structure theorems.

[3] We use this term for the set of elements of a ring commutative with all the elements of the ring.

THEOREM 5. *The ring $\mathfrak{L}$ is simple.*

We recall that a ring is *simple* if it has no proper two-sided ideals. Let $\mathfrak{B} \neq 0$ be a two-sided ideal of $\mathfrak{L} = \Phi'_n$. If $\Sigma E_{ij}\beta'_{ij} = B$ is an element $\neq 0$ in $\mathfrak{B}$, by (2) we obtain that $\beta'_{ij} \epsilon \mathfrak{B}$. At least one of these, say β'_{pq} is $\neq 0$. It follows that $1 = \beta'_{pq}\beta'^{-1}_{pq} \epsilon \mathfrak{B}$. Hence $\mathfrak{B} = \mathfrak{L}$.

THEOREM 6. *The ring $\mathfrak{L}$ is a direct sum of irreducible right (left) ideals.*

The set $E_{kk}\mathfrak{L}$ consisting of the elements $\sum_j E_{kj}\alpha'_{kj}$ is an irreducible right ideal. For if $\mathfrak{J}$ is a right ideal $\neq 0$ contained in $E_{kk}\mathfrak{L}$, let $B = \sum_j E_{kj}\beta'_{kj} \epsilon \mathfrak{J}$ where $\beta'_{kl} \neq 0$. Then $\mathfrak{J}$ contains $BE_{lk}\beta'^{-1}_{kl} = E_{kk}$ and hence all the elements of $E_{kk}\mathfrak{L}$. Evidently $\mathfrak{L} = E_{11}\mathfrak{L} \oplus \cdots \oplus E_{nn}\mathfrak{L}$.

5. Automorphisms and anti-automorphisms of $\mathfrak{L}$. Let F_{ij} be n^2 linear transformations such that

$$F_{ij}F_{kl} = \delta_{jk}F_{il}, \qquad F_{11} + \cdots + F_{nn} = 1. \tag{3}$$

If y is a vector $\neq 0$, there is an F_{pp} such that $yF_{pp} \neq 0$. It follows that the vectors $y_i = yF_{pi}$ form a basis for $\mathfrak{R}$ over Φ. For if $\Sigma y_i\beta_i = 0$, $(\Sigma y_i\beta_i)F_{jp} = \sum_i y(F_{pi}F_{jp})\beta_i = (yF_{pp})\beta_j = 0$ and hence $\beta_j = 0$ for $j = 1, \cdots, n$. Relative to the y_i, we have

$$y_rF_{ij} = yF_{pr}F_{ij} = \delta_{ir}yF_{pj} = \delta_{ir}y_j. \tag{4}$$

If S is the linear transformation such that $x_iS = y_i$, S^{-1} is defined by $y_iS^{-1} = x_i$. From (4) and the definition of the E_{ij} we obtain that $F_{ij} = S^{-1}E_{ij}S$. An important application of this result is the following theorem.

THEOREM 7. *Any automorphism of $\mathfrak{L}$ has the form $\Sigma E_{ij}\alpha'_{ij} \to S^{-1}(\Sigma E_{ij}\alpha'^s_{ij})S$ where $\alpha' \to \alpha'^s$ is an automorphism in Φ'.*

Let G be an automorphism of $\mathfrak{L} = \Phi'_n$. Then the transformations $F_{ij} = E^G_{ij}$ satisfy (3) and hence there is an S in $\mathfrak{L}$ such that $E^G_{ij} = S^{-1}E_{ij}S$. The mapping $A \to SA^GS^{-1} \equiv A^H$ is an automorphism in $\mathfrak{L}$ such that $E^H_{ij} = E_{ij}$. Since Φ' is the complete set of elements commutative with the E_{ij}, it follows that H induces an automorphism s in Φ' and hence $(\Sigma E_{ij}\alpha'_{ij})^H = \Sigma E_{ij}\alpha'^s_{ij}$. Then $A^G = S^{-1}(\Sigma E_{ij}\alpha'^s_{ij})S$.

Now let J be an anti-automorphism in $\mathfrak{L}$. Since the transformations $F_{ij} = E^J_{ji}$ satisfy (3), there exists an S in $\mathfrak{L}$ such that $E^J_{ji} = S^{-1}E_{ij}S$. The correspondence $A \to SA^JS^{-1} \equiv A^K$ is an anti-automorphism sending E_{ij} into E_{ji}. It therefore induces an anti-automorphism t in Φ' and $(\Sigma E_{ij}\alpha'_{ij})^J = S^{-1}(\Sigma E_{ji}\alpha'^t_{ij})S$.

THEOREM 8. *The ring $\mathfrak{L}$ has an anti-automorphism if and only if Φ has an anti-automorphism. Moreover if $\mathfrak{L}$ has an anti-automorphism J, then $A^J = S^{-1}(\Sigma E_{ji}\alpha'^t_{ij})S$ where $S \epsilon \mathfrak{L}$ and $\alpha' \to \alpha'^t$ is an anti-automorphism in Φ'.*[4]

[4] The ring Φ has an anti-automorphism if and only if Φ' has.

If J is involutorial in the sense that $J^2 = 1$, then $E_{ij} = E_{ij}^{J^2} = (S^J S^{-1})E_{ij}S(S^J)^{-1}$. Hence $S^J = \sigma' S$, σ' in Φ'. If $\sigma' \neq -1$, $S^J + S = (\sigma' + 1)S \equiv T$ and $(\sigma' + 1)$ has an inverse in Φ'. Then $(\Sigma E_{ij}\alpha'_{ij})^J = T^{-1}(\Sigma E_{ji}\alpha'^{u}_{ij})T$ where $T^J = T$ and $\alpha'^u = (\sigma' + 1)\alpha'^t(\sigma' + 1)^{-1}$. Thus we may suppose at the start that $A^J = S^{-1}A^K S$ where $S^J = \pm S$, $A^K = \Sigma E_{ji}\alpha'^{t}_{ij}$. Since $A^{K^2} = S(SA^J S^{-1})^J S^{-1} = A$, K is involutorial. Hence t is involutorial. We note finally that $S^J = S^{-1}S^K S = \pm S$ implies that $S^K = \pm S$.

THEOREM 9. *The ring $\mathfrak{L}$ has an involutorial anti-automorphism if and only if Φ has an involutorial anti-automorphism. If $\mathfrak{L}$ has an involutorial anti-automorphism J, then $A^J = S^{-1}(\Sigma E_{ji}\alpha'^{t}_{ij})S$ where $\alpha' \rightarrow \alpha'^t$ is an involutorial anti-automorphism in Φ', and $S \in \mathfrak{L}$ and satisfies the equations $\sigma'_{ij} = \pm\sigma'^{t}_{ji}$.*

We consider now the special case of a commutative Φ and we obtain as a corollary to the above theorems the following

THEOREM 10. *If Φ is a field, then any automorphism in Φ_n that leaves the elements of the center Φ invariant is inner. Any anti-automorphism of Φ_n leaving the elements of Φ invariant has the form $A \rightarrow S^{-1}A'S \equiv A^J$, A' the transpose of A. The anti-automorphism J is involutorial if and only if $S' = \pm S$.*

6. Commuting rings of endomorphisms. Suppose that $n = rs$ and define

$$G_{\alpha\beta} = \sum_{\mu=0}^{s-1} E_{\mu r+\alpha,\mu r+\beta}, \qquad \alpha, \beta = 1, \cdots, r,$$

$$H_{\kappa\lambda} = \sum_{\gamma=1}^{r} E_{(\kappa-1)r+\gamma,(\lambda-1)r+\gamma}, \qquad \kappa, \lambda = 1, \cdots, s.$$

One readily verifies that

$$G_{\alpha\beta}G_{\gamma\delta} = \delta_{\beta\gamma}G_{\alpha\delta}, \qquad G_{11} + \cdots + G_{rr} = 1, \tag{5}$$

$$H_{\kappa\lambda}H_{\mu\nu} = \delta_{\lambda\mu}H_{\kappa\nu}, \qquad H_{11} + \cdots + H_{ss} = 1, \tag{6}$$

$$G_{\alpha\beta}H_{\kappa\lambda} = E_{(\kappa-1)r+\alpha,(\lambda-1)r+\beta} = H_{\kappa\lambda}G_{\alpha\beta}. \tag{7}$$

By (7) every element of $\mathfrak{L}$ has the form $\Sigma G_{\alpha\beta}B_{\alpha\beta}$ where $B_{\alpha\beta}$ is a sum $\Sigma H_{\kappa\lambda}\beta'_{\kappa\lambda}$, β' in Φ', and if $\Sigma G_{\alpha\beta}B_{\alpha\beta} = 0$, $B_{\alpha\beta} = 0$. It follows that $(\Phi'_s)_r \cong \Phi'_{sr}$. Similarly, every element has one and only one expression of the type $\Sigma H_{\kappa\lambda}C_{\kappa\lambda}$ where $C_{\kappa\lambda}$ has the form $\sum G_{\alpha\beta}\gamma'_{\alpha\beta}$, γ' in Φ'. If $A = \sum G_{\alpha\beta}B_{\alpha\beta}$, $B_{\alpha\beta} = \sum_{\gamma} G_{\gamma\alpha}AG_{\beta\gamma}$. Hence the condition that A commute with all $G_{\alpha\beta}$ is that $A = B_{\alpha\alpha} = B_{\beta\beta} = \Sigma H_{\kappa\lambda}\beta'_{\kappa\lambda}$. Similarly, if A commutes with all $H_{\kappa\lambda}$ it has the form $\Sigma G_{\alpha\beta}\gamma'_{\alpha\beta}$.

Now let Φ_r denote the ring of endomorphisms in $\mathfrak{R}$ of the form $\Sigma G_{\alpha\beta}\rho_{\alpha\beta}$, ρ in Φ. A Φ_r-endomorphism is a Φ-endomorphism which commutes with all the $G_{\alpha\beta}$. Hence it is an element of $\mathfrak{L}$ commutative with all the $G_{\alpha\beta}$. It therefore belongs to Φ'_s, the set of endomorphisms of the form $\Sigma H_{\kappa\lambda}\beta'_{\kappa\lambda}$. Because of the symmetry, Φ_r may be characterized as the set of Φ'_s-endomorphisms of $\mathfrak{R}$.

Now suppose that $\mathfrak{R}$ is any commutative group in which there is defined a ring of endomorphisms of the type Φ_r where Φ is a division ring containing the identity endomorphism. Our assumptions are therefore that 1) there are r^2

endomorphisms $G_{\alpha\beta}$ in Φ_r such that (5) holds, 2) $G_{\alpha\beta}\rho = \rho G_{\alpha\beta}$ for any ρ in Φ and 3) every endomorphism of Φ_r has a unique representation in the form $\Sigma G_{\alpha\beta}\rho_{\alpha\beta}$. We suppose also that the ascending chain condition holds for Φ_r-subgroups. Since $\Phi_r \geqq \Phi \geqq 1$, $\mathfrak{R}$ is a vector space over Φ, though it is not clear *a priori* that $\mathfrak{R}$ is finite dimensional.

Let x be an element $\neq 0$ in $\mathfrak{R}$. Since $\Sigma G_{\alpha\alpha} = 1$, there is a $G_{\delta\delta}$ such that $xG_{\delta\delta} \neq 0$. Set $x_\alpha = xG_{\delta\alpha}$. Then these elements are linearly independent over Φ. If $\mathfrak{R}_1$ denotes the set of elements $\Sigma x_\alpha\rho_\alpha$, $\rho \in \Phi$, $\mathfrak{R}_1$ is a Φ_r-subgroup. If $\mathfrak{R}_1 \neq \mathfrak{R}$, let y be a vector not in $\mathfrak{R}_1$. As before, there is a $G_{\epsilon\epsilon}$ such that $yG_{\epsilon\epsilon}$ is not in $\mathfrak{R}_1$ and if $x_{r+\alpha} = yG_{\epsilon\alpha}$, the elements $\Sigma x_{r+\alpha}\rho_\alpha$ form a Φ_r-subgroup $\mathfrak{R}_2$ independent of $\mathfrak{R}_1$ in the sense that $\mathfrak{R}_1 \wedge \mathfrak{R}_2 = 0$. If $\mathfrak{R}_1 + \mathfrak{R}_2 \neq \mathfrak{R}$, we may repeat the process thereby obtaining a chain $\mathfrak{R}_1 < \mathfrak{R}_1 + \mathfrak{R}_2 < \cdots$. By the finiteness assumption this breaks off and we obtain $\mathfrak{R} = \mathfrak{R}_1 \oplus \cdots \oplus \mathfrak{R}_s$. Hence $\mathfrak{R}$ has finite dimensionality $n = rs$ over Φ.

THEOREM 11. *Let $\mathfrak{R}$ be a commutative group and Φ_r a matrix ring of endomorphisms in $\mathfrak{R}$ where Φ is a division ring containing* 1. *If $\mathfrak{R}$ satisfies the ascending chain condition for Φ_r-subgroups, then it has finite dimensionality $n = rs$ over Φ and $\mathfrak{R} = \mathfrak{R}_1 \oplus \cdots \oplus \mathfrak{R}_s$ where the $\mathfrak{R}_i$ are irreducible Φ_r-subgroups.*

The irreducibility of $\mathfrak{R}_i$ is seen as follows: If z is any vector $\neq 0$ in $\mathfrak{R}$, there is a ζ such that $zG_{\zeta 1}, \cdots, zG_{\zeta r}$ are independent over Φ. Hence the set $z\Phi_r = \mathfrak{R}_i$ for any $z \neq 0$ in $\mathfrak{R}_i$.

If we use the basis $x_1, \cdots, x_n$ determined above for $\mathfrak{R}$ over Φ and define the linear transformations E_{ij} as before, we obtain $G_{\alpha\beta} = \sum_{\mu=0}^{s-1} E_{\mu r+\alpha,\mu r+\beta}$. For, $G_{\alpha\beta}$ is linear in $\mathfrak{R}$ over Φ and it has the same effect as $\sum_{\mu=0}^{s-1} E_{\mu r+\alpha,\mu r+\beta}$ on the x_i. Hence we may apply the above discussion to obtain the following

THEOREM 12. *Let Φ_r be a matrix ring of endomorphisms in a commutative group $\mathfrak{R}$ such that the conditions of the preceding theorem hold. Then the ring of endomorphisms commutative with the given endomorphisms has the form Φ'_s where Φ' is a division ring anti-isomorphic to Φ. The original set of endomorphisms Φ_r is the complete set commutative with those in Φ'_s.*

7. Isomorphism of matrix rings. Suppose now that we have a ring that may be regarded as a matrix ring Φ'_n and as a Ψ'_r where Φ' and Ψ' are division rings containing the identity. Then we may suppose that Φ'_n is the ring of linear transformations of an n-dimensional vector space $\mathfrak{R}$ over Φ, Φ anti-isomorphic to Φ'. Let $G_{\alpha\beta}$, $\alpha, \beta = 1, \cdots, r$, be the matrix units of Ψ'_r. The endomorphisms $\Sigma G_{\alpha\beta}\rho_{\alpha\beta}$, ρ in Φ, form a ring Φ_r and we have seen that the dimensionality of $\mathfrak{R}$ over Φ is $rs = n$. Hence $r \leqq n$. By reversing the roles of Φ' and Ψ' we obtain $n \leqq r$ and hence $r = n$. It follows that there is a linear transformation S such that $G_{ij} = S^{-1}E_{ij}S$, E_{ij} the matrix units for Φ'_n. Since the elements ψ' in Ψ' may be characterized as the linear transformations commuting with the G_{ij} and those of Φ', as the linear transformations commuting with the E_{ij}, we have $\Psi' = S^{-1}\Phi'S$.

THEOREM 13. *If $\Phi'_n = \Psi'_r$ where Φ' and Ψ' are division rings, then $r = n$, Φ' and Ψ' are isomorphic, and there exists an S in Φ'_n such that $\Psi' = S^{-1}\Phi' S$ and $G_{ij} = S^{-1}E_{ij}S$ for the corresponding matrix units.*

8. Semi-linear transformations. We shall now discuss a type of transformation of a vector space, first considered by C. Segre, that forms a generalization of the concept of a linear transformation. Let S be an automorphism of the division ring Φ. Then a transformation T of $\mathfrak{R}$ is called a *semi-linear transformation* of $\mathfrak{R}$ over Φ if

$$(8) \qquad (x + y)T = xT + yT, \qquad (x\alpha)T = (xT)\alpha^S$$

for all x, y in $\mathfrak{R}$ and all α in Φ. If $T \neq 0$, S is uniquely determined by T. For then there exists a vector u such that $uT \neq 0$, and if S and S' are automorphisms of Φ for which (8) holds, then $(uT)\alpha^S = (uT)\alpha^{S'}$ for all α. Hence $\alpha^S = \alpha^{S'}$ and $S = S'$. We shall call S the *automorphism* of T.

The condition $\alpha T = T\alpha^S$ evidently implies that the endomorphism T commutes with the set of endomorphisms Φ. If $S = 1$, T, of course, commutes with the individual members of Φ and T is a linear transformation. The commutativity with Φ implies that a semi-linear transformation transforms any subspace $\mathfrak{S}$ of $\mathfrak{R}$ into another subspace. It is also clear that if $\mathfrak{S}_1$ and $\mathfrak{S}_2$ are subspaces such that $\mathfrak{S}_1 \leqq \mathfrak{S}_2$, then the image $\mathfrak{S}_1T \leqq$ the image $\mathfrak{S}_2T$. Since for any two subspaces $\mathfrak{S}_1$, $\mathfrak{S}_2$, $\mathfrak{S}_1 + \mathfrak{S}_2$ may be characterized as the smallest subspace containing $\mathfrak{S}_1$ and $\mathfrak{S}_2$, it follows that if T is a $(1 - 1)$ semi-linear transformation, then $(\mathfrak{S}_1 + \mathfrak{S}_2)T = \mathfrak{S}_1T + \mathfrak{S}_2T$. In a similar manner $(\mathfrak{S}_1 \wedge \mathfrak{S}_2)T = \mathfrak{S}_1T \wedge \mathfrak{S}_2T$. Thus any $(1 - 1)$ semi-linear transformation of a vector space $\mathfrak{R}$ induces a lattice isomorphism in the lattice of subspaces of $\mathfrak{R}$. For this reason the semi-linear transformations are on a par with linear transformations in projective geometry.

If T is an arbitrary semi-linear transformation, we let $\mathfrak{N} = \mathfrak{N}(T)$ denote the space of vectors z such that $zT = 0$ and we suppose that $z_1, \cdots, z_r$ is a basis for this space. We determine a subspace $\mathfrak{S}$ such that $\mathfrak{R} = \mathfrak{N} \oplus \mathfrak{S}$ and let $y_1, \cdots, y_{n-r}$ be a basis for $\mathfrak{S}$. Then it follows directly that $y_1T, \cdots, y_{n-r}T$ is a basis for $\mathfrak{R}T$. Hence if we call the dimensionality of $\mathfrak{R}T$ the *rank* of T and the dimensionality of $\mathfrak{N}$ the *nullity* of T, then we have the following extension of the well-known theorem on linear equations:

$$\text{rank } T + \text{nullity } T = n.$$

If $x_1, \cdots, x_n$ is a basis for $\mathfrak{R}$ over Φ, we may write $x_iT = \Sigma x_j\tau_{ji}$ and call (τ_{ij}) the *matrix* of T relative to this basis. The semi-linear transformation T is determined by its matrix and its automorphism since $(\Sigma x_i\xi_i)T = \Sigma x_j\tau_{ji}\xi_i^S$. In terms of the coordinates $(\xi_1, \cdots, \xi_n)$ of the vector x, we may describe T as the transformation that sends $(\xi_1, \cdots, \xi_n)$ into $(\eta_1, \cdots, \eta_n)$ where $\eta_j = \Sigma\tau_{ji}\xi_i^S$. Now if (τ_{ij}) is an arbitrary matrix and S is any automorphism, then the equation $(\Sigma x_i\xi_i)T = \Sigma x_j\tau_{ji}\xi_i^S$ defines a semi-linear transformation with automorphism S and with matrix (τ_{ij}) relative to the basis $x_1, \cdots, x_n$.

If $y_1, \cdots, y_n$ is a second basis for $\mathfrak{R}$ over Φ and $y_i = \Sigma x_j\beta_{ji}$, a simple computation shows that the matrix of the semi-linear transformation T relative to this basis is $(\beta)^{-1}(\tau)(\beta^S)$ where (τ) is its matrix relative to the x's. Hence the theory of semi-linear transformations corresponds to a theory of matrices with elements in a division ring in which two matrices (τ) and (σ) are regarded as equivalent if there exists a matrix (β) such that $(\sigma) = (\beta)^{-1}(\tau)(\beta^S)$, S, a fixed automorphism.

Now suppose that Γ is a set of semi-linear transformations. If $\mathfrak{S}$ is a proper subspace of $\mathfrak{R}$ invariant under all the T in Γ, we may choose a basis $y_1, \cdots, y_n$ for $\mathfrak{R}$ such that $y_1, \cdots, y_r$ is a basis for $\mathfrak{S}$ $(0 < r < n)$. Then the matrix of T relative to this basis is of the form $\begin{pmatrix} \tau_1 & * \\ 0 & \tau_2 \end{pmatrix}$ where (τ_1) is a matrix of T in $\mathfrak{S}$ and (τ_2) is a matrix of T in the difference space $\mathfrak{R} - \mathfrak{S}$. Conversely, if there exists a basis relative to which the matrices of Γ have this "reduced" form, then $\mathfrak{R}$ is reducible when regarded as a group relative to $\Omega = (\Gamma, \Phi)$ the logical sum of Γ and Φ. In view of the relation between the matrices of a semi-linear transformation, we may state this condition also in the following way: If (τ) is the matrix of T relative to the basis $x_1, \cdots, x_n$ and S is the automorphism of T, then there exists a matrix (β) independent of T such that $(\beta)^{-1}(\tau)(\beta^S) = \begin{pmatrix} \tau_1 & * \\ 0 & \tau_2 \end{pmatrix}$.

Now if $\mathfrak{R} = \mathfrak{R}_s > \mathfrak{R}_{s-1} > \cdots > \mathfrak{R}_1 > 0$ is a composition series for $\mathfrak{R}$ relative to Ω, we choose a basis $y_1, \cdots, y_n$ of $\mathfrak{R}$ over Φ such that $y_1, \cdots, y_{n_1}$ is a basis for $\mathfrak{R}_1$, $y_{n_1+1}, \cdots, y_{n_1+n_2}$ is a basis for $\mathfrak{R}_2$, etc. Then if (β) is the matrix relating the y's to the x's, the matrix of T relative to the y's is

$$(\beta)^{-1}(\tau)(\beta^S) = \begin{bmatrix} \tau_1 & & & * \\ & \tau_2 & & \\ & & \ddots & \\ 0 & & & \tau_s \end{bmatrix} \tag{9}$$

where (τ_i) is a matrix of the semi-linear transformation induced by T in $\mathfrak{R}_i - \mathfrak{R}_{i-1}$ and the blocks below the "diagonal" consist of 0's. The irreducibility of $\mathfrak{R}_i - \mathfrak{R}_{i-1}$ amounts to the following matrix irreducibility: it is impossible to find a matrix (β_i) independent of T such that $(\beta_i)^{-1}(\tau_i)(\beta_i^S)$ has the reduced form $\begin{pmatrix} \tau_{i1} & * \\ 0 & \tau_{i2} \end{pmatrix}$. Conversely if (β) is any matrix for which (9) holds, where the (τ_i) are irreducible, then $(\beta)^{-1}(\tau)(\beta^S)$ arises from a composition series in the way indicated.

In a similar fashion we see that if $\mathfrak{R} = \mathfrak{R}_1 \oplus \cdots \oplus \mathfrak{R}_s$ where $\mathfrak{R}_i \neq 0$ is an Ω-subgroup, there is a matrix (β) such that

$$(\beta)^{-1}(\tau)(\beta^S) = \begin{bmatrix} \tau_1 & & & 0 \\ & \tau_2 & & \\ & & \ddots & \\ 0 & & & \tau_s \end{bmatrix}$$

for all T where, in this case, we have 0's on both sides of the diagonal, and where (τ_i) is the matrix of the induced transformation T in $\mathfrak{N}_i$.

We note the following combinatorial properties of the set of semi-linear transformations. If T_1 and T_2 are semi-linear transformations with automorphisms S_1 and S_2, respectively, then T_1T_2 is a semi-linear transformation with automorphism S_1S_2. If $S_1 = S_2 = S$, then $T_1 + T_2$ is a semi-linear transformation with automorphism S and if T_1 is $(1 - 1)$, T_1^{-1} is a semi-linear transformation with automorphism S_1^{-1}.

Now if (τ_1) and (τ_2) are the matrices, relative to the same basis, of T_1 and T_2, then the matrix of T_1T_2 is $(\tau_2)(\tau_1^{S_2})$. Thus the matrix of T^k is $(\tau)(\tau^S) \cdots (\tau^{S^{k-1}})$. If T is $(1 - 1)$, the matrix of T^{-1} is $(\tau^{S^{-1}})^{-1}$. If T_1 and T_2 have the same automorphism, then the matrix of $T_1 + T_2$ is $(\tau_1) + (\tau_2)$. As a special case of these facts we note that the correspondence between a linear transformation T and its matrix (τ) is an anti-automorphism between the ring of linear transformations $\mathfrak{L}$ and the ring of matrices Φ_n. Since Φ is associated in an invariant manner with $\mathfrak{L}$, this correspondence has certain advantages over the one noted previously between $\mathfrak{L}$ and Φ_n'.

As applications of our computations and of the results of the first chapter we note the following theorems on matrices in Φ_n.

THEOREM 14. *If (ϵ_i) $(i = 1, \cdots, s)$ are matrices $\neq 0$ in Φ_n such that*

$$(\epsilon_i)^2 = (\epsilon_i), \qquad (\epsilon_i)(\epsilon_j) = 0 \quad \text{if} \quad i \neq j, \qquad \Sigma(\epsilon_i) = 1,$$

then there exists a non-singular matrix (β) in Φ_n such that

$$(10) \qquad (\beta)^{-1}(\epsilon_1)(\beta) = \begin{pmatrix} \left.\begin{matrix} 1 & & \\ & \ddots & \\ & & 1 \end{matrix}\right\} n_1 & \\ & \begin{matrix} 0 & & \\ & \ddots & \\ & & 0 \end{matrix} \end{pmatrix}, \qquad (\beta)^{-1}(\epsilon_2)(\beta) = \begin{pmatrix} \left.\begin{matrix} 0 & & \\ & \ddots & \\ & & 0 \end{matrix}\right\} n_1 & & \\ & \left.\begin{matrix} 1 & & \\ & \ddots & \\ & & 1 \end{matrix}\right\} n_2 & \\ & & \begin{matrix} 0 & & \\ & \ddots & \\ & & 0 \end{matrix} \end{pmatrix}, \cdots.$$

This is obtained by using the (ϵ_i) to define linear transformations E_i such that

$$E_i^2 = E_i \neq 0, \qquad E_iE_j = 0 \quad \text{if} \quad i \neq j, \qquad \Sigma E_i = 1.$$

Then $\mathfrak{R} = \mathfrak{R}E_1 \oplus \cdots \oplus \mathfrak{R}E_s$ and so, relative to a suitable basis, we obtain the matrices (10) for $E_1, E_2, \cdots$.

If T is a semi-linear transformation, we may use the Remark following Fitting's lemma to obtain a decomposition $\mathfrak{R} = \mathfrak{N} \oplus \mathfrak{S}$ where $\mathfrak{N}$ and $\mathfrak{S}$ are subspaces invariant under T and T is nilpotent in $\mathfrak{N}$ and non-singular in $\mathfrak{S}$. This implies the following

THEOREM 15. *If (τ) is a matrix in Φ_n and S is an automorphism in Φ, there exists a matrix (β) such that*

$$(\beta)^{-1}(\tau)(\beta^S) = \begin{pmatrix} \nu & 0 \\ 0 & \sigma \end{pmatrix}$$

where $(\nu) \cdots (\nu^{S^k}) = 0$ for sufficiently large k and (σ) is a unit.

CHAPTER 3

NON-COMMUTATIVE PRINCIPAL IDEAL DOMAINS

1. Definitions and examples. In studying a linear transformation or, more generally, a semi-linear transformation T with automorphism S, we are usually interested in the ring of transformations $\Phi[T]$ generated by T and the scalar multiplications $x \to x\alpha$. Evidently $\Phi[T]$ contains the transformations $\alpha_0 + T\alpha_1 + T^2\alpha_2 + \cdots + T^m\alpha_m$. On the other hand, $(T^k\alpha)(T^l\beta) = T^{k+l}\alpha^{S^l}\beta$ and hence the set of polynomials in T is closed under multiplication. It follows that $\Phi[T]$ coincides with the set of these polynomials. Now, it is convenient to introduce a certain ring $\Phi[t, S]$ of polynomials in an indeterminate t. First, let Φ be the abstract division ring isomorphic to the ring of endomorphisms Φ in the vector space $\mathfrak{R}$.[1] Let $\Phi[t, S]$ denote the set of polynomials

$$\alpha_0 + t\alpha_1 + t^2\alpha_2 + \cdots + t^m\alpha_m$$

where t is an indeterminate and the coefficients α_i are in Φ. We define $\alpha_0 + t\alpha_1 + \cdots + t^m\alpha_m = \beta_0 + t\beta_1 + \cdots + t^{m'}\beta_{m'}$ if $\alpha_0 = \beta_0$, $\alpha_1 = \beta_1$, $\cdots$, and we add polynomials according to the rule $(\alpha_0 + t\alpha_1 + \cdots) + (\beta_0 + t\beta_1 + \cdots) = (\alpha_0 + \beta_0) + t(\alpha_1 + \beta_1) + \cdots$. Multiplication is defined by means of the distributive law and

$$(t^k\alpha)(t^l\beta) = t^{k+l}\alpha^{S^l}\beta.$$

It is readily verified that $\Phi[t, S]$ is a ring.

Now, with the polynomial $\alpha_0 + t\alpha_1 + \cdots + t^m\alpha_m$ we may associate the endomorphism $\alpha_0 + T\alpha_1 + \cdots + T^m\alpha_m \in \Phi[T]$. Our correspondence is then a representation of $\Phi[t, S]$ in $\mathfrak{R}$ and $\mathfrak{R}$ is a $\Phi[t, S]$-module.

If $\alpha(t) = \alpha_0 + \cdots + t^m\alpha_m$, $\alpha_m \neq 0$, the *degree* of $\alpha(t)$ is m. We also define the degree of 0 to be $-\infty$ and note that

$$\deg\,[\alpha(t) + \beta(t)] = \max\,(\deg \alpha(t), \deg \beta(t)),$$

$$\deg\,[\alpha(t)\beta(t)] = \deg \alpha(t) + \deg \beta(t).$$

The second equation shows that $\Phi[t, S]$ has no zero-divisors, i.e. $\Phi[t, S]$ is a domain of integrity. It shows also that the only units of this domain are the elements $\neq 0$ in Φ. Now let $\beta(t) = \beta_0 + \cdots + t^{m'}\beta_{m'}$ with $\beta_{m'} \neq 0$ and $m' \leqq m$. Then

$$\alpha(t) - \beta(t)t^{m-m'}(\beta_{m'}^{-1})^{S^{m-m'}}\alpha_m = \alpha_0' + t\alpha_1' + \cdots + t^{m-1}\alpha_{m-1}'.$$

Hence, if we continue this division process, we obtain polynomials $\gamma(t)$ and $\rho(t)$ such that

$$\alpha(t) = \beta(t)\gamma(t) + \rho(t)$$

[1] It is not necessary for our purposes to make any distinction in notation between these two systems.

where $\deg \rho(t) < \deg \beta(t)$. Similarly, we may find a $\gamma_1(t)$ and a $\rho_1(t)$ such that $\alpha(t) = \gamma_1(t)\beta(t) + \rho_1(t)$, $\deg \rho_1(t) < \deg \beta(t)$.

Now let $\mathfrak{J}$ be a right ideal $\neq 0$ in $\Phi[t, S]$. We choose an element $\beta(t) \neq 0$ in $\mathfrak{J}$ having least degree for the non-zero elements of $\mathfrak{J}$. Then if $\alpha(t)$ is any element in $\mathfrak{J}$, $\alpha(t) = \beta(t)\gamma(t) + \rho(t)$ where $\deg \rho(t) < \deg \beta(t)$ and since $\rho(t) = \alpha(t) - \beta(t)\gamma(t) \in \mathfrak{J}$, $\rho(t) = 0$ because of the minimality of the degree of $\beta(t)$. Thus $\alpha(t) = \beta(t)\gamma(t)$ and $\mathfrak{J} = \beta(t)\Phi[t, S]$, the ideal of right multiples of $\beta(t)$. An ideal of this form will be called a *principal right ideal*. Similarly, every left ideal is principal in the sense that it has the form $\Phi[t, S]\beta(t)$. Now we shall call a domain of integrity $\mathfrak{o}$ a *principal ideal domain* if every right ideal is a principal right ideal $a\mathfrak{o}$ and every left ideal is a principal left ideal $\mathfrak{o}a$. Thus $\Phi[t, S]$ is an example of a domain of this type. One may verify that the following are other examples:

1) The ring of integers.

2) Any division ring.

3) The subring of Hamilton's quaternion algebra consisting of quaternions $1\alpha_0 + i\alpha_1 + j\alpha_2 + k\alpha_3$ where the α's are either all rational integers or all halves of odd integers.

4) The ring $\Phi[t, ']$ of differential polynomials. Here the definition is similar to that of $\Phi[t, S]$ with the modification that the rule $\alpha t = t\alpha + \alpha'$ replaces $\alpha t = t\alpha^S$ and $(\alpha + \beta)' = \alpha' + \beta'$, $(\alpha\beta)' = \alpha\beta' + \alpha'\beta$.

In this chapter we consider in some detail the theory of principal ideal domains. The principal applications that we note are to the theory of semi-linear transformations, obtained by specializing $\mathfrak{o}$ to be $\Phi[t, S]$.

2. Elementary properties. Let $\mathfrak{o}$ be a principal ideal domain. If $a\mathfrak{o}$ and $b\mathfrak{o}$ are right ideals $\neq 0$ such that $a\mathfrak{o} \geqq b\mathfrak{o}$, then $b = ac$, or a is a left factor of b. If $a\mathfrak{o} = b\mathfrak{o}$, $au = b$ and $bv = a$. Hence $a = auv$ and $a(1 - uv) = 0$, $uv = 1$. Similarly $vu = 1$ so that u and v are units in $\mathfrak{o}$.[2] Hence a and b are right associates. Similar remarks hold for left ideals. Throughout this chapter there is a complete parallelism between the theory of right ideals and that of left ideals. We shall therefore state the results for right ideals only.

If $a_1\mathfrak{o} \leqq a_2\mathfrak{o} \leqq \cdots$ is a chain of right ideals, the set theoretic sum of these ideals is a right ideal and hence has the form $a\mathfrak{o}$. Since $a \in a_N\mathfrak{o}$ for a suitable N, $a\mathfrak{o} = a_N\mathfrak{o} = a_{N+1}\mathfrak{o} = \cdots$. Now suppose that $a_1\mathfrak{o} \geqq a_2\mathfrak{o} \geqq \cdots$ is a descending chain and that all of the $a_i\mathfrak{o}$ contain a fixed element $b \neq 0$. Then $b = a_ib_i$, $a_i = a_{i-1}c_{i-1}$ and hence $b = a_{i-1}(c_{i-1}b_i) = a_{i-1}b_{i-1}$. It follows that $b_{i-1} = c_{i-1}b_i$ and $\mathfrak{o}b_1 \leqq \mathfrak{o}b_2 \leqq \cdots$ so that $\mathfrak{o}b_N = \mathfrak{o}b_{N+1} = \cdots$ for N sufficiently large. Thus $c_N, c_{N+1}, \cdots$ are units, and $a_N\mathfrak{o} = a_{N+1}\mathfrak{o} = \cdots$. We note next that the descending chain condition holds only if $\mathfrak{o}$ is a division ring. For suppose that a is an element $\neq 0$ in $\mathfrak{o}$ and consider the chain $a\mathfrak{o} \geqq a^2\mathfrak{o} \geqq \cdots$. Let k be an integer such that $a^k\mathfrak{o} = a^{k+1}\mathfrak{o}$. Then $a^{k+1} = a^ku$ where u is a unit; hence $a = u$ is a unit.

[2] If $\mathfrak{o}$ is any domain of integrity and $uv = 1$ in $\mathfrak{o}$, then $(1 - vu)v = 0$; hence $vu = 1$ also.

THEOREM 1. *The ascending chain condition holds for right ideals of* $\mathfrak{o}$. *Any descending chain of right ideals having an intersection* $\neq 0$ *contains only a finite number of distinct ideals. If the descending chain condition holds without restriction,* $\mathfrak{o}$ *is a division ring.*

Let a and b be elements $\neq 0$ and consider the ideal $a\mathfrak{o} + b\mathfrak{o}$ of elements $ax + by$, x and y arbitrary in $\mathfrak{o}$. This is the smallest ideal containing $a\mathfrak{o}$ and $b\mathfrak{o}$. Now let $a\mathfrak{o} + b\mathfrak{o} = d\mathfrak{o}$. Then $d = ap + bq$ is a highest common left factor of a and b, i.e. d is a left factor of a and b and any left factor of a and b is a left factor of d. The element d is determined to within a unit right factor. Any one of the determinations of d will be denoted by (a, b). Now write $a = da_1$, $b = db_1$. Then $a(1 - pa_1) = da_1 - apa_1 = bqa_1$ and similarly, $b(1 - qb_1) = apb_1$. Since either p or q is $\neq 0$, this proves that the intersection $a\mathfrak{o} \wedge b\mathfrak{o} = m\mathfrak{o} \neq 0$. The element m is a least common right multiple of a and b in the obvious sense: m is a common right multiple of a and b and m is a right multiple of any common right multiple of a and b. We denote m by $[a, b]$ and note that it is determined to within a unit right factor.

THEOREM 2. *Any two elements a and b, $\neq 0$, have a highest common left factor* (a, b) *and a least common right multiple* $[a, b]$ *determined to within unit right factors.*

The existence of common right multiples $\neq 0$ enables us to use the ordinary construction of fractions to obtain a quotient division ring for $\mathfrak{o}$. For this purpose we consider the pairs (a, b) with $b \neq 0$. Define $(a, b) \sim (c, d)$ if, for $m = bd_1 = db_1$, we have $ad_1 = cb_1$. This relation is symmetric, reflexive and transitive. Let a/b denote the set of pairs $(c, d) \sim (a, b)$. We define $a/b + c/d \equiv (ad_1 + cb_1)/m$. If $c \neq 0$, let $n = bc_2 = cb_2$ and define $(a/b)(c/d) \equiv ac_2/db_2$. For $c = 0$ we set $(a/b)(0/d) = 0/d$. It is readily seen that these functions are single valued and that the sets a/b called (right) *fractions* form a division ring Φ relative to these functions as addition and multiplication. The division ring Φ contains a subring $\bar{\mathfrak{o}}$ whose elements $a/1$ are in isomorphic correspondence with the elements of $\mathfrak{o}$. Thus if we replace $\mathfrak{o}$ by $\bar{\mathfrak{o}}$, we may suppose that the domain $\mathfrak{o}$ is a subring of a division ring. The element $a/b = (a/1)(b/1)^{-1}$ so that Φ is the smallest division subring of Φ containing $\mathfrak{o}$.

3. Finitely generated $\mathfrak{o}$-modules. We suppose that $\mathfrak{o}$ is an arbitrary ring with an identity and that $\mathfrak{M}$ is an $\mathfrak{o}$-module in which $x1 = x$ for all x. We recall the defining properties of the product:

$$(x + y)a = xa + ya$$

$$x(a + b) = xa + xb$$

$$x(ab) = (xa)b$$

for all x, y in $\mathfrak{M}$ and all a, b in $\mathfrak{o}$. We say that $\mathfrak{M}$ is *finitely generated* if there exist n *generators* $x_1, \cdots, x_n$ in $\mathfrak{M}$ such that every element in $\mathfrak{M}$ may be expressed in the form $\Sigma x_i a_i$, a_i in $\mathfrak{o}$. If the ascending chain condition holds for the submodules of $\mathfrak{M}$, it is readily seen that $\mathfrak{M}$ is finitely generated.

Now suppose that $\mathfrak{N}$ is a submodule of $\mathfrak{M}$ and let $\mathfrak{I}_i \equiv \mathfrak{I}_i(\mathfrak{N})$ denote the set of elements a_i that occur as multipliers of elements of the form $x_i a_i + x_{i+1}a_{i+1} + \cdots + x_n a_n$ in $\mathfrak{N}$. Then $\mathfrak{I}_i$ is a right ideal. Evidently if $\mathfrak{N} \leqq \mathfrak{P}$, a second submodule, then $\mathfrak{I}_i(\mathfrak{N}) \leqq \mathfrak{I}_i(\mathfrak{P})$. On the other hand, it follows readily that if $\mathfrak{N} \leqq \mathfrak{P}$ and $\mathfrak{I}_i(\mathfrak{N}) = \mathfrak{I}_i(\mathfrak{P})$ for $i = 1, \cdots, n$, then $\mathfrak{N} = \mathfrak{P}$. This remark enables us to prove

THEOREM 3. *If $\mathfrak{o}$ is a ring that satisfies the ascending (descending) chain condition for right ideals, then any finitely generated $\mathfrak{o}$-module $\mathfrak{M}$ satisfies the ascending (descending) chain condition for submodules.*

For let $\mathfrak{M}_1 \leqq \mathfrak{M}_2 \leqq \cdots$ be an ascending chain of submodules and let $\mathfrak{I}_i^{(k)} = \mathfrak{I}_i(\mathfrak{M}_k)$. Then $\mathfrak{I}_i^{(1)} \leqq \mathfrak{I}_i^{(2)} \leqq \cdots$ and hence there is an integer N such that $\mathfrak{I}_i^{(N)} = \mathfrak{I}_i^{(N+1)} = \cdots$ for all i. This implies that $\mathfrak{M}_N = \mathfrak{M}_{N+1} = \cdots$. The descending chain condition may be treated in a similar manner.

If the elements of $\mathfrak{M}$ are expressible in one and only one way in the form $\Sigma x_i a_i$, then $\mathfrak{M}$ is called a *free* module with the *basis* $x_1, \cdots, x_n$. An equivalent condition is that the x_i be generators of $\mathfrak{M}$ and that $\Sigma x_i d_i = 0$ only when all the $d_i = 0$. As in the case of division rings discussed in Chapter 2, we may construct, for any ring $\mathfrak{o}$, a free module having a prescribed number of base elements. The theorem on the invariance of the number of base elements is not, however, true without restriction on $\mathfrak{o}$. Thus it may be possible to choose elements $y_1, \cdots, y_m$, $m < n$, in $\mathfrak{M}$ such that every $x = \Sigma y_k b_k$ for suitable b_k in $\mathfrak{o}$. We shall now show that the invariance theorem holds under either one of the following assumptions:

1) $\mathfrak{o}$ is a subring of a division ring.
2) The ascending chain condition holds for the right ideals of $\mathfrak{o}$.

For suppose that $\mathfrak{M}$ has $m < n$ generators $y_k = \Sigma x_i a_{ik}$.[3] Then each $x_i = \Sigma y_k b_{ki}$ and hence $x_i = \Sigma x_j a_{jk} b_{ki}$. By the uniqueness assumption, $\Sigma a_{jk} b_{ki} = \delta_{ji}$ so that

$$(a)(b) \equiv \begin{pmatrix} a_{11} & \cdots & a_{1m} & 0 & \cdots & 0 \\ \cdot & \cdots & \cdot & \cdot & \cdots & \cdot \\ \cdot & \cdots & \cdot & \cdot & \cdots & \cdot \\ a_{m1} & \cdots & a_{mm} & 0 & \cdots & 0 \\ a_{m+11} & \cdots & a_{m+1m} & 0 & \cdots & 0 \\ \cdot & \cdots & \cdot & \cdot & \cdots & \cdot \\ \cdot & \cdots & \cdot & \cdot & \cdots & \cdot \\ a_{n1} & \cdots & a_{nm} & 0 & \cdots & 0 \end{pmatrix} \begin{pmatrix} b_{11} & \cdots & b_{1m} & b_{1m+1} & \cdots & b_{1n} \\ \cdot & \cdots & \cdot & \cdot & \cdots & \cdot \\ \cdot & \cdots & \cdot & \cdot & \cdots & \cdot \\ b_{m1} & \cdots & b_{mm} & b_{mm+1} & \cdots & b_{mn} \\ 0 & \cdots & 0 & 0 & \cdots & 0 \\ \cdot & \cdots & \cdot & \cdot & \cdots & \cdot \\ \cdot & \cdots & \cdot & \cdot & \cdots & \cdot \\ 0 & \cdots & 0 & 0 & \cdots & 0 \end{pmatrix} = 1.$$

Since $(b)(a) \neq 1$, this is impossible when 1) holds, as is evident from Theorem 1, Chapter 2. We note next that the mapping $\Sigma x_i a_i \rightarrow \Sigma y_k a_k$ is an $\mathfrak{o}$-endomorphism A such that $\mathfrak{M}A = \mathfrak{M}$. On the other hand $\mathfrak{Z}_A$, the set of elements mapped into 0 by A, includes $x_{m+1}, \cdots, x_n$ and so $\mathfrak{Z}_A \neq 0$. This is excluded when 2) holds by Theorem 7, Chapter 1.[4]

[3] By symmetry it suffices to consider this case.

[4] The last result is due to C. J. Everett. A fuller discussion of these questions is contained in his paper [3].

Now let $\mathfrak{F}$ be a free $\mathfrak{o}$-module with the basis $e_1, \cdots, e_n$ and let $\mathfrak{M}$ be any $\mathfrak{o}$-module with n generators $x_1, \cdots, x_n$. The correspondence $\Sigma e_i a_i \rightarrow \Sigma x_i a_i$ is an $\mathfrak{o}$-homomorphism between $\mathfrak{F}$ and $\mathfrak{M}$. Hence $\mathfrak{M}$ is isomorphic to $\mathfrak{F} - \mathfrak{N}$, $\mathfrak{N}$ the set of elements mapped into 0 by the homomorphism. We shall use this result later to obtain the structure of finitely generated $\mathfrak{o}$-modules for any principal ideal domain $\mathfrak{o}$.

An $\mathfrak{o}$-module that is generated by a single element is called *cyclic*. If we regard $\mathfrak{o}$ as a module relative to the ordinary multiplication xa, we see that $\mathfrak{o}$ is a free cyclic module since $a = 1a$. A right ideal $\mathfrak{J}$ of $\mathfrak{o}$ is an $\mathfrak{o}$-submodule. $\mathfrak{J}$ is cyclic if and only if it is principal, and $\mathfrak{J} = a\mathfrak{o}$ is free if and only if a is not a left zero divisor. Now if $\mathfrak{M}$ is any cyclic $\mathfrak{o}$-module and x is a generator of $\mathfrak{M}$, the correspondence between a in $\mathfrak{o}$ and xa in $\mathfrak{M}$ is an $\mathfrak{o}$-homomorphism. Thus in this case $\mathfrak{M}$ is isomorphic to the difference module $\mathfrak{o} - \mathfrak{J}$ where $\mathfrak{J}$ is the right ideal of elements b such that $xb = 0$. The ideal $\mathfrak{J}$ is called the *order* of x.

4. Cyclic $\mathfrak{o}$-modules. From now on $\mathfrak{o}$ will denote a principal ideal domain. If $\mathfrak{M}$ is an $\mathfrak{o}$-module and x is an element of $\mathfrak{M}$, we shall say that x has *finite order* if its order $\mathfrak{J} \neq 0$. Suppose that x and y have finite orders $\mathfrak{J}_1$ and $\mathfrak{J}_2$. Then $\mathfrak{J}_3 = \mathfrak{J}_1 \wedge \mathfrak{J}_2 \neq 0$ and since $(x + y)b = 0$ for all b in $\mathfrak{J}_3$, the element $x + y$ has finite order. Next if a is any element $\neq 0$ in $\mathfrak{o}$, $a\mathfrak{o} \wedge \mathfrak{J}_1 = \mathfrak{J}_4 \neq 0$ and if b is an element $\neq 0$ in $\mathfrak{J}_4$, $b = ac$ where $c \neq 0$. Then $(xa)c = xb = 0$. Thus the totality of elements of finite order is a submodule of $\mathfrak{M}$.

We consider now a cyclic $\mathfrak{o}$-module $\mathfrak{M}$ whose generator has finite order and hence, without loss of generality, we may suppose that $\mathfrak{M} = \mathfrak{o} - a\mathfrak{o}$, $a \neq 0$. Any submodule of $\mathfrak{o} - a\mathfrak{o}$ has the form $b\mathfrak{o} - a\mathfrak{o}$ where $b\mathfrak{o} \geqq a\mathfrak{o}$ and hence $a = bc$. The submodule $b\mathfrak{o} - a\mathfrak{o}$ is cyclic since it is generated by the coset of the element b. Since the order of the coset $b + a\mathfrak{o}$ is $c\mathfrak{o}$, $b\mathfrak{o} - a\mathfrak{o}$ is $\mathfrak{o}$-isomorphic to $\mathfrak{o} - c\mathfrak{o}$. By the Second Isomorphism Theorem, $(\mathfrak{o} - a\mathfrak{o}) - (b\mathfrak{o} - a\mathfrak{o})$ is $\mathfrak{o}$-isomorphic to $\mathfrak{o} - b\mathfrak{o}$. Thus with a factorization $a = bc$ of an element $a \neq 0$ we may associate a chain of $\mathfrak{o}$-modules $\mathfrak{o} - a\mathfrak{o} \geqq b\mathfrak{o} - a\mathfrak{o} \geqq a\mathfrak{o} - a\mathfrak{o} = 0$ whose difference modules are respectively $\mathfrak{o} - b\mathfrak{o}$ and $\mathfrak{o} - c\mathfrak{o}$, and conversely.

We seek a condition on a and b, $\neq 0$, that $\mathfrak{o} - a\mathfrak{o}$ and $\mathfrak{o} - b\mathfrak{o}$ be $\mathfrak{o}$-isomorphic. Let 1_a be the coset containing the element 1 in $\mathfrak{o} - a\mathfrak{o}$. By an isomorphism this is mapped into a coset u_b in $\mathfrak{o} - b\mathfrak{o}$. Then $1_a c$ corresponds to $u_b c$. Since $0 \rightarrow 0$ is an isomorphism, $u_b a = b\mathfrak{o}$. If u is any element of u_b, $ua = bv = m$. Since 1_b, the coset containing 1 in $\mathfrak{o} - b\mathfrak{o}$, has the form $u_b c$ for a suitable c, we have $uc = 1 + bq$. Hence the highest common left factor (u, b) of u and b is 1. Since $ua_1 \in b\mathfrak{o}$ only if $a_1 = ac_1$, m is the least common right multiple $[u, b]$ of u and b. Following Ore, we shall say that a and b are *right similar* if there is a u in $\mathfrak{o}$ such that $(u, b) = 1$ and $a = u^{-1}[u, b]$, or $ua\mathfrak{o} = u\mathfrak{o} \wedge b\mathfrak{o}$ and $u\mathfrak{o} + b\mathfrak{o} = \mathfrak{o}$. We have, therefore, shown that $\mathfrak{o} - a\mathfrak{o}$ and $\mathfrak{o} - b\mathfrak{o}$ are $\mathfrak{o}$-isomorphic only when a and b are right similar. The converse also holds, since, as is seen by retracing the above steps, $1_a c \rightarrow u_b c$ is an isomorphism. Now the condition $m = ua = bv = [u, b]$ implies that a and v have no common right factor, i.e. $\mathfrak{o}a + \mathfrak{o}v = \mathfrak{o}$, and $(u, b) = 1$ implies that m is a least common left multiple of a and v. Thus

a and b are *left similar* in the obvious sense. Because of the equivalence of left and of right similarity we shall refer to this property simply as *similarity*. If we consider $\mathfrak{o}$ as a left module relative to left multiplication, we obtain

THEOREM 4. *The $\mathfrak{o}$ modules $\mathfrak{o} - a\mathfrak{o}$ and $\mathfrak{o} - b\mathfrak{o}$ $(a, b \neq 0)$ are isomorphic if and only if the left modules $\mathfrak{o} - \mathfrak{o}a$ and $\mathfrak{o} - \mathfrak{o}b$ are isomorphic. For either of these conditions to hold it is necessary and sufficient that a and b be similar.*

We note that ua, and hence uav, are similar to a if u and v are units. If $\mathfrak{o}$ is commutative and $m = ua = bv$, $up + bq = 1$; then $a = aup + abq = b(vp + aq)$ so that b is a factor of a. Similarly a is a factor of b. Hence, in this case, a and b are similar if and only if they differ by units.

Let a be an element $\neq 0$ and not a unit. Then $\mathfrak{o} > a\mathfrak{o} > 0$. Since the chain conditions hold, the $\mathfrak{o}$-module $\mathfrak{o} - a\mathfrak{o}$ has a composition series. Such a series corresponds to a chain of ideals $\mathfrak{o} = a_0\mathfrak{o} > a_1\mathfrak{o} > a_2\mathfrak{o} > \cdots > a_n\mathfrak{o} = a\mathfrak{o}$ such that $(a_i\mathfrak{o} - a\mathfrak{o}) - (a_{i+1}\mathfrak{o} - a\mathfrak{o})$, and hence $a_i\mathfrak{o} - a_{i+1}\mathfrak{o}$, is irreducible. If $a_{i+1} = a_ib_{i+1}$, $a_0 = 1$, then $a_i\mathfrak{o} - a_{i+1}\mathfrak{o}$ is isomorphic to $\mathfrak{o} - b_{i+1}\mathfrak{o}$. Hence $a = b_1b_2 \cdots b_n$ where the b_i are irreducible in the sense that they are neither 0 nor units and they have no proper factors. Conversely if $a = b_1b_2 \cdots b_n$, b_i irreducible, we obtain a composition series $\mathfrak{o} - a\mathfrak{o} > b_1\mathfrak{o} - a\mathfrak{o} > b_1b_2\mathfrak{o} - a\mathfrak{o} > \cdots > 0$. Thus we may apply the Jordan-Hölder theorem to obtain the following

THEOREM 5. *Any element $a \neq 0$ and not a unit may be written as $b_1 \cdots b_n$, b_i irreducible. If $a = c_1 \cdots c_m$ where the c_j are irreducible then $m = n$ and the b's and the c's may be paired into similar pairs.*

The number n of irreducible factors b_i in $a = b_1 \cdots b_n$ will be called the *length* of a. It is also the length of a composition series for $\mathfrak{o} - a\mathfrak{o}$. Let b be a second element $\neq 0$ and not a unit and suppose first that $(a, b) = 1$. Then $a^{-1}[a, b] = b'$ is similar to b. Hence length b' = length b. Now let $(a, b) = d$, $a = da_1$, $b = db_1$. Then $(a_1, b_1) = 1$ and length $[a_1, b_1]$ = length a_1 + length $a_1^{-1}[a_1, b_1]$ = length a_1 + length b_1. Since $[a, b] = d[a_1, b_1]$, length $[a, b]$ = length d + length a_1 + length b_1 and length $[a, b]$ + length d = length a_1 + length b_1 + 2 length d = length ab we have

THEOREM 6. *If a, b are $\neq 0$ and not units, then length $[a, b]$ + length (a, b) = length ab.*

A proper direct decomposition of $\mathfrak{o} - a\mathfrak{o}$ is associated with a set of ideals $a_i\mathfrak{o}$ such that $\mathfrak{o} > a_i\mathfrak{o} > a\mathfrak{o}$, $a_1\mathfrak{o} + \cdots + a_n\mathfrak{o} = \mathfrak{o}$ and $a_i\mathfrak{o} \wedge (a_1\mathfrak{o} + \cdots + a_{i-1}\mathfrak{o} + a_{i+1}\mathfrak{o} + \cdots + a_n\mathfrak{o}) = a\mathfrak{o}$. Thus the elements a_i are proper factors of a, the highest common left factor $(a_1, \cdots, a_n) = 1$ and the least common right multiple $[a_i, (a_1, \cdots, a_{i-1}, a_{i+1}, \cdots, a_n)] = a$. If $a = a_ib_i$, $a_i\mathfrak{o} - a\mathfrak{o}$ is $\mathfrak{o}$-isomorphic to $\mathfrak{o} - b_i\mathfrak{o}$. The condition that $a_i\mathfrak{o} - a\mathfrak{o}$ be indecomposable is that no proper divisors b_i', b_i'' of b_i exist such that $[b_i', b_i''] = b_i$ and $(b_i', b_i'') = 1$, or that no proper divisors a_i', a_i'' of a exist such that $[a_i', a_i''] = a$, $(a_i', a_i'') = a_i$. This of course affords an interpretation of the Krull-Schmidt theorem. A more usual interpretation is obtained by making use of the following general lattice theoretic argument.

Let $\mathfrak{M}$ be a group relative to a set Ω of endomorphisms containing the inner automorphisms and suppose that $\mathfrak{M} = \mathfrak{M}_1 \oplus \cdots \oplus \mathfrak{M}_n$, i.e.

$$\begin{aligned} &\mathfrak{M} = \mathfrak{M}_1 + \cdots + \mathfrak{M}_n, \\ &\mathfrak{M}_i \wedge (\mathfrak{M}_1 + \cdots + \mathfrak{M}_{i-1} + \mathfrak{M}_{i+1} + \cdots + \mathfrak{M}_n) = 0. \end{aligned} \tag{1}$$

Set $\mathfrak{N}_i = \mathfrak{M}_1 + \cdots + \mathfrak{M}_{i-1} + \mathfrak{M}_{i+1} + \cdots + \mathfrak{M}_n$. Then by repeated application of Dedekind's distributive law we obtain $(\mathfrak{N}_1 \wedge \cdots \wedge \mathfrak{N}_{i-1} \wedge \mathfrak{N}_{i+1} \wedge \cdots \wedge \mathfrak{N}_n) = \mathfrak{M}_i$. Hence

$$\begin{aligned} &0 = \mathfrak{N}_1 \wedge \cdots \wedge \mathfrak{N}_n, \\ &\mathfrak{N}_i + (\mathfrak{N}_1 \wedge \cdots \wedge \mathfrak{N}_{i-1} \wedge \mathfrak{N}_{i+1} \wedge \cdots \wedge \mathfrak{N}_n) = \mathfrak{M}. \end{aligned} \tag{2}$$

Conversely, if we have a set of $\mathfrak{N}_i$ satisfying these conditions, we may define $\mathfrak{M}_i = (\mathfrak{N}_1 \wedge \cdots \wedge \mathfrak{N}_{i-1} \wedge \mathfrak{N}_{i+1} \wedge \cdots \wedge \mathfrak{N}_n)$ and obtain $\mathfrak{N}_i = \mathfrak{M}_1 + \cdots + \mathfrak{M}_{i-1} + \mathfrak{M}_{i+1} + \cdots + \mathfrak{M}_n$ and $\mathfrak{M} = \mathfrak{M}_1 \oplus \cdots \oplus \mathfrak{M}_n$. Thus we have a complete dualism between the two types of decompositions. We note also that $\mathfrak{M} - \mathfrak{N}_i$ is Ω-isomorphic to $\mathfrak{M}_i$. Hence we have the following dual of the Krull-Schmidt theorem:

THEOREM 7. *Let $\mathfrak{M}$ be an Ω-group such that Ω includes all the inner automorphisms and both chain conditions hold. Suppose that $\mathfrak{N}_1, \cdots, \mathfrak{N}_n$ and $\mathfrak{N}'_1, \cdots, \mathfrak{N}'_{n'}$ are two sets of Ω-subgroups $\neq \mathfrak{M}$ satisfying* (2) *and such that $\mathfrak{M} - \mathfrak{N}_i$, $\mathfrak{M} - \mathfrak{N}'_i$ are indecomposable. Then $n = n'$ and there is an Ω-automorphism H in $\mathfrak{M}$ and a suitable ordering of the $\mathfrak{N}'_i$ such that $\mathfrak{N}_i H = \mathfrak{N}'_i$. In particular, $\mathfrak{M} - \mathfrak{N}_i$ and $\mathfrak{M} - \mathfrak{N}'_i$ are Ω-isomorphic.*

We return now to $\mathfrak{o}$ and shall call an element a *indecomposable* if a is neither 0 nor a unit and $\mathfrak{o} - a\mathfrak{o}$ is indecomposable. The latter condition holds if and only if there are no proper factors a' and a'' of a such that $a = [a', a'']$ and $(a', a'') = 1$. If a has such a decomposition and $a = a'b'' = a''b'$, a is a least common left multiple of b' and b'' and these elements have no common right factors other than units. It follows that $\mathfrak{o} - a\mathfrak{o}$ is indecomposable if and only if $\mathfrak{o} - \mathfrak{o}a$ is indecomposable. The dual of the Krull-Schmidt theorem implies

THEOREM 8. *An element $a \neq 0$ and not a unit may be written as $[c_1, \cdots, c_n]$ where the c_i are indecomposable and $(c_i, [c_1, \cdots, c_{i-1}, c_{i+1}, \cdots, c_n]) = 1$. If we have a second decomposition $a = [d_1, \cdots, d_m]$ of this type, then $n = m$ and the c's and the d's may be paired into similar pairs.*

Polynomial domains. Let $\mathfrak{o}$ be the polynomial domain $\Phi[t, S]$. If $a = t^m + t^{m-1}\alpha_1 + \cdots + \alpha_m$, $m > 0$, and d is any element of the domain then $d = aq + r$ where $\deg r < \deg a$. Hence in each coset of $a\Phi[t, S]$ there is an element of degree $< m$. As is readily seen, this element is uniquely determined. It follows that any coset of $a\Phi[t, S]$ may be written in one and only one way in the form $\{1\}\xi_1 + \{t\}\xi_2 + \cdots + \{t^{m-1}\}\xi_m$ where $\{t^k\}$ is the coset containing t^k and $\xi_i \in \Phi$. Thus if we regard the $\Phi[t, S]$-module $\Phi[t, S] - a\Phi[t, S]$ as a Φ-module, we see that its dimensionality is the degree of a.

As a consequence of this we obtain the result that similar polynomials have the same degree. The degrees of the irreducible polynomials that occur in a factorization of any a are therefore invariants of a. Suppose that a is similar to b, say $a = u^{-1}[u, b]$ with $(u, b) = 1$. If $[u, b] = ua = bv$, let

$$u = bq_1 + u_1, \qquad v = q_2a + v_1$$

where deg $u_1 <$ deg b and deg $v_1 <$ deg a. Then

$$b(q_1 - q_2)a = bv_1 - u_1a.$$

Unless $q_1 = q_2$ the degree of the left hand side is $\geqq$ deg a + deg b while the degree of the right hand side is $<$ deg a + deg b. Hence $q_1 = q_2$ and $bv_1 = u_1a$. The pair b, u_1 have no common left factors other than units and v_1, a have no common right factors other than units. Hence $a = u_1^{-1}[u_1, b]$ where deg $u_1 <$ deg b. For example, if $b = t - \beta$, β in Φ, then we may take $u_1 = \sigma$ in Φ and $t - \beta = \sigma\sigma^{-1}(t - \beta) = [\sigma, t - \beta]$. Hence the elements similar to $t - \beta$ are right associates of polynomials of the form $\sigma^{-1}(t - \beta)$ or $t - \sigma^{-1}\beta\sigma^S$.

Now let $\Phi = R(i, j)$, the quaternion algebra over a real closed field: The elements of Φ are $\alpha_0 + i\alpha_1 + j\alpha_2 + ij\alpha_3$ where $i^2 = j^2 = -1$ and $ij = -ji$. Assume $S = 1$. If $a(t) = a_0 + ta_1 + t^2a_2 + \cdots + t^ma_m$, a_i in Φ, we define $\overline{a(t)} = \bar{a}_0 + t\bar{a}_1 + t^2\bar{a}_2 + \cdots + t^m\bar{a}_m$ where $\overline{\alpha_0 + i\alpha_1 + j\alpha_2 + ij\alpha_3} = \alpha_0 - i\alpha_1 - j\alpha_2 - ij\alpha_3$. One readily verifies that

$$\overline{a(t) + b(t)} = \overline{a(t)} + \overline{b(t)}, \quad \overline{a(t)b(t)} = \overline{b(t)}\,\overline{a(t)} \tag{3}$$

and

$$a(t) + \overline{a(t)}, \qquad a(t)\overline{a(t)} = \overline{a(t)}a(t) \tag{4}$$

have coefficients in R. Thus $a(t)\overline{a(t)}$ may be factored into linear factors in $R(i)[t]$. Hence the irreducible factors of $a(t)$ in $\Phi[t]$ are linear and the only irreducible polynomials in $\Phi[t]$ are the linear ones. As in the commutative case we may use the identity

$$t^k - r^k = (t - r)(t^{k-1} + t^{k-2}r + \cdots + r^{k-1})$$

to prove that the remainder obtained by dividing $a(t) = a_0 + ta_1 + \cdots + t^ma_m$ on the left by $t - r$ is $a_0 + ra_1 + \cdots + r^ma_m$. Hence $t - r$ is an exact divisor of $a(t)$ if and only if r is a left hand root of $a(t)$ in the sense that $a_0 + ra_1 + \cdots + r^ma_m = 0$. Thus we have proved that every polynomial of degree >0 has a left hand root, and, in a similar fashion, we see that these polynomials have right hand roots $(a_0 + a_1r + \cdots + a_mr^m = 0)$. In this sense Φ is *algebraically closed*.

Next let $\Phi = R(i)$, where $i^2 = -1$ and R is a real closed field. Suppose that S is the automorphism $\alpha_0 + i\alpha_1 \to \alpha_0 - i\alpha_1$. If $a(t) = a_0 + ta_1 + \cdots + t^ma_m$, a_i in Φ, is in $\Phi[t, S]$, we define $\overline{a(t)} = \bar{a}_0 - ta_1 + t^2\bar{a}_2 - t^3a_3 + \cdots$ (or, $\bar{t} = -t$, $t^i a_i = a_i^{S^i} t^i$). Then (3) is valid and $a(t)\overline{a(t)} = \overline{a(t)}a(t) = \alpha(t^2)$, a polynomial in t^2 with coefficients in R. We may factor $\alpha(t^2)$ into factors of the form $t^2 - a$ where $a \in \Phi$; and these are irreducible in $\Phi[t, s]$ unless $a = b\bar{b}$, i.e. unless a is real and non-negative. Hence any $a(t)$ has linear or quadratic factors and our result gives a special form to which every irreducible polynomial is similar.

5. Two-sided ideals. Any two-sided ideal $\mathfrak{J}$ has the form $a\mathfrak{o} = \mathfrak{o}a'$. It follows that $a = ua'$, $a' = av$ and $a = uav$, $a' = ua'v$. Since $ua \in \mathfrak{J}$, $ua = au'$ and $a = au'v$, $u'v = 1$. Hence v is a unit. Similarly, u is a unit and $a\mathfrak{o} = \mathfrak{o}a$, $a'\mathfrak{o} = \mathfrak{o}a'$, i.e. any right generator is a left generator and vice versa. We shall denote generators of two-sided ideals by a^*, b^*, $\cdots$. These elements are characterized by the property that given any x there is an x' ($\hat{x}$) such that $xa^* = a^*x'$ ($a^*x = \hat{x}a^*$). This shows, of course, that the correspondence $x \rightarrow x'$ ($x \rightarrow \hat{x}$) is $(1 - 1)$ and hence it is an automorphism in $\mathfrak{o}$.

If $\mathfrak{J}_1$ and $\mathfrak{J}_2$ are two-sided ideals, then so are $\mathfrak{J}_1 + \mathfrak{J}_2$, $\mathfrak{J}_1 \wedge \mathfrak{J}_2$ and the product $\mathfrak{J}_1\mathfrak{J}_2$, defined as the set of sums $\Sigma y_1 y_2$, y_i in $\mathfrak{J}_i$.[5] If $\mathfrak{J}_1 = a^*\mathfrak{o}$, $\mathfrak{J}_2 = b^*\mathfrak{o}$, then $\mathfrak{J}_1\mathfrak{J}_2 = (a^*\mathfrak{o})(b^*\mathfrak{o}) = a^*(\mathfrak{o}b^*)\mathfrak{o} = a^*(b^*\mathfrak{o})\mathfrak{o} = a^*b^*\mathfrak{o}$. Evidently $\mathfrak{J}_1 \wedge \mathfrak{J}_2 \geqq \mathfrak{J}_1\mathfrak{J}_2$. Now suppose that $\mathfrak{J}_2 \geqq \mathfrak{J}_1 \neq 0$; then $a^* = b^*c$ and if $x \in \mathfrak{o}$, there is an x' and an $\bar{x}$ such that $xa^* = a^*x'$ and $xb^* = b^*\bar{x}$. Hence $b^*\bar{x}c = xa^* = a^*x' = b^*cx'$ and $\bar{x}c = cx'$. Since $\bar{x}$ is arbitrary, $c = c^*$ generates a two-sided ideal $c^*\mathfrak{o}$ such that $a^*\mathfrak{o} = (b^*\mathfrak{o})(c^*\mathfrak{o})$. Evidently $c^*\mathfrak{o} \geqq a^*\mathfrak{o}$.

LEMMA 1. *If $\mathfrak{J}_1$ and $\mathfrak{J}_2$ are two-sided ideals $\neq 0$, the condition that $\mathfrak{J}_2 \geqq \mathfrak{J}_1$ is $\mathfrak{J}_1 = \mathfrak{J}_2\mathfrak{J}_3$ where $\mathfrak{J}_3$ is a two-sided ideal containing $\mathfrak{J}_1$.*

By a maximal two-sided ideal $p^*\mathfrak{o}$ we shall mean a two-sided ideal $\neq \mathfrak{o}$ which is contained in no two-sided ideal $\neq \mathfrak{o}$ and $\neq p^*\mathfrak{o}$. In a similar fashion we define a maximal right ideal $p\mathfrak{o}$. Thus $p\mathfrak{o}$ is maximal if and only if p is irreducible.

Now let $p_1^*\mathfrak{o}$ be a maximal two-sided ideal containing $a^*\mathfrak{o} \neq 0, \mathfrak{o}$. Such ideals exist since $\mathfrak{o} - a^*\mathfrak{o}$ satisfies the chain conditions. We have $a^*\mathfrak{o} = (p_1^*\mathfrak{o})(a_1^*\mathfrak{o})$ where $a_1^*\mathfrak{o} \neq a^*\mathfrak{o}$ since $p_1^*\mathfrak{o} \neq \mathfrak{o}$. If $a_1^*\mathfrak{o} = \mathfrak{o}$, $a^*\mathfrak{o} = p_1^*\mathfrak{o}$. Otherwise $a_1^*\mathfrak{o} = (p_2^*\mathfrak{o})(a_2^*\mathfrak{o})$ where $\mathfrak{o} \geqq a_2^*\mathfrak{o} > a_1^*\mathfrak{o}$. Continuing this process we obtain the following

LEMMA 2. *Any two-sided ideal $a^*\mathfrak{o} \neq 0, \neq \mathfrak{o}$ may be factored as $(p_1^*\mathfrak{o})(p_2^*\mathfrak{o}) \cdots (p_k^*\mathfrak{o})$ where the $p^*\mathfrak{o}$ are maximal (or unfactorable) two-sided ideals.*

Suppose that $p^*\mathfrak{o}$ is maximal and contains (or is a divisor of) $(a^*\mathfrak{o})(b^*\mathfrak{o})$. If $p^*\mathfrak{o} \ngeqq a^*\mathfrak{o}$, $p^*\mathfrak{o} + a^*\mathfrak{o} = \mathfrak{o}$, and hence $b^*\mathfrak{o} = \mathfrak{o}b^*\mathfrak{o} = (p^*\mathfrak{o} + a^*\mathfrak{o})b^*\mathfrak{o} = (p^*\mathfrak{o})(b^*\mathfrak{o}) + (a^*\mathfrak{o})(b^*\mathfrak{o}) \leqq p^*\mathfrak{o}$.

LEMMA 3. *If $p^*\mathfrak{o}$ is maximal and is a divisor of $(a^*\mathfrak{o})(b^*\mathfrak{o})$, then $p^*\mathfrak{o}$ is a divisor of either $a^*\mathfrak{o}$ or of $b^*\mathfrak{o}$.*

Let $p^*\mathfrak{o}$ and $q^*\mathfrak{o}$ be maximal two-sided ideals. If $p^*\mathfrak{o} = q^*\mathfrak{o}$, evidently $(p^*\mathfrak{o})(q^*\mathfrak{o}) = (q^*\mathfrak{o})(p^*\mathfrak{o})$. Now suppose $p^*\mathfrak{o} \neq q^*\mathfrak{o}$. The ideal $(p^*\mathfrak{o} \wedge q^*\mathfrak{o}) \leqq p^*\mathfrak{o}$ so that $(p^*\mathfrak{o} \wedge q^*\mathfrak{o}) = (p^*\mathfrak{o})(q_1^*\mathfrak{o})$. Now $q^*\mathfrak{o} \geqq (p^*\mathfrak{o})(q_1^*\mathfrak{o})$ and since $q^*\mathfrak{o} \ngeqq p^*\mathfrak{o}$, $q^*\mathfrak{o} \geqq q_1^*\mathfrak{o}$. Hence $(p^*\mathfrak{o})(q^*\mathfrak{o}) \geqq (p^*\mathfrak{o} \wedge q^*\mathfrak{o})$. Since the reverse inequality holds, $(p^*\mathfrak{o})(q^*\mathfrak{o}) = (p^*\mathfrak{o} \wedge q^*\mathfrak{o})$. By symmetry we have

[5] In general if $\mathfrak{A}$ and $\mathfrak{B}$ are subrings of a ring, $\mathfrak{A}\mathfrak{B}$ is defined as the set of elements Σab, a in $\mathfrak{A}$ and b in $\mathfrak{B}$. The following rules hold: $\mathfrak{A}(\mathfrak{B}\mathfrak{C}) = (\mathfrak{A}\mathfrak{B})\mathfrak{C}$, $\mathfrak{A}(\mathfrak{B} + \mathfrak{C}) = \mathfrak{A}\mathfrak{B} + \mathfrak{A}\mathfrak{C}$, $(\mathfrak{B} + \mathfrak{C})\mathfrak{A} = \mathfrak{B}\mathfrak{A} + \mathfrak{C}\mathfrak{A}$.

LEMMA 4. *If $p^*\mathfrak{o}$, $q^*\mathfrak{o}$ are maximal two-sided ideals, then $(p^*\mathfrak{o})(q^*\mathfrak{o}) = (q^*\mathfrak{o})(p^*\mathfrak{o})$.*

These lemmas yield, as in the commutative case,

THEOREM 9. *The two-sided ideals of $\mathfrak{o}$ form a commutative multiplicative system. Any two-sided ideal $\neq 0$, $\neq \mathfrak{o}$ has one and only one factorization as a product of maximal two sided ideals.*

It follows from this theorem that if $a^*\mathfrak{o} = (p_1^*\mathfrak{o})^{e_1} \cdots (p_s^*\mathfrak{o})^{e_s}$ where $p_i^*\mathfrak{o}$ is maximal and $p_i^*\mathfrak{o} \neq p_j^*\mathfrak{o}$ if $i \neq j$, then any two-sided ideal containing $a^*\mathfrak{o}$ has the form $(p_1^*\mathfrak{o})^{f_1} \cdots (p_s^*\mathfrak{o})^{f_s}$ with $f_i \leqq e_i$. Hence if $a^*\mathfrak{o} = (p_1^*\mathfrak{o})^{e_1} \cdots (p_s^*\mathfrak{o})^{e_s}$, $b^*\mathfrak{o} = (p_1^*\mathfrak{o})^{f_1} \cdots (p_s^*\mathfrak{o})^{f_s}$ where $e_i \geqq 0$ and $f_i \geqq 0$, then $a^*\mathfrak{o} + b^*\mathfrak{o} = (p_1^*\mathfrak{o})^{h_1} \cdots (p_s^*\mathfrak{o})^{h_s}$ where $h_i = \min(e_i, f_i)$, and $a^*\mathfrak{o} \wedge b^*\mathfrak{o} = (p_1^*\mathfrak{o})^{g_1} \cdots (p_s^*\mathfrak{o})^{g_s}$ where $g_i = \max(e_i, f_i)$. If $a^*\mathfrak{o} + b^*\mathfrak{o} = \mathfrak{o}$, $a^*\mathfrak{o} \wedge b^*\mathfrak{o} = (a^*\mathfrak{o})(b^*\mathfrak{o})$. Thus a necessary and sufficient condition that $a^*\mathfrak{o} = (p^*\mathfrak{o})^e$, $p^*\mathfrak{o}$ maximal, is that it be impossible to write $a^*\mathfrak{o} = (b^*\mathfrak{o} \wedge c^*\mathfrak{o})$ where $b^*\mathfrak{o}$, $c^*\mathfrak{o}$ are proper divisors such that $b^*\mathfrak{o} + c^*\mathfrak{o} = \mathfrak{o}$.

Two-sided ideal in $\Phi[t, s]$. Let $a^* = t^n + t^{n-1}\alpha_1 + \cdots + t^{n-k}\alpha_k$, $\alpha_k \neq 0$, generate a two-sided ideal in $\Phi[t, S]$. Then since t^{n-k} generates a two-sided ideal, this is true also for $t^k + t^{k-1}\alpha_1 + \cdots + \alpha_k$. Hence we may suppose that $k = n$ and $\alpha_n \neq 0$. If ξ is any element in Φ, there is a ξ' in $\Phi[t, S]$ such that

$$\xi(t^n + t^{n-1}\alpha_1 + \cdots + \alpha_n) = (t^n + t^{n-1}\alpha_1 + \cdots + \alpha_n)\xi'.$$

Hence deg $\xi' = 0$, or ξ' is in Φ. Then $\xi' = \alpha_n^{-1}\xi\alpha_n$. If $n \neq 0$, $\xi^{S^n} = \xi'$ so that $\xi^{S^n} = \alpha_n^{-1}\xi\alpha_n$. Thus we see that if no power of S other than $S^0 = 1$ is an inner automorphism, the only elements a^* are $t^k\alpha$ and the two-sided ideals are $t^k\mathfrak{o} = \mathfrak{o}t^k$, $k = 0, 1, 2, \cdots$.

Now suppose that $S^r \in \mathfrak{K}$, $r > 0$, where $\mathfrak{K}$ is the group of inner automorphisms of Φ, and let S^r be the least positive power having this property. Accordingly let $\xi^{S^r} = \mu^{-1}\xi\mu$ for all ξ. Then if $S^n \in \mathfrak{K}$, n is a multiple of r. If $a^* = t^n + t^{n_2}\beta_2 + t^{n_3}\beta_3 + \cdots + \beta_s$, $\beta_i \neq 0$, $n > n_2 > \cdots$ and $\xi a^* = a^*\xi'$ where, of necessity, $\xi' = \beta_s^{-1}\xi\beta_s$, then S^n, S^{n_2}, $\cdots$ are inner. Hence $a^* = t^{mr} + t^{(m-1)r}\gamma_1 + \cdots + \gamma_m$, $\gamma_m \neq 0$ and $\gamma_i\xi' = \mu^{-(m-i)}\xi\mu^{(m-i)}\gamma_i$ for all ξ. Since $ta^* = a^*t'$, $t' = t$ and hence $\gamma_i^S = \gamma_i$. Conversely, the conditions

$$(5) \qquad \gamma_i\xi' = \mu^{-(m-i)}\xi\mu^{(m-i)}\gamma_i, \qquad \gamma_i^S = \gamma_i$$

imply that $xa^* = a^*x'$ for $x = \xi$ in Φ and $x = t$. It follows that this holds for all x in $\Phi[t, S]$. The general form of a generator of a two-sided ideal is therefore $t^k a^*\gamma$ where a^* is as indicated.

6. Bounded ideals. A right ideal $a\mathfrak{o}$ will be called *bounded* if it contains a two-sided ideal $\neq 0$. The join of all two-sided ideals contained in $a\mathfrak{o}$ is then a two-sided ideal called the *bound* $\mathfrak{J} = a^*\mathfrak{o} = \mathfrak{o}a^*$ of $a\mathfrak{o}$. If $z \in \mathfrak{J}$, $xz \in a\mathfrak{o}$ for any x in $\mathfrak{o}$ and hence $z \in \mathfrak{J}'$, the ideal of annihilators of the difference module $\mathfrak{o} - a\mathfrak{o}$. Hence if $a\mathfrak{o}$ is bounded, $\mathfrak{J}' \neq 0$. Conversely if $\mathfrak{J}' \neq 0$, $1\mathfrak{J}' \leqq a\mathfrak{o}$, and $a\mathfrak{o}$ is bounded with bound $\mathfrak{J} \geqq \mathfrak{J}'$. Thus $\mathfrak{J} = \mathfrak{J}'$. This characterization of the bound implies that if a and b are similar and $a\mathfrak{o}$ is bounded, then $b\mathfrak{o}$ is bounded and they have the same bounds. In particular if $a\mathfrak{o} = a^*\mathfrak{o}$ is a two-sided ideal, then $a\mathfrak{o} = b\mathfrak{o}$.

A second characterization of boundedness and bounds is obtained as follows: Let b be an element similar to some right factor of a and let $\mathfrak{J}' = \Delta b\mathfrak{o}$ be the intersection of all $b\mathfrak{o}$ of this type. Suppose that $\mathfrak{J}' \neq 0$. If x is any element of $\mathfrak{o}$, let $(x, a) = e$, $x = ex_1$, $a = ea_1$ so that $(x_1, a_1) = 1$. Let $m_1 = [x_1, a_1] = x_1a_2 = a_1x_2$. Then a_2 is similar to the right divisor a_1 of a. Hence if $d \in \mathfrak{J}'$, $d = a_2d'$ and $xd = ex_1a_2d' = ea_1x_2d' = ax_2d' \in a\mathfrak{o}$. This implies that $a\mathfrak{o}$ is bounded with bound $\mathfrak{J} \geqq \mathfrak{J}'$. On the other hand, let $a\mathfrak{o}$ be bounded with bound $\mathfrak{J}$. Then if b is similar to a right factor of a, $\mathfrak{o} - b\mathfrak{o}$ is $\mathfrak{o}$-isomorphic to a submodule of $\mathfrak{o} - a\mathfrak{o}$ and hence if $d \in \mathfrak{J}$, $d = 1d \in b\mathfrak{o}$. Since $b\mathfrak{o}$ is arbitrary, $d \in \Delta b\mathfrak{o} = \mathfrak{J}'$ and so $\mathfrak{J} \leqq \mathfrak{J}'$. Hence $\mathfrak{J} = \mathfrak{J}'$.

THEOREM 10. *The following conditions are equivalent*: 1) *$a\mathfrak{o}$ is bounded;* 2) *there exist elements $z \neq 0$ such that $xz \in a\mathfrak{o}$ for all x;* 3), *$\Delta b\mathfrak{o}$, the intersection of all $b\mathfrak{o}$ where b is similar to a right factor of a, is $\neq 0$. If these conditions hold, the bound of $a\mathfrak{o}$ is the set of elements z satisfying* 2), *or, the set $\Delta b\mathfrak{o}$ of* 3).

COROLLARY. *If a and b are similar and $a\mathfrak{o}$ is bounded, then $b\mathfrak{o}$ is bounded and has the same bound as $a\mathfrak{o}$.*

Similar definitions hold for left ideals. Now if $a\mathfrak{o} \geqq a^*\mathfrak{o}$, consider $\mathfrak{o}a$ and let $\mathfrak{o}a + \mathfrak{o}a^* = \mathfrak{o}d$. Then $d = ka + la^*$. Since $a^* = aa_1$, we have $da_1 = kaa_1 + la^*a_1 = kaa_1 + la_1'a^* = (ka + la_1'a)a_1$. Hence $d = ua$ where $u = k + la_1'$. Then $\mathfrak{o}d \leqq \mathfrak{o}a$ and $\mathfrak{o}a^* = a^*\mathfrak{o} \leqq \mathfrak{o}a$. Thus $\mathfrak{o}a$ is bounded and its bound is the same as that of $a\mathfrak{o}$.

THEOREM 11. *If $a\mathfrak{o}$ is bounded with bound $a^*\mathfrak{o} = \mathfrak{o}a^*$, then $\mathfrak{o}a$ is bounded and has the bound $a^*\mathfrak{o}$.*

If $a\mathfrak{o}$ and $b\mathfrak{o}$ are bounded with bounds $a^*\mathfrak{o}$ and $b^*\mathfrak{o}$ respectively, then $a^*\mathfrak{o} \wedge b^*\mathfrak{o}$ is a two-sided ideal $\neq 0$. Hence $a\mathfrak{o} \wedge b\mathfrak{o}$ is bounded and evidently its bound is $a^*\mathfrak{o} \wedge b^*\mathfrak{o}$. It follows also from the definition of the bound that if $b\mathfrak{o} \geqq a\mathfrak{o}$, say $a = bc$, and $a\mathfrak{o}$ is bounded with bound $a^*\mathfrak{o}$, then $b\mathfrak{o}$ is bounded with bound $b^*\mathfrak{o} \geqq a^*\mathfrak{o}$. Similarly $\mathfrak{o}c$, and hence $c\mathfrak{o}$, is bounded with bound containing $a^*\mathfrak{o}$. If we combine these two facts, we see that if $a\mathfrak{o} = bcd\mathfrak{o}$ is bounded, then $c\mathfrak{o}$ is bounded and its bound contains $a^*\mathfrak{o}$.

THEOREM 12. *If $a\mathfrak{o} = bcd\mathfrak{o}$ is bounded with bound $a^*\mathfrak{o}$, then $c\mathfrak{o}$ is bounded and has the bound $c^*\mathfrak{o}$ containing $a^*\mathfrak{o}$.*

Now let p be irreducible. Then $p\mathfrak{o}$ is a maximal right ideal. Suppose that $p\mathfrak{o} \geqq (a^*\mathfrak{o})(b^*\mathfrak{o})$. If $p\mathfrak{o} \ngeqq a^*\mathfrak{o}$, then $p\mathfrak{o} + a^*\mathfrak{o} = \mathfrak{o}$ and hence $b^*\mathfrak{o} = (p\mathfrak{o})(b^*\mathfrak{o}) + (a^*\mathfrak{o})(b^*\mathfrak{o}) \leqq p\mathfrak{o}$. If $p\mathfrak{o}$ is bounded, it follows that its bound $p^*\mathfrak{o}$ is a maximal two-sided ideal.

Now let q be indecomposable and let $q\mathfrak{o}$ be bounded with bound $q^*\mathfrak{o}$. Suppose that $q^*\mathfrak{o} = \overset{*}{q_1}\mathfrak{o} \wedge \overset{*}{q_2}\mathfrak{o}$, $\overset{*}{q_1}\mathfrak{o} + \overset{*}{q_2}\mathfrak{o} = \mathfrak{o}$. Set $q_1\mathfrak{o} = \overset{*}{q_1}\mathfrak{o} + q\mathfrak{o}$, $q_2\mathfrak{o} = \overset{*}{q_2}\mathfrak{o} + q\mathfrak{o}$. Then $q_1\mathfrak{o} + q_2\mathfrak{o} = \mathfrak{o}$ and $\overset{*}{q_1}\mathfrak{o}\overset{*}{q_2}\mathfrak{o} = \overset{*}{q_2}\mathfrak{o}\overset{*}{q_1}\mathfrak{o} = q^*\mathfrak{o}$. If $x \in \overset{*}{q_1}\mathfrak{o} + q\mathfrak{o}$, $x(\overset{*}{q_2}\mathfrak{o}) \leqq q\mathfrak{o}$ and if $x \in \overset{*}{q_2}\mathfrak{o} + q\mathfrak{o}$ also, $x(\overset{*}{q_1}\mathfrak{o}) \leqq q\mathfrak{o}$. Hence $x\mathfrak{o} \leqq q\mathfrak{o}$ and $x \in q\mathfrak{o}$. Thus $q_1\mathfrak{o} \wedge$

$q_2\mathfrak{o} = q\mathfrak{o}$. By the indecomposability of $q\mathfrak{o}$ we have either $q_1\mathfrak{o} = q\mathfrak{o}$ or $q_2\mathfrak{o} = q\mathfrak{o}$. Accordingly $q\mathfrak{o} \geqq q_1^*\mathfrak{o}$ or $q\mathfrak{o} \geqq q_2^*\mathfrak{o}$. Since $q^*\mathfrak{o}$ is the bound of $q\mathfrak{o}$, either $q_1^*\mathfrak{o} = q^*\mathfrak{o}$ or $q_2^*\mathfrak{o} = q^*\mathfrak{o}$. It follows that $q^*\mathfrak{o}$ cannot be factored as a product of proper two-sided factors which are relatively prime, i.e $q^*\mathfrak{o}$ is a power of a maximal two-sided ideal.

THEOREM 13. *If p is irreducible and $p\mathfrak{o}$ is bounded, then its bound $p^*\mathfrak{o}$ is a maximal two-sided ideal. If q is indecomposable and $q\mathfrak{o}$ is bounded, then its bound $q^*\mathfrak{o}$ is a power of a maximal two-sided ideal.*

An element a is a *total divisor* of $b \neq 0$ if there is a two-sided ideal $\mathfrak{J}$ such that $a\mathfrak{o} \geqq \mathfrak{J} \geqq b\mathfrak{o}$. Thus $a\mathfrak{o}$ is bounded with bound $a^*\mathfrak{o}$ containing $b\mathfrak{o}$. Since we have seen that $\mathfrak{o}a \geqq a^*\mathfrak{o}$ and since $\mathfrak{o}a^* = a^*\mathfrak{o} \geqq \mathfrak{o}b$ is evident, we also have the result that $\mathfrak{o}a \geqq \mathfrak{o}a^* \geqq \mathfrak{o}b$. An equivalent condition that is more symmetric is due to Teichmüller, namely, $(a\mathfrak{o} \wedge \mathfrak{o}a) \geqq \mathfrak{o}b\mathfrak{o}$. For if $(a\mathfrak{o} \wedge \mathfrak{o}a) \geqq \mathfrak{o}b\mathfrak{o}$, $a\mathfrak{o}$ contains the two-sided ideal $\mathfrak{o}b\mathfrak{o}$ which contains $b\mathfrak{o}$. Conversely if $a\mathfrak{o} \geqq a^*\mathfrak{o} \geqq b\mathfrak{o}$, $a^*\mathfrak{o} \geqq \mathfrak{o}b\mathfrak{o}$ and $a\mathfrak{o} \geqq \mathfrak{o}b\mathfrak{o}$. Similarly, $\mathfrak{o}a \geqq \mathfrak{o}b\mathfrak{o}$ and so $(a\mathfrak{o} \wedge \mathfrak{o}a) \geqq \mathfrak{o}b\mathfrak{o}$. The notion of total divisibility is a similarity invariant as is seen in the following theorem.

THEOREM 14. *If a is a total divisor of b, and a' is similar to a and b' is similar to b, then a' is a total divisor of b'.*

We have seen that if $a\mathfrak{o}$ is bounded and a' is similar to a, then $a'\mathfrak{o}$ is bounded and has the same bound $a^*\mathfrak{o}$ as $a\mathfrak{o}$. Hence if a is a total divisor of b, then a' is a total divisor of b also. Now suppose that $b' = u^{-1}[u, b]$ where $(u, b) = 1$. Then $u\mathfrak{o} + b\mathfrak{o} = \mathfrak{o}$, and if $a\mathfrak{o} \geqq a^*\mathfrak{o} \geqq b\mathfrak{o}$, then $u\mathfrak{o} + a^*\mathfrak{o} = \mathfrak{o}$. Thus $u^{-1}[u, a^*]$ is similar to a^* and since $a^*\mathfrak{o}$ is a two-sided ideal, $u^{-1}[u, a^*]\mathfrak{o} = a^*\mathfrak{o}$ and $ua^* = [u, a^*]$. Since $(u\mathfrak{o} \wedge a^*\mathfrak{o}) \geqq (u\mathfrak{o} \wedge b\mathfrak{o})$, we may write $ub' = [u, b] = [u, a^*]c = ua^*c$. Hence $b' = a^*c$ and $a^*\mathfrak{o} \geqq b'\mathfrak{o}$ so that a is a total divisor of b'. It follows that a' also is a total divisor of b'.

Bounded elements of $\Phi[t, S]$. Let Γ be the center of Φ and suppose that $(\Phi:\Gamma) = m(<\infty)$ and that a power of S not S^0 is inner. Let S^r be the smallest positive power having this property, where $\xi^{S^r} = \mu^{-1}\xi\mu$ for all ξ. Then S induces an automorphism in Γ and $S^r = 1$ in Γ. If S^t is the smallest positive power of the induced automorphism which is the identity then $t = r$. For, as we shall prove later (Chapter 5, **9**), if S^t leaves the elements of Γ invariant, then S^t is inner. Hence $t \geqq r$ and since $t \leqq r$ is evident, $t = r$. If Γ_0 is the subfield of elements invariant under S, then from the Galois theory of fields, $(\Gamma:\Gamma_0) = r$.[6] Hence $(\Phi:\Gamma_0) = mr$.

Since S and S^r commute and $\xi^{S^r} = \mu^{-1}\xi\mu$, $(\mu^S)^{-1}\xi\mu^S = \mu^{-1}\xi\mu$. Hence $\mu^S = \delta\mu$ where $\delta \in \Gamma$. It follows that $\delta\delta^S \cdots \delta^{S^{r-1}} = 1$. Then, as we show in **9**, $\delta = \eta(\eta^S)^{-1}$ where $\eta \in \Gamma$. By replacing μ by $\eta\mu$ and changing the notation, we may suppose that $\mu^S = \mu$.

Now suppose that $a^* = t^{mr}\gamma_0 + t^{(m-1)r}\gamma_1 + \cdots + \gamma_m$, where $\gamma_m = 1$, generates a two-sided ideal. This may be written as $u^m\delta_0 + u^{m-1}\delta_1 + \cdots + \delta_m$,

[6] See Chapter 4, **19**.

$\delta_m = 1$ where $u = t^r\mu^{-1}$, $u^2 = t^{2r}\mu^{-2}$, $\cdots$. The conditions on a^* are that $\delta_i \in \Gamma_0$, and the general form of an element that generates a two-sided ideal is

$$t^k(u^m\delta_0 + u^{m-1}\delta_1 + \cdots + \delta_m)\gamma, \qquad \delta_i \text{ in } \Gamma_0 .$$

If $a(t)$ is a polynomial of degree h let

$$u^i = a(t)q_i(t) + r_i(t), \qquad i = 0, 1, \cdots, mrh,$$

where $\deg r_i(t) < h$. Since the polynomials of degree $< h$ form a space of dimensionality $mrh = N$ over Γ_0, there exist elements $\delta_0, \delta_1, \cdots, \delta_N$ in Γ_0 such that $\Sigma r_i(t)\delta_i = 0$. Hence $\Sigma u^i\delta_i \equiv a^*(t) = a(t)q(t)$ where $q(t) = \Sigma q_i(t)\delta_i$, and $a(t)\Phi[t, S] \geqq a^*(t)\Phi[t, S]$.

THEOREM 15. *If Φ has finite dimensionality over its center and S^r, $0 < r < \infty$, is inner, then every ideal in $\Phi[t, S]$ is bounded.*

7. Matrices with elements in $\mathfrak{o}$. If U and $V \in \mathfrak{o}_n$ the ring of $n \times n$ matrices with elements in $\mathfrak{o}$ and $UV = 1$, then $VU = 1$ also. This is an immediate consequence of the fact that $\mathfrak{o}$ may be embedded in a division ring. Thus U is a unit in $\mathfrak{o}_n$. If A and B are any two $n \times r$ matrices (n rows, r columns) with elements in $\mathfrak{o}$ and $B = UAV$ where U and V are units in $\mathfrak{o}_n$ and $\mathfrak{o}_r$ respectively, then A and B are *associates*. We shall consider in this section the problem of selecting a canonical form among the associates of a given matrix. This will be applied in the next section to obtain the structure of an arbitrary $\mathfrak{o}$-module.

Let a and b be elements $\neq 0$ in $\mathfrak{o}$ and $a\mathfrak{o} + b\mathfrak{o} = d\mathfrak{o}$, $(a\mathfrak{o} \wedge b\mathfrak{o}) = m\mathfrak{o}$. Then there are elements p, q, r, s such that $ap + bq = d$, $ar + bs = 0$ where $m = ar = -bs \neq 0$ and $\mathfrak{o}r + \mathfrak{o}s = \mathfrak{o}$. If $a = da_1$, $b = db_1$ and c_1, d_1 are elements such that $c_1r + d_1s = 1$, we set $u = c_1p + d_1q$ and we may verify that

$$\begin{pmatrix} a_1 & b_1 \\ c_1 - ua_1 & d_1 - ub_1 \end{pmatrix}\begin{pmatrix} p & r \\ q & s \end{pmatrix} = \begin{pmatrix} 1 & 0 \\ 0 & 1 \end{pmatrix}.$$

Hence the matrix $\begin{pmatrix} p & r \\ q & s \end{pmatrix}$ is a unit in $\mathfrak{o}_2$ and

$$\begin{array}{ccc}
 & \begin{array}{ccccc} & i & & j & \end{array} & \\
V = & \left[\begin{array}{ccccc}
\begin{smallmatrix} 1 & & \\ & \ddots & \\ & & 1 \end{smallmatrix} & \vdots & & \vdots & \\
\cdots & p & \cdots & r & \cdots \\
 & \vdots & \begin{smallmatrix} 1 & & \\ & \ddots & \\ & & 1 \end{smallmatrix} & \vdots & \\
\cdots & q & \cdots & s & \cdots \\
 & & & & \begin{smallmatrix} 1 & & \\ & \ddots & \\ & & 1 \end{smallmatrix}
\end{array}\right] & \begin{array}{c} \\ i \\ \\ j \\ \\ \end{array}
\end{array}$$

is a unit in $\mathfrak{o}_r$. If A has the i-th row $(c_1, \cdots, c_{i-1}, a, c_{i+1}, \cdots, c_{j-1}, b, c_{j+1}, \cdots, c_r)$, then the i-th row of AV is $(c_1, \cdots, c_{i-1}, d, c_{i+1}, \cdots, c_{j-1}, 0, c_{j+1}, \cdots, c_r)$. A similar result holds for the columns of A.

We note next that the following "elementary" transformations may be performed by multiplying A on the left and on the right by units:

I. Adding to the i-th column the j-th column multiplied by q on the right ($i \neq j$). This is done by multiplying A on the right by $(1 + e_{ji}q)$. To add to the i-th row the j-th row multiplied on the left by q, form $(1 + e_{ij}q)A$.

II. Interchanging the i-th and the j-th columns (rows): Form $A(1 + e_{ij} + e_{ji} - e_{ii} - e_{jj})$ (or $(1 + e_{ij} + e_{ji} - e_{ii} - e_{jj})A$).

III. Multiplying the i-th column (row) on the right (left) by a unit u: Form $A(1 + (u - 1)e_{ii})$ (or $(1 + (u - 1)e_{ii})A$).

If $A \neq 0$, let $a_{pq} \neq 0$ be an element of A whose length is least for the non-zero elements of A. By performing operations of type II, we obtain an associate $B = (b_{ij})$ for which $b_{11} \neq 0$ has the smallest length. If b_{11} is not a left factor of one of the b_{1i}, a suitable associate BU has in place of b_{11} the element $d \neq 0$ whose length is less than that of b_{11}. Similarly, if b_{11} is not a right factor of every b_{i1}, b_{11} may be replaced by an element of smaller length. After a finite number of these replacements, we obtain an associate C of A for which the element c_{11} is $\neq 0$ and is a left factor of every c_{1i} and a right factor of every c_{i1}. If $c_{1i} = c_{11}q_i$, we multiply successively the first column on the right by $-q_i$ and add to the 2nd, 3rd, $\cdots$, r-th columns. This leaves the first column unaltered and replaces c_{1i}, $i > 1$, by 0. If we use a similar procedure on the rows, we obtain an associate D of A such that

$$D = \begin{pmatrix} d_1 & 0 & \cdots & 0 \\ 0 & d_{22} & \cdots & d_{2r} \\ \cdot & \cdot & & \cdot \\ 0 & d_{n2} & \cdots & d_{nr} \end{pmatrix}, \qquad d_1 \neq 0.$$

The same process applied to the matrix (d_{ij}) and repeated to submatrices shows that A has an associate in diagonal form $\{d_1, \cdots, d_s, 0, \cdots, 0\}$, $d_i \neq 0$.

We wish to show that we may suppose that each d_i is a total divisor of d_j for $j > i$. If d_i is a left factor of bd_j for every b, $d_i\mathfrak{o} \geqq \mathfrak{o}d_j\mathfrak{o}$ and, as we have seen, d_i is a total divisor of d_j. Now suppose that there is a $b \neq 0$ such that d_1 is not a left factor of bd_2. Add the second row multiplied on the left by b to the first. The corner of the resulting matrix is

$$D_2 = \begin{pmatrix} d_1 & bd_2 \\ 0 & d_2 \end{pmatrix}.$$

This has an associate

$$D_2V_2 = \begin{pmatrix} d'_{11} & 0 \\ d'_{12} & d'_{22} \end{pmatrix}$$

where d'_{11} is a highest common left factor of d_1 and bd_2 and hence has length less than that of d_1. This matrix may be diagonalized to a form in which the element in the (1, 1) position has smaller length than d_1. Repeated applications of this process will yield an associate $\{e_1, \cdots, e_s, 0, \cdots, 0\}$ of A in which each e_i is a left factor of every be_j with $j > i$. Hence

THEOREM 16. *Any rectangular matrix with elements in* $\mathfrak{o}$ *has an associate* $\{e_1, \cdots, e_s, 0, \cdots, 0\}$ *in diagonal form where each* e_i *is a total divisor of* e_j, $j > i$.

We may replace e_i by $u_i e_i v_i$, u_i and v_i units, and obtain another diagonal matrix having the same properties as $\{e_1, \cdots, e_s, 0, \cdots, 0\}$. If $\mathfrak{o}$ is a division ring, we may, therefore, suppose that $e_i = 1$. Hence we have the

COROLLARY. *If* $\mathfrak{o}$ *is a division ring, any rectangular matrix with elements in* $\mathfrak{o}$ *has an associate of the form* $\{1, \cdots, 1, 0, \cdots, 0\}$.

We consider next the special case where $\mathfrak{o}$ is commutative. Let h_i denote the highest common factor of the i-rowed minors of A. Since the columns of any AV are linear combinations of those of A, h_i is a divisor of the i-rowed minors of AV. Similarly h_i is a factor of the i-rowed minors of any matrix UA. Hence if U and V are units, h_i is a highest common factor of the i-rowed minors of UAV. Now if U and V are chosen so that $UAV = \{e_1, \cdots, e_s, 0, \cdots, 0\}$ where e_i is a factor of e_j for $j > i$, it is evident that $h_i = e_1 \cdots e_i$ and so $e_i = h_i h_{i-1}^{-1}$. This enables us to compute directly the normal form $\{e_1, \cdots, e_s, 0, \cdots, 0\}$ of A. It shows also that the e_i are uniquely determined except for unit multipliers. In **11** we shall show that in the general case, the e_i are determined in the sense of similarity by the matrix A.

8. The structure of finitely generated $\mathfrak{o}$-modules. We have seen that any finitely generated $\mathfrak{o}$-module has the form $\mathfrak{F} - \mathfrak{N}$ where $\mathfrak{F}$ is a free module with the basis $x_1, \cdots, x_n$ and $\mathfrak{N}$ is a submodule. We consider first the structure of $\mathfrak{N}$ in the following

THEOREM 17. *If* $\mathfrak{o}$ *is a principal ideal ring and* $\mathfrak{F}$ *is a free* $\mathfrak{o}$*-module, then any submodule* $\mathfrak{N}$ *of* $\mathfrak{F}$ *is free. The number of elements in a basis for* $\mathfrak{N}$ *is* $\leqq$ *that for* $\mathfrak{F}$.

Let $\mathfrak{N}$ be a submodule of $\mathfrak{F}$ and suppose that it is a submodule of $(x_1, \cdots, x_n)$, $\cdots$, of $(x_{n_1}, \cdots, x_n)$ but not of $(x_{n_1+1}, \cdots, x_n)$ where in general $(y_1, \cdots, y_r)$ denotes the $\mathfrak{o}$-module generated by the y_i. The multipliers of x_{n_1} of the elements y in $\mathfrak{N}$ form a right ideal $b_{n_1}\mathfrak{o} \neq 0$. Thus there is an element $y_1 = x_{n_1} b_{n_1} + \sum_{j > n_1} x_j b_j$ in $\mathfrak{N}$ and if $z = x_{n_1} d_{n_1} + \Sigma x_j d_j$ is any element in $\mathfrak{N}$, we have $d_{n_1} = b_{n_1} k$. Hence $z - y_1 k \in (x_{n_1+1}, \cdots, x_n)$. Consider next the $\mathfrak{o}$-module $\mathfrak{N}_1 = \mathfrak{N} \wedge (x_{n_1+1}, \cdots, x_n)$. Treating it in a similar fashion, we obtain an $n_2 > n_1$ such that $\mathfrak{N}_1 \leqq (x_j, \cdots, x_n)$ for $j \leqq n_2$ but $\mathfrak{N}_1 \nleqq (x_{n_2+1}, \cdots, x_n)$. Hence there is a $y_2 = x_{n_2} b_{n_2} + \sum_{j > n_2} x_j b_j$ such that for each z in $\mathfrak{N}_1$ there is a k in $\mathfrak{o}$ such that $z - y_2 k \in (x_{n_2+1}, \cdots, x_n)$. If we continue this process, we obtain $r \leqq n$ elements $y_1, \cdots, y_r$ in $\mathfrak{N}$ where $y_i = x_{n_i} b_{n_i} + \sum_{j > n_i} x_j b_{ji}$, $b_{n_i} \neq 0$, $n_1 < n_2 < \cdots$, such that each z in $\mathfrak{N}$ has the form $\Sigma y_i k_i$. This expression is unique, as is evident from the form of the y's.

Next we may replace the basis x_i of $\mathfrak{F}$ by $\bar{x}_i = \Sigma x_j u_{ji}$ where (u) is a unit in $\mathfrak{o}_n$. Likewise the elements $y_k = \Sigma x_i b_{ik}$ may be replaced by elements $\bar{y}_k = \Sigma y_l v_{lk}$, (v) a unit in $\mathfrak{o}_r$. Then we obtain $\bar{y}_k = \Sigma \bar{x}_i e_{ik}$ where $(e) = (u)^{-1}(b)(v)$

is an associate of (b). It follows from Theorem 16 that, for a suitable choice of the $\bar{x}$ and the $\bar{y}$, we have $\bar{y}_k = \bar{x}_k e_k$, where $e_k \neq 0$ if $k = 1, \cdots, s$, $e_k = 0$ if $k > s$ and each e_k is a total divisor of e_l for $l > k$. We return to the original notation and write x and y in place of $\bar{x}$ and $\bar{y}$.

Consider now the difference module $\mathfrak{F} - \mathfrak{N}$. It is generated by the cosets $\{x_i\}$ containing x_i. If $\{x_1\}c_1 + \cdots + \{x_n\}c_n = 0$, $x_1c_1 + \cdots + x_nc_n \in \mathfrak{N}$ and hence $c_j \in e_j\mathfrak{o}$ if $j = 1, \cdots, s$ and $c_j = 0$ if $j > s$. Since $x_je_j \in \mathfrak{N}$ for $j \leqq s$, $\mathfrak{F} - \mathfrak{N}$ is a direct sum of the cyclic modules $\{x_i\}$. The cyclic modules $\{x_1\}, \cdots, \{x_s\}$ are finite and $\{x_{s+1}\}, \cdots, \{x_n\}$ are infinite. The j-th of these $(j \leqq s)$ is isomorphic to $(\mathfrak{o} - e_j\mathfrak{o})$ and if e_j is a unit, we may delete the corresponding $\{x_j\}$. If $a_1, \cdots, a_s$ are any elements in $\mathfrak{o}$, there exist k_i such that $a_ik_i = e_sb$ and if k is a common multiple of the k_i, then $a_ik = e_sc_i$. Hence $(\{x_1\}a_1 + \cdots + \{x_s\}a_s)k = 0$. It follows that the module $\mathfrak{P} - \mathfrak{N}$ of cosets $\{x_1\}a_1 + \cdots + \{x_s\}a_s$ may be characterized as the totality of elements of $\mathfrak{F} - \mathfrak{N}$ that have finite order. The difference module $(\mathfrak{F} - \mathfrak{N}) - (\mathfrak{P} - \mathfrak{N})$ is a free module of dimensionality $n - s$. Evidently this number is an invariant of $\mathfrak{F} - \mathfrak{N}$. If we make use of the fact that any finitely generated $\mathfrak{o}$-module is $\mathfrak{o}$-isomorphic to an $\mathfrak{F} - \mathfrak{N}$, we obtain the following theorems.

THEOREM 18. *Any finitely generated $\mathfrak{o}$-module is a direct sum of its submodule of elements of finite order and of a free $\mathfrak{o}$-module.*

THEOREM 19. *Any finitely generated $\mathfrak{o}$-module is a direct sum of cyclic $\mathfrak{o}$-modules. The orders $e_i\mathfrak{o} \neq 0$ may be chosen so that e_i is a total divisor of e_j if $j > i$.*[7]

For the further study of finitely generated $\mathfrak{o}$-modules we shall restrict ourselves to modules having only elements of finite order. Thus $s = n$ in the above notation. As a consequence of Theorem 19, an indecomposable $\mathfrak{o}$-module is cyclic with order $q_i\mathfrak{o}$, q_i indecomposable. Any module is a direct sum of modules of the form $\mathfrak{o} - q_i\mathfrak{o}$. By the Krull-Schmidt theorem the q_i are determined in the sense of similarity. We shall call these elements *elementary divisors* of the module.

9. Bounded indecomposable elements. We have seen that if $\mathfrak{o} - q\mathfrak{o}$ is indecomposable and $q\mathfrak{o}$ is bounded, its bound $q^*\mathfrak{o}$ is a power $(p^*\mathfrak{o})^e$ of a maximal two-sided ideal $(p^*\mathfrak{o})$. If $q = p_1 \cdots p_f$ is a factorization of q into irreducible elements p_i, the ideals $p_i\mathfrak{o}$ are bounded and have bounds containing $q^*\mathfrak{o}$. Since $p_i\mathfrak{o}$ is maximal, its bound is a maximal two-sided ideal. Hence this bound is $p^*\mathfrak{o}$. Now let $p_1, \cdots, p_f$ be arbitrary irreducible elements having the property that the bound of $p_i\mathfrak{o}$ is $p^*\mathfrak{o}$. Suppose that $p_{i+1} \cdots p_f\mathfrak{o} \geqq (p^*\mathfrak{o})^{f-i}$. Then $p_ip_{i+1} \cdots p_f\mathfrak{o} \geqq p_i(p^*\mathfrak{o})^{f-i} = (p_i\mathfrak{o})(p^*\mathfrak{o})^{f-i} \geqq (p^*\mathfrak{o})(p^*\mathfrak{o})^{f-i} = (p^*\mathfrak{o})^{f-i+1}$. Thus we have proved that $p_1 \cdots p_f\mathfrak{o}$ is bounded with bound $(p^*\mathfrak{o})^e$, $e \leqq f$. Evidently this implies that the bound of $p_1 \cdots p_k\mathfrak{o} \wedge p_{k+1} \cdots p_f\mathfrak{o}$ is $(p^*\mathfrak{o})^e$ with $e \leqq \min(k, f - k)$.

We now form a direct sum $\mathfrak{N}_h$ of h cyclic modules each $\mathfrak{o}$-isomorphic to $\mathfrak{o} - q\mathfrak{o}$,

[7] The ordinary theory of finitely generated commutative groups is obtained from Theorems 18 and 19 by specializing $\mathfrak{o}$ to be the ring of rational integers.

$q = p_1 \cdots p_f$ indecomposable, and we suppose that this module is cyclic. Then $\mathfrak{N}_h$ is $\mathfrak{o}$-isomorphic to $\mathfrak{o} - q_h\mathfrak{o}$. The bound of $q_h\mathfrak{o}$ is $(p^*\mathfrak{o})^e$ and the length of q_h is fh. Thus $fh \leqq ek$ if k is the length of p^*. Now consider $\mathfrak{N}_{h+1}$, a direct sum of $h + 1$ cyclic modules isomorphic to $\mathfrak{o} - q\mathfrak{o}$. We assert that either $\mathfrak{N}_{h+1}$ is cyclic or $q_h\mathfrak{o} = (p^*\mathfrak{o})^e$. For, if $\mathfrak{N}_{h+1}$ is not cyclic, it is a direct sum of $s > 1$ cyclic modules whose orders are $e_i\mathfrak{o}$, where e_i is a total divisor of e_j if $j > i$. By the Krull-Schmidt theorem the indecomposable parts of e_i are similar to q and hence the length of $e_1 \geqq$ length of q and the bound of $e_1\mathfrak{o}$ is $(p^*\mathfrak{o})^e$. Then the length of $e_2 \geqq ek \geqq fh$ the length of q_h. Since

$$\text{length } q_h + \text{length } q \geqq \text{length } e_1 + \text{length } e_2,$$

we see that length $e_2 =$ length $q_h =$ length $(p^*)^e$. Hence $q_h\mathfrak{o} = (p^*\mathfrak{o})^e$. If $\mathfrak{N}_{h+1}$ is cyclic, we form $\mathfrak{N}_{h+2}$ and repeat the process. Since the lengths of q_h, $q_{h+1}, \cdots$ form an increasing sequence bounded by the length of $(p^*)^e$, we obtain an integer k' such that $\mathfrak{N}_{k'}$ is cyclic but $\mathfrak{N}_{k'+1}$ is not. Then $q_{k'}\mathfrak{o} = (p^*\mathfrak{o})^e$ and $\mathfrak{o} - (p^*\mathfrak{o})^e$ is decomposable as a direct sum of k' modules isomorphic to $\mathfrak{o} - q\mathfrak{o}$. This proves the important

THEOREM 20. *If $\mathfrak{o} - q\mathfrak{o}$ is indecomposable and $q\mathfrak{o}$ is bounded with bound $(p^*\mathfrak{o})^e$, $p^*\mathfrak{o}$ maximal, then $\mathfrak{o} - (p^*\mathfrak{o})^e$ is decomposable into a direct sum of k' modules $\mathfrak{o}$-isomorphic to $\mathfrak{o} - q\mathfrak{o}$. A necessary and sufficient condition that the indecomposable modules $\mathfrak{o} - q\mathfrak{o}$ and $\mathfrak{o} - r\mathfrak{o}$, with $q\mathfrak{o}$ and $r\mathfrak{o}$ bounded, be $\mathfrak{o}$-isomorphic is that they have the same bounds.*

COROLLARY. *If $p_1\mathfrak{o} \geqq p^*\mathfrak{o}$, $p_2\mathfrak{o} \geqq p^*\mathfrak{o}$ where the p_i are irreducible, then p_1 and p_2 are similar.*

For $p_i\mathfrak{o}$ has the bound $p^*\mathfrak{o}$ and $\mathfrak{o} - p_i\mathfrak{o}$ is indecomposable. In particular the factors p_i of q are all similar.

Let $p^*\mathfrak{o}$ be an arbitrary maximal two-sided ideal $\neq \mathfrak{o}$, 0 and $p\mathfrak{o} \neq \mathfrak{o}$ a maximal right ideal containing $p^*\mathfrak{o}$. If $p_1, \cdots, p_h$ are similar to p, it follows as above that $p_1 \cdots p_h\mathfrak{o}$ has the bound $(p^*\mathfrak{o})^{h'}$ with $h' \leqq h$. Suppose that we have already determined elements $p_1, \cdots, p_h$ such that $p_1 \cdots p_h\mathfrak{o}$ has the bound $(p^*\mathfrak{o})^h$. Then there exists an element p_{h+1} such that the bound of $p_1 \cdots p_{h+1}\mathfrak{o}$ is $(p^*\mathfrak{o})^{h+1}$. For otherwise, for every p' similar to p we have that $p_1 \cdots p_hp'\mathfrak{o} \geqq (p^*\mathfrak{o})^h$. Since the intersection $\Delta p'\mathfrak{o} = p^*\mathfrak{o}$, $\Delta p_1 \cdots p_hp'\mathfrak{o} = p_1 \cdots p_h(p^*\mathfrak{o})$ and it contains $(p^*\mathfrak{o})^h$. It follows that $p_1 \cdots p_h\mathfrak{o} \geqq (p^*\mathfrak{o})^{h-1}$ contrary to the choice of $p_1, \cdots, p_h$. Thus for every integer e there exist p_i, $i = 1, \cdots, e$, such that the bound of $p_1 \cdots p_e\mathfrak{o}$ is $(p^*\mathfrak{o})^e$. Then $p_1 \cdots p_e\mathfrak{o}$ is indecomposable since otherwise its bound would be $(p^*\mathfrak{o})^{e'}$ with $e' < e$. By the preceding theorem we obtain

THEOREM 21. *Let $q = p_1 \cdots p_e$ where p_i is irreducible and $p_i\mathfrak{o}$ has the bound $p^*\mathfrak{o}$. Then a necessary and sufficient condition that q be indecomposable is that the bound of $q\mathfrak{o}$ be $(p^*\mathfrak{o})^e$.*

A comparison of lengths shows that k', the number of indecomposable components in a direct decomposition of $\mathfrak{o} - (p^*\mathfrak{o})^e$, is the same as k, the length of p^* We shall call this number the *capacity* of $p^*\mathfrak{o}$. Our criterion for indecomposability has a number of important consequences which we now note.

THEOREM 22. *If* $q = rst$ *is indecomposable and* $q\mathfrak{o}$ *is bounded then* s *is indecomposable, i.e. any submodule and any difference module of an indecomposable* $\mathfrak{o} - q\mathfrak{o}$, *with* $q\mathfrak{o}$ *bounded, is indecomposable.*

Suppose that $r = p_1 \cdots p_k$, $s = p_{k+1} \cdots p_l$, $t = p_{l+1} \cdots p_e$ where the p_i are irreducible. Let $p^*\mathfrak{o}$ be the bound of $p_i\mathfrak{o}$. Then the bound of $q\mathfrak{o}$ is $(p^*\mathfrak{o})^e$. If $s\mathfrak{o}$ is decomposable, $(s\mathfrak{o}) \geqq (p^*\mathfrak{o})^{l-k-1}$. We have seen that $t\mathfrak{o} \geqq (p^*\mathfrak{o})^{e-l}$ and so $st\mathfrak{o} \geqq s(p^*\mathfrak{o})^{e-l} = (s\mathfrak{o})(p^*\mathfrak{o})^{e-l} \geqq (p^*\mathfrak{o})^{e-k-1}$. Similarly, $rst\mathfrak{o} \geqq (p^*\mathfrak{o})^{e-1}$ contrary to the fact that $(p^*\mathfrak{o})^e$ is the bound of $rst\mathfrak{o}$.

THEOREM 23. *If* $q_1\mathfrak{o}$, $q_2\mathfrak{o} \geqq q\mathfrak{o}$, *a bounded ideal, and* q *is indecomposable, then either* $q_1\mathfrak{o} \geqq q_2\mathfrak{o}$ *or* $q_2\mathfrak{o} \geqq q_1\mathfrak{o}$.

If $q_1\mathfrak{o} \wedge q_2\mathfrak{o} = q_3\mathfrak{o}$, $q_3\mathfrak{o} \geqq q\mathfrak{o}$ and hence q_3 is indecomposable. If the bound of $q_i\mathfrak{o}$ is $(p^*\mathfrak{o})^{e_i}$, $i = 1, 2$, that of $q_3\mathfrak{o}$ is $(p^*\mathfrak{o})^{e_3}$, $e_3 = \max(e_1, e_2)$. Hence the length of q_3 = maximum length of (q_1, q_2), say = length of q_1. Then $q_3\mathfrak{o} = q_1\mathfrak{o} \leqq q_2\mathfrak{o}$.

This theorem readily implies the following

THEOREM 24. *If* q *is indecomposable and* $q\mathfrak{o}$ *is bounded, then* $\mathfrak{o} - q\mathfrak{o}$ *has only one composition series.*

Now let $b\mathfrak{o}$ be any bounded ideal with bound of the form $(p^*\mathfrak{o})^e$, $p^*\mathfrak{o}$ maximal. Suppose that $b = [q_1, \cdots, q_\lambda]$ is a direct decomposition of b into indecomposable elements where the bound of $q_i\mathfrak{o}$ is $(p^*\mathfrak{o})^{e_i}$ and $e_1 \geqq \cdots \geqq e_\lambda \geqq 1$. Evidently $e = e_1$. We assert that $\lambda \leqq k$, the capacity of $p^*\mathfrak{o}$. For suppose that $\lambda > k$. If $q_i'\mathfrak{o} \geqq q_i\mathfrak{o}$, $[q_1', q_2', \cdots]$ is a direct decomposition of this element since we clearly have $q_i'\mathfrak{o} + (q_1'\mathfrak{o} \wedge \cdots \wedge q_{i-1}'\mathfrak{o} \wedge q_{i+1}'\mathfrak{o} \wedge \cdots) = \mathfrak{o}$. We choose the divisors $q_i'\mathfrak{o}$ of $q_i\mathfrak{o}$ to have the length e_k for $i = 1, \cdots, k$ and form $q' = [q_1', \cdots, q_k']$, or $q_1'\mathfrak{o} \wedge \cdots \wedge q_k'\mathfrak{o} = q'\mathfrak{o}$. Since $q'\mathfrak{o} \geqq (p^*\mathfrak{o})^{e_k}$ and the latter is decomposable into k indecomposable ideals of length e_k, we have $q'\mathfrak{o} = (p^*\mathfrak{o})^{e_k}$. Thus $(p^*\mathfrak{o})^{e_k} \geqq (q_1\mathfrak{o} \wedge \cdots \wedge q_k\mathfrak{o})$. On the other hand, $q_{k+1}\mathfrak{o}, \cdots, q_\lambda\mathfrak{o}$ all contain $(p^*\mathfrak{o})^{e_k}$ and this contradicts $(q_1\mathfrak{o} \wedge \cdots \wedge q_k\mathfrak{o}) + q_{k+1}\mathfrak{o} = \mathfrak{o}$.

THEOREM 25. *If* $b\mathfrak{o}$ *has the bound* $(p^*\mathfrak{o})^e$, $p^*\mathfrak{o}$ *maximal with capacity* k, *then a direct decomposition of* b *has at most* k *terms.*

Applications of the polynomial case. Suppose that $\mathfrak{o} = \Phi[t]$, i.e. $S = 1$. The two-sided ideals of this domain are generated by polynomials whose coefficients are in the center Γ of Φ. Let ρ be an element of Φ which is algebraic over Γ in the sense that it is a root of a polynomial $\alpha(t)$ in $\Gamma[t]$. Then $\alpha(t)$ is divisible by $t - \rho$. If $\alpha(t)$ has least degree for polynomials in $\Gamma[t]$ having the root ρ, $\alpha(t)$ is irreducible in $\Gamma[t]$. Hence if σ is a second element in Φ such that $\alpha(\sigma) = 0$, the corollary to Theorem 20 implies that $t - \rho$ and $t - \sigma$ are similar and so $\sigma = \beta^{-1}\rho\beta$. Since our hypothesis that ρ and σ satisfy the same irreducible equation is equivalent to the assumption that $\xi(\rho) \to \xi(\sigma)$, $\xi(t)$ in $\Gamma[t]$, is an isomorphism between $\Gamma(\rho)$ and $\Gamma(\sigma)$ over Γ,[8] we have proved

THEOREM 26. *Let* Φ *be a division ring with center* Γ *and let* $\Gamma(\rho)$ *and* $\Gamma(\sigma)$ *be isomorphic subfields of* Φ *which are algebraic over* Γ. *Then any isomorphism between* $\Gamma(\rho)$ *and* $\Gamma(\sigma)$ *over* Γ *may be extended to an inner automorphism in* Φ.

[8] i.e. an isomorphism leaving the elements of Γ invariant.

We consider next $\Phi[t, S]$ where $S^r = 1$ for $r > 0$, and no smaller power of S is an inner automorphism. If we use the form, determined on p. 38, of the elements which generate two-sided ideals, we see that $t^r - \gamma$ generates a maximal two-sided ideal if γ is any element $\neq 0$ in the center Γ and $\gamma^S = \gamma$. A necessary and sufficient condition that $t^r - \gamma$ be divisible by $t - \rho$ is that $\gamma = N(\rho) \equiv \rho\rho^S \cdots \rho^{S^{r-1}}$. For, since γ commutes with ρ, $\gamma = \rho\rho^S \cdots \rho^{S^{r-1}}$ implies that $\gamma = \rho^S \cdots \rho^{S^{r-1}}\rho = N(\rho^S)$ and conversely. Since

$$\begin{aligned}(t^r - \gamma) &= (t - \rho)(t^{r-1} + t^{r-2}\rho^{S^{r-1}} + \cdots + \rho^S \cdots \rho^{S^{r-1}}) + (N(\rho) - \gamma) \\ &= (t^{r-1} + t^{r-2}\rho^{S^{r-1}} + \cdots + \rho^S \cdots \rho^{S^{r-1}})(t - \rho) + (N(\rho^S) - \gamma),\end{aligned}$$

our assertion is evident. Since $t^r - \gamma$ generates a maximal ideal, any two irreducible factors of $t^r - \gamma$ are similar. Moreover, $t - \rho$ and $t - \sigma$ are similar if and only if $\sigma = \beta^{-1}\rho\beta^S$. Hence we have

THEOREM 27. *Let Φ be a division ring and S an automorphism in Φ such that $S^r = 1, 0 < r < \infty$, and no smaller power of S is inner. If γ is in the center of Φ, $\gamma^S = \gamma$ and ρ and σ are elements of Φ such that $N(\rho) = \gamma = N(\sigma)$, then there exists an element β in Φ such that $\sigma = \beta^{-1}\rho\beta^S$.*

COROLLARY. *A necessary and sufficient condition that $N(\sigma) = 1$ is that $\sigma = \beta^{-1}\beta^S$.*

We remark that the conditions on S amount to the statement that S generates a finite group $\mathfrak{G}$ of outer automorphisms, i.e. all the automorphisms $\neq 1$ in $\mathfrak{G}$ are outer. Now suppose again that $\gamma \in \Gamma$, $\gamma^S = \gamma$ and let $r = r_1r_2$. Let $\gamma^{r_1} = N(\rho)$, ρ in Φ. Then $t^r - \gamma^{r_1} = (t^{r_2} - \gamma)q(t)$ is divisible by $t - \rho$. Since the irreducible factors of $t^r - \gamma^{r_1}$ are all similar, $t^{r_2} - \gamma$ is divisible on the left by a suitable $t - \sigma$. Since

$$t^{r_2} - \gamma = (t - \sigma)(t^{r_2-1} + \cdots + \sigma^S \cdots \sigma^{S^{r_2-1}}) + (\sigma\sigma^S \cdots \sigma^{S^{r_2-1}} - \gamma),$$

$\gamma = \sigma\sigma^S \cdots \sigma^{S^{r_2-1}}$. Since $\gamma \in \Gamma$, we have $\gamma = \sigma^S \cdots \sigma^{S^{r_2-1}}\sigma$ and since $\gamma^S = \gamma$, $\gamma = \sigma^S \cdots \sigma^{S^{r_2-1}}\sigma^{S^{r_2}}$. Thus $\sigma^{S^{r_2}} = \sigma$.

THEOREM 28. *Suppose that Φ, S and γ are as in the preceding theorem. If $r = r_1r_2$ and γ^{r_1} is the norm of an element in Φ, then $\gamma = \sigma\sigma^S \cdots \sigma^{S^{r_2-1}}$ where $\sigma^{S^{r_2}} = \sigma$.*

10. Bounded $\mathfrak{o}$-modules. An $\mathfrak{o}$-module $\mathfrak{M}$ is *bounded* if there exist elements $b \neq 0$ in $\mathfrak{o}$ such that $xb = 0$ for all x in $\mathfrak{M}$. The totality of these b's is then a two-sided ideal $\mathfrak{B} \neq 0$ which we call the *bound* of $\mathfrak{M}$. This is in agreement with the previous definitions given in the cyclic case. It is readily seen that a necessary and sufficient condition that $\mathfrak{M}$ be bounded is that the orders of any set of generators y_i be bounded right ideals. The bound of the order of any x in $\mathfrak{M}$ is a divisor of $\mathfrak{B}$.

If $q_1, \cdots, q_u$ are the elementary divisors of $\mathfrak{M}$, we have seen that these elements are determined up to similarity by $\mathfrak{M}$. On the other hand, any two indecomposable elements q are similar if and only if they have the same bounds. Hence we have the following fundamental

THEOREM 29. *The bounds of the elementary divisors of a bounded* $\mathfrak{o}$*-module are invariant: They are independent of the particular decomposition.*

Let $\mathfrak{B} = (p_1^*\mathfrak{o})^{f_1} \cdots (p_r^*\mathfrak{o})^{f_r}$ be a factorization of the bound of $\mathfrak{M}$ into distinct maximal ideals $p_i^*\mathfrak{o}$. Set $\mathfrak{B}_i = \mathfrak{B}(p_i^*\mathfrak{o})^{-f_i}$ and $\mathfrak{M}^{(i)} = \mathfrak{M}\mathfrak{B}_i$ the subset of finite sums of elements xb_i, b_i in $\mathfrak{B}_i$. Since $\mathfrak{B}_i$ is a right ideal, $\mathfrak{M}^{(i)}$ is a submodule and since $\mathfrak{B}_1 + \cdots + \mathfrak{B}_r = \mathfrak{o}$, $\mathfrak{M} = \mathfrak{M}^{(1)} + \cdots + \mathfrak{M}^{(r)}$. The elements of $\mathfrak{M}^{(i)}$ satisfy the equation $x_i(p_i^*\mathfrak{o})^{f_i} = 0$ so that the bound of $\mathfrak{M}^{(i)}$ is $(p^*\mathfrak{o})^{f_i'}$ with $f_i' \leqq f_i$. It follows that $\mathfrak{M}^{(i)} \wedge (\mathfrak{M}^{(1)} + \cdots + \mathfrak{M}^{(i-1)} + \mathfrak{M}^{(i+1)} + \cdots + \mathfrak{M}^{(r)}) = 0$ and hence $\mathfrak{M} = \mathfrak{M}^{(1)} \oplus \cdots \oplus \mathfrak{M}^{(r)}$. Moreover, since $xp_1^{*f_1'} \cdots p_r^{*f_r'} = 0$ for any x, we must have $f_i' = f_i$. Suppose that y is an element of $\mathfrak{M}$ satisfying the equation $y(p_j^*\mathfrak{o})^k = 0$ for some k. Then if we write $y = y_1 + \cdots + y_r$, y_i in $\mathfrak{M}^{(i)}$, we have $y_i(p_j^*\mathfrak{o})^k = 0$ and since $(p_j^*\mathfrak{o})^k + (p_i^*\mathfrak{o})^{f_i} = \mathfrak{o}$, $y_i = 0$ if $i \neq j$. Thus $\mathfrak{M}^{(j)}$ may be characterized as the totality of elements y_j such that $y_j(p_j^*\mathfrak{o})^k = 0$ for some k. We note also that if $\mathfrak{N}$ is any submodule of $\mathfrak{M}$, then $\mathfrak{N} = \mathfrak{N}^{(1)} \oplus \cdots \oplus \mathfrak{N}^{(r)}$ where $\mathfrak{N}^{(i)} = \mathfrak{M}^{(i)} \wedge \mathfrak{N}$.

We now restrict our attention to the case $r = 1$, or $\mathfrak{B} = (p^*\mathfrak{o})^f$. In this case let $\mathfrak{M} = \mathfrak{M}_1 \oplus \cdots \oplus \mathfrak{M}_u$ be a decomposition of $\mathfrak{M}$ into indecomposable $\mathfrak{o}$-modules, and let $(p^*\mathfrak{o})^{e_i}$ be the bound of the cyclic module $\mathfrak{M}_i$ where $e_1 \geqq \cdots \geqq e_u$. Evidently $e_1 = f$.

Consider first an indecomposable $\mathfrak{o}$-module $\mathfrak{N}$ with bound $(p^*\mathfrak{o})^g$. If x is a generator of $\mathfrak{N}$ and $q\mathfrak{o}$ is its order, $q = p_1 \cdots p_g$, p_i irreducible, then $xp_1 \cdots p_{g-1} = y$ is $\neq 0$ and $y(p^*\mathfrak{o}) \doteq 0$. Thus the submodule $\mathfrak{N}_0$ of elements y_0 such that $y_0(p^*\mathfrak{o}) = 0$ is $\neq 0$. Since $\mathfrak{N}_0 \leqq \mathfrak{N}$, it is indecomposable and since its bound is $p^*\mathfrak{o}$, $\mathfrak{N}_0$ is irreducible. We note also that the submodule $\mathfrak{N}(p^*\mathfrak{o})^j$ is indecomposable and its elementary divisor has the length max $(0, g - j)$.

In the general case where $\mathfrak{M} = \mathfrak{M}_1 \oplus \cdots \oplus \mathfrak{M}_u$, if $y(p^*\mathfrak{o}) = 0$ and $y = y_1 + \cdots + y_u$, y_i in $\mathfrak{M}_i$, then $y_i(p^*\mathfrak{o}) = 0$. Hence u may be characterized as the length of the submodule $\mathfrak{M}_0$ of elements y_0 such that $y_0(p^*\mathfrak{o}) = 0$. We have also $\mathfrak{M}(p^*\mathfrak{o})^j = \mathfrak{M}_1(p^*\mathfrak{o})^j \oplus \cdots \oplus \mathfrak{M}_u(p^*\mathfrak{o})^j$ and hence the number $\delta(j, \mathfrak{M})$ of bounds $(p^*\mathfrak{o})^{e_i}$ with exponents $e_i > j$ is the length of the intersection $\mathfrak{M}(p^*\mathfrak{o})^j \wedge \mathfrak{M}_0$.

If $\mathfrak{N}$ is a submodule of $\mathfrak{M}$, $\mathfrak{N}(p^*\mathfrak{o})^j \leqq \mathfrak{M}(p^*\mathfrak{o})^j$ and hence $\mathfrak{N}(p^*\mathfrak{o})^j \wedge \mathfrak{N}_0 = \mathfrak{N}(p^*\mathfrak{o})^j \wedge \mathfrak{M}_0 \leqq \mathfrak{M}(p^*\mathfrak{o})^j \wedge \mathfrak{M}_0$. It follows that $\delta(j, \mathfrak{N}) \leqq \delta(j, \mathfrak{M})$ and therefore if $(p^*\mathfrak{o})^{g_1}$, $(p^*\mathfrak{o})^{g_2}$, $\cdots$ are the bounds of the elementary divisors of $\mathfrak{N}$ and $g_1 \geqq g_2 \geqq \cdots$, then we have $e_i \geqq g_i$. If we apply this and the decomposition $\mathfrak{M} = \mathfrak{M}^{(1)} \oplus \cdots \oplus \mathfrak{M}^{(r)}$ noted above to the case where $\mathfrak{M}$ is cyclic, we obtain the "necessity" part of the following

THEOREM 30. *Suppose that* $a = [q_{11}, \cdots, q_{1u_1}; \cdots; \cdots, q_{ru_r}]$ *is a direct decomposition of a into indecomposable elements* q_{ij} *where the bound of* $q_{ij}\mathfrak{o}$ *is* $(p_i^*\mathfrak{o})^{e_{ij}}$ *and* $e_{i1} \geqq e_{i2} \geqq \cdots$. *Similarly, let* $b = [s_{11}, \cdots, s_{1u_1}; \cdots; \cdots, s_{ru_r}]$ *where the bound of* $s_{ij}\mathfrak{o}$ *is* $(p_i^*\mathfrak{o})^{g_{ij}}$ *and* $g_{i1} \geqq g_{i2} \geqq \cdots \geqq 0$. *Then a necessary and sufficient condition that b be similar to a factor of a is that* $e_{ij} \geqq g_{ij}$.

To prove the sufficiency we observe that there exists a divisor q_{ij}' of q_{ij} with bound $(p_i^*\mathfrak{o})^{g_{ij}}$. For, we obtain such an element by taking the product of the first g_{ij} irreducible factors of q_{ij}. It follows that $a' = [q_{11}', \cdots, q_{1u_1}'; \cdots;$

$\cdots, q'_{ru_r}]$ is a factor of a. Since, by Theorems 20 and 21, q'_{ij} is similar to s_{ij}, a' is similar to b.

11. The invariant factors. Let $\mathfrak{M} = \mathfrak{M}_1 \oplus \cdots \oplus \mathfrak{M}_s$ be a decomposition of the finite $\mathfrak{o}$-module $\mathfrak{M}$ as a direct sum of modules $\mathfrak{o}$-isomorphic to $\mathfrak{o} - e_i\mathfrak{o}$ where e_i is a total divisor of e_j, $j > i$. We wish to show that the e_i are invariants in the sense of similarity.

Suppose first that $\mathfrak{M}$ is bounded. We obtain a decomposition of $\mathfrak{M}$ into indecomposable $\mathfrak{o}$-modules by decomposing the e_i into indecomposable elements q_{ijl} where the bound of $q_{ijl}\mathfrak{o}$ is $(p_j^*\mathfrak{o})^{h_{ijl}}$, $p_j^*\mathfrak{o}$ a maximal two-sided ideal. Let $p^*\mathfrak{o}$ be one of the $p_j^*\mathfrak{o}$'s, k its capacity and $q_1, \cdots, q_u$ the indecomposable parts of the e's for which the bounds of $q_i\mathfrak{o}$ have the form $(p^*\mathfrak{o})^{h_i}$. We recall that if the bound of $q_i\mathfrak{o}$ is $(p^*\mathfrak{o})^{h_i}$, then the length of q_i is h_i. If q is one of the q_i and q is an indecomposable part of e_r, then e_{r+1} is divisible by $(p^*)^h$. Hence by Theorem 30, e_{r+1} contains at least k q's whose lengths h_i are $\geqq h$. Since k is the capacity of $p^*\mathfrak{o}$, these are all of the q's corresponding to p^* that occur in the decomposition of e_{r+1}. Thus we may arrange the q's in a sequence $q_1, \cdots, q_k$; $q_{k+1}, \cdots, q_{2k}; \cdots; q_{tk+1}, \cdots, q_{tk+m}$ so that their lengths form a non-increasing sequence and $q_1, \cdots, q_k$ are indecomposable parts of e_s, $q_{k+1}, \cdots, q_{2k}$ the indecomposable parts of e_{s-1}, etc.

If we have a second decomposition of $\mathfrak{M}$ as $\mathfrak{M}'_1 \oplus \cdots \oplus \mathfrak{M}'_{s'}$ where $\mathfrak{M}'_i$ is $\mathfrak{o}$-isomorphic to $\mathfrak{o} - e'_i\mathfrak{o}$ and e'_i is a total divisor of e'_j for $j > i$, then we may arrange the indecomposable elements q' corresponding to p^* in the same way. By the Krull-Schmidt theorem, q_i and q'_i are similar and their number is the same. Thus the indecomposable parts of e_{s-j} and of e'_{s-j} may be paired into similar pairs and so e_{s-j} and e'_{s-j} are similar and $s = s'$.

Now let $\mathfrak{M}$ be arbitrary and let $\mathfrak{A} = a^*\mathfrak{o}$ be the bound of $e_{s-1}\mathfrak{o}$ and $\mathfrak{B} = b^*\mathfrak{o}$ that of $e'_{s'-1}\mathfrak{o}$. Let $\mathfrak{N}_s$ be the submodule of $\mathfrak{M}_s$ of elements y such that $y\mathfrak{A}\mathfrak{B} = y\mathfrak{B}\mathfrak{A} = 0$. If $e_s = ca^*$ and y_s is a generator of $\mathfrak{M}_s$ of order $e_s\mathfrak{o}$, then $y_sc \in \mathfrak{N}_s$, and its order is $a^*\mathfrak{o}$. Hence if z_s is a generator of $\mathfrak{N}_s$, its order has the form $d_s\bar{a}\mathfrak{o}$ where $\bar{a}$ is similar to a^*. Since $a^*\mathfrak{o}$ is two-sided, $\bar{a}$ differs from a^* by a unit so that the order of z_s is $d_sa^*\mathfrak{o} = a^*\bar{d}_s\mathfrak{o}$. Now suppose that $\mathfrak{N}$ is the submodule of $\mathfrak{M}$ consisting of the elements y such that $y\mathfrak{A}\mathfrak{B} = 0$. Evidently $\mathfrak{N}$ is bounded, and $\mathfrak{N} = \mathfrak{M}_1 \oplus \cdots \oplus \mathfrak{M}_{s-1} \oplus \mathfrak{N}_s$ is a decomposition of $\mathfrak{N}$ into cyclic modules whose orders are bounded and where the bound of each order is a divisor of the next order. Similarly, we have $\mathfrak{N} = \mathfrak{M}'_1 \oplus \cdots \oplus \mathfrak{M}'_{s'-1} \oplus \mathfrak{N}'_{s'}$ where $\mathfrak{N}'_{s'} \leqq \mathfrak{M}'_{s'}$. Hence by our result in the bounded case, $s' = s$ and e_i and e'_i are similar for $i = 1, \cdots, s - 1$. It follows then by the Krull-Schmidt theorem applied to $\mathfrak{M}$ that e_s and e'_s are similar. We shall call the elements e_i the *invariant factors* of the module. Hence we have

THEOREM 31 (Nakayama). *The invariant factors of an $\mathfrak{o}$-module $\mathfrak{M}$ are determined to within similarity.*

12. The theory of a single semi-linear transformation. We have seen that if T is a semi-linear transformation with automorphism S acting in a vector space

$\mathfrak{R}$ over Φ, we may regard $\mathfrak{R}$ as an $\mathfrak{o} = \Phi[t, S]$-module by defining $x\alpha(t) = x\alpha(T)$. If T_1 and T_2 are semi-linear transformations in $\mathfrak{R}$ having the same automorphism S, then the $\Phi[t, S]$-modules determined by them are isomorphic if and only if there exists a linear transformation A such that $T_2 = A^{-1}T_1A$. For if A is a $\Phi[t, S]$-isomorphism, A is linear since $\xi A = A\xi$ for all ξ in Φ, and $T_1A = AT_2$ so that $T_2 = A^{-1}T_1A$. Conversely if these conditions hold for an automorphism A, then $\alpha(T_1)A = A\alpha(T_2)$ for all $\alpha(t)$ and A is an isomorphism. If the matrices of T_1, T_2 and A relative to a basis $x_1, \cdots, x_n$ are respectively (τ_1), (τ_2) and (α), then the condition $T_2 = A^{-1}T_1A$ is equivalent to $(\tau_2) = (\alpha)(\tau_1)(\alpha^S)^{-1}$ $(= (\beta)^{-1}(\tau_1)(\beta^S), (\beta) = (\alpha)^{-1})$.

Consider now a fixed T. Then if x is any vector, there is a vector xT^m in the sequence $x, xT, \cdots$ which is a linear combination of the xT^i, $i < m$, say $xT^m = x\beta_m + \cdots + xT^{m-1}\beta_1$. Then $x(T^m - T^{m-1}\beta_1 - \cdots - \beta_m) = 0$ so that every element of the $\Phi[t, S]$-module $\mathfrak{R}$ has finite order. It follows from the general theory that $\mathfrak{R} = \mathfrak{R}_1 \oplus \cdots \oplus \mathfrak{R}_s$ where $\mathfrak{R}_i$ is cyclic with generator u_i whose order is $e_i\Phi[t, S]$, $e_i = e_i(t)$ a total divisor of $e_j(t)$ for $j > i$. If the degree of the invariant factor $e_i(t)$ is n_i, the vectors $u_i, \cdots, u_iT^{n_i-1}$ form a basis for $\mathfrak{R}_i$ over Φ. Hence $u_1, \cdots, u_1T^{n_1-1}; \cdots; \cdots, u_sT^{n_s-1}$ is a basis for $\mathfrak{R}$ over Φ, and relative to this basis the matrix of T is

$$\begin{pmatrix} \tau^{(1)} & & & & \\ & \tau^{(2)} & & & \\ & & \cdot & & \\ & & & \cdot & \\ & & & & \tau^{(s)} \end{pmatrix}$$

where

$$\tau^{(i)} = \begin{pmatrix} 0 & \cdot & \cdot & \cdot & \cdot & \beta_{n_i}^{(i)} \\ 1 & 0 & \cdot & \cdot & \cdot & \beta_{n_i-1}^{(i)} \\ 0 & 1 & \cdot & & & \cdot \\ \cdot & & \cdot & \cdot & & \cdot \\ \cdot & & & \cdot & 0 & \cdot \\ 0 & \cdot & \cdot & 0 & 1 & \beta_1^{(i)} \end{pmatrix},$$

if $e_i(t) = t^{n_i} - t^{n_i-1}\beta_1^{(i)} - \cdots - \beta_{n_i}^{(i)}$. If (α) is any matrix in Φ_n there exists a matrix (ρ) such that $(\rho)^{-1}(\alpha)(\rho^S)$ has this form. Similarly, we may obtain a canonical form for (α) corresponding to the decomposition of $\mathfrak{R}$ into indecomposable $\mathfrak{o}$-modules.

As an illustration we consider the case of a linear transformation T acting in $\mathfrak{R}$ and Φ where $\Phi = R(i, j)$ is the quaternion algebra over a real closed field. We have seen that the irreducible polynomials in $\mathfrak{o} = \Phi[t]$ are linear. The bound $p^*(t)$ of $p(t) = (t - \alpha)$ is $t - \alpha$ if $\alpha \in R$. Otherwise it is $N(t - \alpha) = (t - \alpha)(t - \bar{\alpha})$. We obtain in this way all the irreducible polynomials (with leading coefficient 1) in $R[t]$. Now consider $(t - \alpha)^e\mathfrak{o}$. Its bound is $(p^*)^e\mathfrak{o}$. For, otherwise, $(t - \alpha)^e$ would be a divisor of $(p^*)^f$ with $f < e$. This is clearly impossible if $p^* = (t - \alpha)$. In the other case we obtain

$$(t - \alpha)^e q(t) = N(t)^f = (t - \alpha)^f (t - \bar{\alpha})^f.$$

Since $\bar{\alpha} \in R(\alpha)$, $q(t)$ has coefficients in $R(\alpha)$ and since this ring is commutative, we obtain a contradiction to the unique factorization theorem in $R(\alpha)[t]$. It follows now that $(t - \alpha)^e$ is indecomposable and that every indecomposable element is similar to one of this form. Hence if $\mathfrak{S}$ is an indecomposable subspace of $\mathfrak{R}$, it is generated by a vector y whose order is $(t - \alpha)^e \mathfrak{o}$. If we use the basis $y_k = y(T - \alpha)^{k-1}$, $k = 1, \cdots, e$, for $\mathfrak{S}$, we obtain the matrix.

$$\begin{pmatrix} \alpha & & & \\ 1 & \alpha & & \\ & \ddots & \ddots & \\ & & 1 & \alpha \end{pmatrix} \tag{6}$$

Two such matrices are similar if and only if their diagonal elements α_1 and α_2 are similar and the condition for this is that α_1 and α_2 satisfy the same irreducible polynomial in $R[t]$. Any matrix is similar to one having blocks of the form (6) strung down the main diagonal.

We return to the general case and the decomposition $\mathfrak{R} = \mathfrak{R}_1 \oplus \cdots \oplus \mathfrak{R}_s$ where the $\mathfrak{R}_i$ are cyclic with orders $e_i(t)$ the invariant factors. In order to determine the $e_i(t)$ we choose a basis $x_1, \cdots, x_f$ for $\mathfrak{R}$ over Φ and write $x_i T = \Sigma x_j \tau_{ji}$. Then $\mathfrak{R}$ is $\Phi[t, S]$-isomorphic to the difference module of the free $\Phi[t, S]$-module $\mathfrak{F}$, whose basis is $e_1, \cdots, e_n$, with respect to a submodule $\mathfrak{N}$ containing the elements $f_i = e_i t - \Sigma e_j \tau_{ji}$. We assert that the f's form a basis for $\mathfrak{N}$. For if f is any element in $\mathfrak{N}$, we may choose polynomials $\varphi_1(t), \cdots, \varphi_f(t)$ so that $f - \Sigma f_i \varphi_i(t) = \Sigma e_i \beta_i$, β_i in Φ. Then $\Sigma x_i \beta_i = 0$ and so $\beta_i = 0$. Thus $f = \Sigma f_i \varphi_i(t)$. Now suppose that $\Sigma f_i \varphi_i(t) = 0$. Then $\Sigma e_i[t\varphi_i(t) - \Sigma \tau_{ij} \varphi_j(t)] = 0$ and $t\varphi_i(t) = \Sigma \tau_{ij} \varphi_j(t)$, $i = 1, \cdots, n$. If any $\varphi_i(t) \neq 0$ and $\varphi_k(t)$ is one of these polynomials of maximum degree, the equation $t\varphi_k(t) = \Sigma \tau_{kj} \varphi_j(t)$ is impossible. Hence $\varphi_i(t) = 0$ for all i. From this result we see that the $e_i(t)$ are the diagonal elements in the normal form of the matrix $1t - (\tau)$, expressing the f's in terms of the e's.

In the case where Φ is commutative and T is linear, $\Phi[t]$ is commutative and $e_i(t) = h_i(t) h_{i-1}(t)^{-1}$ where $h_0(t) = 1$ and $h_i(t)$ is the highest common factor of the i-rowed minors of $1t - (\tau)$. The last invariant factor $e_s(t) = \mu(t)$ has the property that $x\mu(T) = 0$ for all x. For we have $u_i \mu(T) = 0$ if u_i is a generator of $\mathfrak{R}_i$ and since any $x = \Sigma u_i \xi_i(T)$, $x\mu(T) = 0$. Since $\mu(t)\, \Phi[t, S]$ is the order of u_s, $\mu(t)$ is the polynomial of least degree with leading coefficient 1 having T or, the matrix (τ) of T, as a root. Since the other invariant factors are factors of $\mu(t)$ and the characteristic polynomial $f(t) = \det(1t - (\tau)) = \Pi e_i(t)$, $f(t)$ and $\mu(t)$ have the same irreducible factors in $\Phi[t]$, differing at most in the multiplicities of these factors. This is the well-known

THEOREM 32 (Frobenius). *Let* (τ) *be a matrix in* Φ_n, Φ *a field, and let* $\mu(t)$ *be the last invariant factor of the matrix* $1t - (\tau)$ *in* $(\Phi[t])_n$. *Then* 1) $\mu((\tau)) = 0$, 2) $\mu(t)$ *is a factor of any polynomial* $\gamma(t)$ *having the property that* $\gamma((\tau)) = 0$ *and* 3) $\mu(t)$ *and the characteristic polynomial* det $(1t - (\tau))$ *have the same irreducible factors in* $\Phi[t]$.

Suppose now that Φ is a division ring such that $(\Phi:\Gamma) = m < \infty$ for Γ the center of Φ and let S be an automorphism in Φ such that S^r, $0 < r < \infty$, is inner but no smaller positive power of S is inner. If Γ_0 is the subfield of Γ consisting of the elements invariant under S, $(\Gamma:\Gamma_0) = r$ and hence $(\Phi:\Gamma_0) = mr$. We have seen that $\xi^{S^r} = \mu^{-1}\xi\mu$ where $\mu^s = \mu$. The two-sided ideals of $\mathfrak{o} = \Phi[t, S]$ are generated by elements of the form $t^j(u^h + u^{h-1}\gamma_1 + \cdots + \gamma_h)$ where $u = t^r\mu^{-1}$ and $\gamma_i \in \Gamma_0$. Every ideal in $\mathfrak{o}$ is bounded.

If $\mathfrak{o}_1 = \Gamma_0[u]$, it is evident that any $\mathfrak{o}$-module is an $\mathfrak{o}_1$-module and if two $\mathfrak{o}$-modules are $\mathfrak{o}$-isomorphic, then they are $\mathfrak{o}_1$-isomorphic. We wish to prove the converse of the latter result for modules that contain no elements of order $t\mathfrak{o}$. For this purpose we consider first an indecomposable module $\mathfrak{R}_1$ of this type. Then $\mathfrak{R}_1$ has the bound $(p^*)^e\mathfrak{o}$ where $p^* = u^h + u^{h-1}\gamma_1 + \cdots + \gamma_h$, γ_i in Γ_0 and $\gamma_h \neq 0$. We have seen that $\mathfrak{R}_1$ may be embedded in a cyclic module $\mathfrak{R}$ whose generator has the order $(p^*)^e\mathfrak{o}$ and that $\mathfrak{R} = \mathfrak{R}_1 \oplus \cdots \oplus \mathfrak{R}_k$ where the $\mathfrak{R}_i$ are $\mathfrak{o}$-isomorphic indecomposable $\mathfrak{o}$-modules and k is the capacity of $p^*\mathfrak{o}$. We may obtain a decomposition of $\mathfrak{R}$ into indecomposable $\mathfrak{o}$-modules by decomposing the $\mathfrak{R}_i$. This yields kl indecomposable $\mathfrak{o}_1$-components for $\mathfrak{R}$. On the other hand, we may also use the following procedure: Let $\rho_1, \cdots, \rho_{mr}$ be a basis for Φ over Γ_0. Since the vectors zt^ju^l; $j = 0, \cdots, r-1$, $l = 0, \cdots, h-1$, form a basis for $\mathfrak{R}$ over Φ, the vectors $zt^ju^l\rho_i$, $i = 1, \cdots, mr$ form a basis for $\mathfrak{R}$ over Γ_0. Hence $\mathfrak{R}$ is a direct sum of the mr^2 cyclic $\mathfrak{o}_1$-modules whose generators are $zt^j\rho_i$. The orders of these modules are $(p^*)^e\mathfrak{o}_1$ and, since $p^*\mathfrak{o}_1$ is maximal in the commutative ring $\mathfrak{o}_1$, they are indecomposable. It follows that $mr^2 = kl$ and $\mathfrak{R}_1$ is a direct sum of $\dfrac{mr^2}{k}$ indecomposable $\mathfrak{o}_1$-modules whose orders are $(p^*)^e\mathfrak{o}_1$.

Now let $\mathfrak{R}_1$ and $\bar{\mathfrak{R}}_1$ be two indecomposable modules having no elements of order $t\mathfrak{o}$ and suppose that $\mathfrak{R}_1$ and $\bar{\mathfrak{R}}_1$ are $\mathfrak{o}_1$-isomorphic. Then if $\bar{p}^*$, $\bar{e}$, $\bar{k}$ have the same significance for $\bar{\mathfrak{R}}_1$ as p^*, e, k have for $\mathfrak{R}_1$, we evidently have $k = \bar{k}$ and $(p^*)^e = (\bar{p}^*)^{\bar{e}}$. Thus the bounds of $\mathfrak{R}_1$ and $\bar{\mathfrak{R}}_1$ are the same and so $\mathfrak{R}_1$ and $\bar{\mathfrak{R}}_1$ are $\mathfrak{o}$-isomorphic. If we use the Krull-Schmidt theorem, we may extend this special case to the following

THEOREM 33. *Suppose that* $\mathfrak{R}$ and $\bar{\mathfrak{R}}$ *are* $\Phi[t, S]$-*modules having only elements of finite order but no element of order* $t\Phi[t, S]$. *Then a necessary and sufficient condition that* $\mathfrak{R}$ *and* $\bar{\mathfrak{R}}$ *be* $\Phi[t, S]$-*isomorphic is that they be* $\Gamma_0[u]$-*isomorphic.*

COROLLARY. *Under the assumptions of the theorem,* $\mathfrak{R}$ *and* $\bar{\mathfrak{R}}$ *are* $\Phi[t, S]$-*isomorphic if and only if they are* $\Phi[u]$-*isomorphic.*

Now let T be a semi-linear transformation in $\mathfrak{R}$ over Φ where Φ and S have the above form. The condition that no vector in $\mathfrak{R}$ has order $t\Phi[t, S]$ is the same as the assumption that T is $(1 - 1)$. Our results therefore give conditions for similarity of $(1-1)$ semi-linear transformations T_1 and T_2 having the same automorphism S. Thus the corollary states that T_1 and T_2 are similar if and only if the linear transformations $U_1 = T_1^r\mu^{-1}$ and $U_2 = T_2^r\mu^{-1}$ are similar. If we recall the connection with matrices, we obtain

THEOREM 34. *Let Φ be a division ring such that $(\Phi:\Gamma) = m < \infty$ where Γ is the center and let S be an automorphism in Φ such that $\xi^{S^r} = \mu^{-1}\xi\mu$, $0 < r < \infty$, $\mu^S = \mu$ and so smaller power of S is inner. If (τ_1) and (τ_2) are non-singular matrices (i.e. units) in Φ_n, then a necessary and sufficient condition that there exists a non-singular matrix (β) such that $(\tau_2) = (\beta)^{-1}(\tau_1)(\beta^S)$ is that $N(\tau_1)\mu^{-1} = (\tau_1)(\tau_1^S) \cdots (\tau_1^{S^{r-1}})\mu^{-1}$ and $N(\tau_2)\mu^{-1}$ be similar in the usual sense.*

Another interesting case of the above theorem is obtained by taking $r = 1$ and $\mu = 1$. The result is the theorem that two linear transformations in $\mathfrak{R}$ over Φ are similar if and only if they are similar as transformations in $\mathfrak{R}$ over Γ, Γ the center. As is readily shown, it is not necessary to assume in this case that T_1 and T_2 are $(1-1)$.

CHAPTER 4

STRUCTURE OF RINGS OF ENDOMORPHISMS AND OF ABSTRACT RINGS

1. The general problem. Special cases. We consider an arbitrary commutative group $\mathfrak{M}$ and a fixed set Ω of endomorphisms $\alpha, \beta, \cdots$ acting in $\mathfrak{M}$. Let $\mathfrak{A}$ be the set of Ω-endomorphisms, i.e. the set of endomorphisms that commute with every endomorphism in Ω. Then $\mathfrak{A}$ is a subring containing the identity endomorphism of the ring of endomorphisms of $\mathfrak{M}$. In this chapter we impose various conditions on the lattice of Ω-subgroups of $\mathfrak{M}$ and investigate the restrictions that these imply for $\mathfrak{A}$. These results will be applied to obtain the structure of abstract rings and finally we shall give some applications to the theory of projective representations of groups and to the Galois theory of division rings.

Examples. 1) $\mathfrak{M}$ a finite commutative group and Ω vacuous.

2) $\mathfrak{M}$ a vector space over a division ring $\Omega = \Phi$. Here $\mathfrak{A}$ is the ring of linear transformations. We have seen that $\mathfrak{A} = \Phi'_n$, Φ' anti-isomorphic to Φ, or, $\mathfrak{A}$ is anti-isomorphic to Φ_n. We recall also that $\mathfrak{A}$ is simple.

3) $\mathfrak{M}$ a vector space over Φ, Ω the logical sum of Φ and a set of semi-linear transformations $T_1, T_2, \cdots$. In this case $\mathfrak{A}$ consists of the linear transformations commutative with $T_1, T_2, \cdots$. It follows that if $(\tau_1), (\tau_2), \cdots; S_1, S_2, \cdots$ are, respectively, the matrices relative to a fixed basis and the automorphisms of $T_1, T_2, \cdots$, then $\mathfrak{A}$ is anti-isomorphic to the subring of Φ_n of matrices (α) such that $(\alpha)(\tau_i) = (\tau_i)(\alpha^{S_i})$.

We wish to show now that any ring $\mathfrak{A}$ with an identity is essentially the ring of Ω-endomorphisms of a certain commutative group $\mathfrak{M}$. The group $\mathfrak{M}$ is the additive group of $\mathfrak{A}$. We have seen that the right multiplication $x \rightarrow xa = xa_r$ is an endomorphism of $\mathfrak{M}$ and that the totality of these endomorphisms is a subring $\mathfrak{A}_r$ of the ring of endomorphisms of $\mathfrak{M}$. The ring $\mathfrak{A}_r$ is isomorphic to $\mathfrak{A}$. Similarly we have defined the left multiplication a_l by $xa_l \equiv ax$ and we have shown that their totality is a ring $\mathfrak{A}_l$ anti-isomorphic to $\mathfrak{A}$.

Now the associative law evidently implies that if $a_r \in \mathfrak{A}_r$ and $b_l \in \mathfrak{A}_l$, then $a_r b_l = b_l a_r$. On the other hand, suppose that B is a single-valued transformation in $\mathfrak{M}$ commutative with all the a_r and let $1B = b$. Then $xB = (1x)B = (1B)x_r = bx_r = bx$. Thus $B = b_l$, and $\mathfrak{A}_l$ is the set of $\mathfrak{A}_r$-endomorphisms. Similarly, $\mathfrak{A}_r$ is the set of $\mathfrak{A}_l$-endomorphisms. The isomorphism between $\mathfrak{A}_r$ and $\mathfrak{A}$ will therefore enable us to apply the theory of rings of Ω-endomorphisms to the theory of abstract rings. We state these fundamental results in

THEOREM 1. *Any ring $\mathfrak{A}$ with an identity is isomorphic to $\mathfrak{A}_r$, the ring of its right multiplications and is anti-isomorphic to $\mathfrak{A}_l$, the ring of its left multiplications. $\mathfrak{A}_r$ is the ring of $\mathfrak{A}_l$-endomorphisms acting in the additive group $\mathfrak{M}$ of $\mathfrak{A}$ and $\mathfrak{A}_l$ is the ring of $\mathfrak{A}_r$-endomorphisms of $\mathfrak{M}$.*

The $\mathfrak{A}_l$-($\mathfrak{A}_r$-) subgroups of the additive group are the left (right) ideals. The $(\mathfrak{A}_l, \mathfrak{A}_r)$-subgroups are the two-sided ideals. We note also that $\mathfrak{A}_r \wedge \mathfrak{A}_l$ consists of the endomorphisms $c_r = c_l$, c in the center $\mathfrak{C}$. For if $a_r = b_l$, $1a_r = 1b_l$ and $a = b$. Then $ax = xa$ for all x and so $a = c$ is in $\mathfrak{C}$.

2. Algebras over a field. In a similar fashion our results will apply to the theory of algebras (hypercomplex systems). These are defined as follows: If Φ is an abstract field, a set $\mathfrak{A}$ is called an *algebra* over Φ if

1. $\mathfrak{A}$ is a ring.
2. The additive group of $\mathfrak{A}$ is a Φ-module and $x1 = x$ for any x in $\mathfrak{A}$ and 1, the identity of Φ.
3. $a_r\alpha = \alpha a_r$, $a_l\alpha = \alpha a_l$ for all a in $\mathfrak{A}$ and all α in Φ.

The last condition may also be written in the form: $(ab)\alpha = (a\alpha)b = a(b\alpha)$ for all a, b in $\mathfrak{A}$ and all α in Φ. Since Φ is a field, the ring of endomorphisms corresponding to Φ is isomorphic to Φ. Hence if we wish to fix our attention on a particular algebra, we may adopt the point of view of Chapter 2 and regard the set of endomorphisms, rather than the abstract field, as fundamental. In the present chapter we shall follow this line and, in fact, the field properties of Φ will play no role. Thus we may equally well study a ring $\mathfrak{A}$ relative to an arbitrary set Φ of endomorphisms $\alpha, \beta, \cdots$ which commute with the left and the right multiplications. This includes the case of ordinary rings, obtained by taking Φ to be vacuous, as well as that of algebras, obtained by taking Φ to be a field. $\mathfrak{A}$ will be called a *Φ-ring*. We shall be concerned with Φ-subrings and with Φ-ideals of $\mathfrak{A}$.

If $\mathfrak{A}$ has an identity 1, then $x\alpha = x(1\alpha) = (1\alpha)x$, so that the endomorphism α is the right and the left multiplication corresponding to the element 1α. It follows that the element 1α is in the center of $\mathfrak{A}$. Any ideal of the ring $\mathfrak{A}$ is necessarily a Φ-ideal. Hence in the statements of many of the important structure theorems, we could, without loss of generality, omit any reference to the set of operators Φ.

If we wish to compare different algebras $\mathfrak{A}_1$ and $\mathfrak{A}_2$, it is natural to suppose that the field Φ is the same for both algebras. Thus we say that $\mathfrak{A}_1$ and $\mathfrak{A}_2$ are *isomorphic* if there is a $(1 - 1)$ correspondence $a_1 \rightarrow a_2$ between them that is both an isomorphism of the rings $\mathfrak{A}_1$ and $\mathfrak{A}_2$ and a Φ-isomorphism of the additive groups: If $a_1 \rightarrow a_2$ and $b_1 \rightarrow b_2$,

$$a_1 + b_1 \rightarrow a_2 + b_2, \qquad a_1\alpha \rightarrow a_2\alpha, \qquad a_1b_1 \rightarrow a_2b_2 .$$

The correspondence is an *isomorphism*. *Homomorphisms*, *automorphisms*, *anti-automorphisms*, etc., are defined in a similar manner.

We consider now some methods of constructing algebras of finite dimensionality. Evidently Φ_n is such an algebra if $a\alpha$ is taken to be the product of a by the diagonal matrix $\{\alpha, \cdots, \alpha\}$. The ring $\mathfrak{L}$ of linear transformations in an n-dimensional vector space $\mathfrak{M}$ over Φ is also an algebra if $A\alpha$ is defined as the product of A with the scalar multiplication α. As we have seen, if $x_1, \cdots, x_n$ is a basis for $\mathfrak{M}$ over Φ and $x_iA = \Sigma x_j\alpha_{ji}$, then the correspondence between A

and the matrix (α_{ij}) is an anti-isomorphism between the algebra $\mathfrak{L}$ and the algebra Φ_n. Now let $\mathfrak{A}$ be an arbitrary algebra. Then the multiplications $x \to xa$ and $x \to ax$ are linear transformations in $\mathfrak{A}$ regarded as a vector space over Φ. Since $x(a\alpha) = (xa)\alpha$ and $(a\alpha)x = (ax)\alpha$, $(a\alpha)_r = a_r\alpha$ and $(a\alpha)_l = a_l\alpha$. It follows from these equations that if $\mathfrak{A}$ has an identity, the correspondence $a \to a_r$ is an algebra isomorphism between $\mathfrak{A}$ and the subalgebra $\mathfrak{A}_r$ of $\mathfrak{L}$. If we combine this correspondence with one of the anti-isomorphisms between $\mathfrak{L}$ and Φ_n, we obtain an anti-isomorphism between $\mathfrak{A}$ and a subalgebra of Φ_n. Similarly we may combine the correspondence $a \to a_l$ with one of the anti-isomorphisms between $\mathfrak{L}$ and Φ_n and obtain an isomorphism between $\mathfrak{A}$ and a subalgebra of Φ_n. To be explicit, let $x_1, \cdots, x_n$ be a basis for $\mathfrak{A}$ over Φ and let $x_ia = \Sigma x_j\rho_{ji}(a)$ and $ax_i = \Sigma x_j\lambda_{ji}(a)$. Then if $(\rho_{ij}(a)) = R(a)$ and $(\lambda_{ij}(a)) = L(a)$, the correspondence $a \to R(a)$ is an anti-isomorphism and the correspondence $a \to L(a)$ is an isomorphism between $\mathfrak{A}$ and subalgebras of Φ_n. It may be remarked that we may also combine the anti-isomorphism $a \to R(a)$ with the anti-isomorphism $(\alpha) \to (\alpha)'$, $(\alpha)'$ the transposed matrix, and obtain a second isomorphism between $\mathfrak{A}$ and a subalgebra of Φ_n.

If $\mathfrak{A}$ does not have an identity, we form the vector space $\mathfrak{B} = \mathfrak{A} + (x_0)$ and we define $(x_0\alpha + a)(x_0\beta + b) = x_0(\alpha\beta) + a\beta + b\alpha + ab$ for a, b in $\mathfrak{A}$. Then $\mathfrak{B}$ is an algebra with the identity x_0, and $\mathfrak{A}$ is contained as a subalgebra of $\mathfrak{B}$. Hence $\mathfrak{B}$ and, *a fortiori* $\mathfrak{A}$, is isomorphic to an algebra of matrices. We have therefore proved

THEOREM 2. *Any algebra with a finite basis is isomorphic to a subalgebra of a matrix algebra.*

This theorem gives one general method of obtaining algebras of finite dimensionality. A second general procedure is the following. Let $\mathfrak{A}$ be a vector space over Φ with the basis $x_1, \cdots, x_n$. We choose n^3 elements γ_{ijk} in Φ and set $x_ix_j = \sum_\alpha x_\alpha\gamma_{\alpha ij}$. In order that this definition lead to an associative algebra it is necessary that the γ's satisfy the associativity equations $\sum_\alpha \gamma_{l\alpha k}\gamma_{\alpha ij} = \sum_\alpha \gamma_{li\alpha}\gamma_{\alpha jk}$. Then if we define $(\Sigma x_i\xi_i)(\Sigma x_j\eta_j) = \Sigma x_\alpha\gamma_{\alpha ij}\xi_i\eta_j$, we obtain $(x_ix_j)x_k = x_i(x_jx_k)$ and hence $(xy)z = x(yz)$ for all x, y, z. Evidently $(xy)\alpha = (x\alpha)y = x(y\alpha)$ and the distributive laws hold. Hence $\mathfrak{A}$ is an algebra.

Examples. 1) Let the basis be x_1, x_2, x_3, x_4 where $x_1x_i = x_i = x_ix_1$, $x_2^2 = 1\alpha$, $x_3^2 = 1\beta$, $x_4^2 = -1\alpha\beta$, $x_2x_3 = -x_3x_2 = x_4$, $x_3x_4 = -x_4x_3 = -x_2\beta$, $x_4x_2 = -x_2x_4 = -x_3\alpha$.

2) Suppose that $\mathfrak{G}$ is a finite group with elements $1, s, \cdots, u$. We put these in $(1 - 1)$ correspondence with the elements of a basis of a vector space and denote the vector corresponding to s by x_s. Then if we define $x_sx_t = x_{st}$, the associativity equations are satisfied. Hence we obtain an algebra, the *group algebra* of $\mathfrak{G}$ over Φ.

3) Let $\mathfrak{A}$ be the difference algebra $\Phi[t] - (\nu(t))$ where $\Phi[t]$ is the ordinary polynomial domain and $(\nu(t))$ is the principal ideal generated by $\nu(t) = t^n - t^{n-1}\beta_1 - \cdots - \beta^n$. Then $\mathfrak{A}$ has the basis 1, $x = \{t\}$ the coset containing t

and $x^2, \cdots, x^{n-1}$. The multiplication table is deducible from the relation $x^n = x^{n-1}\beta_1 + \cdots + 1\beta_n$ and the associative law.

3. Previous results. We now begin the discussion of the general problem formulated in **1**: A commutative group $\mathfrak{M}$ and a set of endomorphisms Ω in $\mathfrak{M}$ are given, what can be said about the structure of $\mathfrak{A}$, the ring of Ω-endomorphisms?

We have seen that if $\mathfrak{N}$ is an Ω-subgroup of $\mathfrak{M}$ and $A \in \mathfrak{A}$, then $\mathfrak{N}A$ is an Ω-subgroup. The set of elements mapped into $\mathfrak{N}$ by A is also an Ω-subgroup. Corresponding to a direct decomposition of $\mathfrak{M}$ into $\mathfrak{M}_1 \oplus \cdots \oplus \mathfrak{M}_u$ where the $\mathfrak{M}_i$ are Ω-subgroups, we have a decomposition

$$(1) \quad 1 = E_1 + \cdots + E_u, \qquad E_iE_j = 0 \text{ if } i \neq j, \qquad E_i^2 = E_i$$

where the E_i are the projections on the $\mathfrak{M}_i$. Conversely, if the E's are given such that (1) holds, then $\mathfrak{M} = \mathfrak{M}E_1 \oplus \cdots \oplus \mathfrak{M}E_u$. The Ω-group $\mathfrak{M}$ is indecomposable if and only if 1 is a primitive idempotent element of $\mathfrak{A}$.

By a *completely primary ring* $\mathfrak{A}$ we shall understand a ring that contains a nil-ideal $\mathfrak{R}$ (i.e. an ideal all of whose elements are nilpotent) such that $\mathfrak{A} - \mathfrak{R}$ is a division ring. If b is any element not in $\mathfrak{R}$, there is a c such that $bc \equiv 1(\mathfrak{R})$ or, $bc = 1 + z$, z in $\mathfrak{R}$. If $z^m = 0$, we have $(1 + z)(1 - z + z^2 - \cdots \pm z^{m-1}) = 1$ and hence b has the inverse $c(1 - z + z^2 - \cdots)$. Thus $\mathfrak{R}$ may be characterized as the totality of singular elements (non-units) of $\mathfrak{A}$ and $\mathfrak{R}$ is therefore uniquely determined. Fitting's lemma (**5**, Chapter 1) yields the following

THEOREM 3. *If $\mathfrak{M}$ is an indecomposable Ω-group and satisfies both chain conditions, then $\mathfrak{A}$, the ring of Ω-endomorphisms, is completely primary.*

By the lemma any A in $\mathfrak{A}$ is either an automorphism or is nilpotent. The latter case occurs when either $\mathfrak{M}A < \mathfrak{M}$ or when there are elements $z \neq 0$ such that $zA = 0$. (These two conditions are equivalent.) Let $\mathfrak{R}$ be the totality of endomorphisms that are not automorphisms. If $B \in \mathfrak{R}$ and A is arbitrary, then AB and BA are in $\mathfrak{R}$. Suppose that $B_1 + B_2 = A$ is an automorphism. Then $C_1 + C_2 = 1$ where $C_i = B_iA^{-1} \in \mathfrak{R}$. Since C_2 is nilpotent, $C_2^r = 0$ for some r and hence $C_1(1 + C_2 + C_2^2 + \cdots + C_2^{r-1}) = 1 = (1 + C_2 + \cdots + C_2^{r-1})C_1$, and C_1 is not in $\mathfrak{R}$. This contradiction proves that $\mathfrak{R}$ is an ideal. If $A + \mathfrak{R}$ is a coset $\neq \mathfrak{R}$, A is an automorphism and hence $(A + \mathfrak{R})(A^{-1} + \mathfrak{R}) = 1 + \mathfrak{R}$. Thus $\mathfrak{A} - \mathfrak{R}$ is a division ring.

An important related result is the following

THEOREM 4 (Schur's lemma). *If $\mathfrak{M}$ is an irreducible Ω-group, then $\mathfrak{A}$ is a division ring.*

If $A \neq 0$ is in $\mathfrak{A}$, $\mathfrak{M}A = \mathfrak{M}$ and the set of elements z such that $zA = 0$ consists of 0 alone. Thus A is an automorphism and hence it has an inverse in $\mathfrak{A}$.

4. Matrix rings.

LEMMA. *Let $\mathfrak{A}$ be an arbitrary ring with an identity and let e_{ij}, $i, j = 1, \cdots, u$, be a set of elements of $\mathfrak{A}$ which satisfy*

$$(2) \qquad 1 = e_{11} + \cdots + e_{uu}, \qquad e_{ij}e_{kl} = \delta_{jk}e_{il}.$$

Then $\mathfrak{A} = \mathfrak{B}_u$ *where* $\mathfrak{B}$ *is the subring of* $\mathfrak{A}$ *consisting of the elements which commute with the* e_{ij}. *The ring* $\mathfrak{B}$ *is isomorphic to* $e_{ii}\mathfrak{A}e_{ii}$.

If $a \in \mathfrak{A}$, we readily verify that $a_{ij} = \sum_p e_{pi}ae_{jp}$ is in $\mathfrak{B}$ and $a = \Sigma e_{ij}a_{ij}$. On the other hand, if a_{ij} are arbitrary elements in $\mathfrak{B}$ and $\Sigma e_{ij}a_{ij} = 0$, then $a_{ij} = \Sigma e_{pi}(\Sigma e_{ij}a_{ij})e_{jp} = 0$. Hence $\mathfrak{A} = \mathfrak{B}_u$. If $a = \Sigma e_{ij}a_{ij}$ is arbitrary, then $e_{ii}ae_{ii} = e_{ii}a_{ii}$. The correspondence between $e_{ii}ae_{ii}$ and a_{ii} in $\mathfrak{B}$ is an isomorphism.

LEMMA. *If* $\mathfrak{M} = \mathfrak{M}_1 \oplus \mathfrak{M}_2$ *and correspondingly,* $1 = E_1 + E_2$, E_i *the projection on* $\mathfrak{M}_i$, *then the ring* $\mathfrak{A}_1$ *of* Ω-*endomorphisms of* $\mathfrak{M}_1$ *is isomorphic to* $E_1\mathfrak{A}E_1$, $\mathfrak{A}$, *the ring of* Ω-*endomorphisms of* $\mathfrak{M}$.

If $A \in \mathfrak{A}$, E_1AE_1 induces an Ω-endomorphism B in $\mathfrak{M}_1$ and maps $\mathfrak{M}_2$ into 0. Hence if $B = 0$, $E_1AE_1 = 0$, and the correspondence between E_1AE_1 and B is an isomorphism between $E_1\mathfrak{A}E_1$ and a subring $\bar{\mathfrak{A}}_1$ of $\mathfrak{A}_1$. On the other hand, if $B \in \mathfrak{A}_1$, $E_1BE_1 = E_1B$ is an element $E_1(E_1BE_1)E_1$ of $E_1\mathfrak{A}E_1$ whose induced effect in $\mathfrak{M}_1$ is B. Hence $\bar{\mathfrak{A}}_1 = \mathfrak{A}_1$.

THEOREM 5. *If* $\mathfrak{M} = \mathfrak{M}_1 \oplus \cdots \oplus \mathfrak{M}_u$ *where the* $\mathfrak{M}_i$ *are* Ω-*isomorphic, then* $\mathfrak{A} = \mathfrak{B}_u$ *where* $\mathfrak{B}$ *is isomorphic to the ring of* Ω-*endomorphisms of one of the* $\mathfrak{M}_i$.

Let E_i be the projection determined by the decomposition and B_{1i} a fixed Ω-isomorphism between $\mathfrak{M}_1$ and $\mathfrak{M}_i$, $i \neq 1$. Set $E_{ii} = E_i$, $E_{1i} = E_{11}B_{1i}E_{ii}$, $E_{i1} = E_{ii}B_{1i}^{-1}E_{11}$ and $E_{ij} = E_{i1}E_{1j}$ if $i \neq j, i \neq 1, j \neq 1$. Then we readily verify that $E_{ij}E_{kl} = \delta_{jk}E_{il}$ for all i, j, k, l. The theorem is therefore an immediate consequence of the above lemmas.

5. Completely reducible groups. We suppose that $\mathfrak{M}$ is a completely reducible Ω-group satisfying one (and hence both) of the chain conditions. Then $\mathfrak{M} = \mathfrak{M}_1 \oplus \cdots \oplus \mathfrak{M}_u$ where the $\mathfrak{M}_i$ are irreducible. We choose the notation so that $\mathfrak{M}_1, \cdots, \mathfrak{M}_{n_1}$ are Ω-isomorphic, $\mathfrak{M}_{n_1+1}, \cdots, \mathfrak{M}_{n_1+n_2}$ are Ω-isomorphic but not Ω-isomorphic to $\mathfrak{M}_1$, etc.

Now if $\mathfrak{N}_1$ and $\mathfrak{N}_2$ are any irreducible Ω-subgroups of $\mathfrak{M}$ and B is an Ω-homomorphism between $\mathfrak{N}_1$ and a part of $\mathfrak{N}_2$, it is clear that either $B = 0$ or B is an Ω-isomorphism between $\mathfrak{N}_1$ and the whole of $\mathfrak{N}_2$. If $1 = E_1 + E_2 + \cdots + E_u$ is the decomposition of 1 into projections corresponding to the decomposition $\mathfrak{M} = \mathfrak{M}_1 \oplus \cdots \oplus \mathfrak{M}_u$ and A is any Ω-endomorphism, then E_iAE_j induces an Ω-homomorphism between $\mathfrak{M}_i$ and a part of $\mathfrak{M}_j$. Hence if i is in the range $n_1 + \cdots + n_{p-1} + 1, \cdots, n_1 + \cdots + n_p$ and j is in another range $n_1 + \cdots + n_{q-1} + 1, \cdots, n_1 + \cdots + n_q$, then E_iAE_j maps $\mathfrak{M}_i$ into 0. Since E_iAE_j maps all the other $\mathfrak{M}_k$ into 0, we have $E_iAE_j = 0$. Thus if we set $E^{(1)} = E_1 + \cdots + E_{n_1}$, $E^{(2)} = E_{n_1+1} + \cdots + E_{n_1+n_2}, \cdots, E^{(t)} = E_{n_1+\cdots+n_{t-1}+1} + \cdots + E_{n_1+\cdots+n_t}$, we obtain $A = \Sigma E_iAE_j = E^{(1)}AE^{(1)} + \cdots + E^{(t)}AE^{(t)}$. Since $(E^{(p)}AE^{(p)})(E^{(q)}BE^{(q)}) = 0$ if $p \neq q$, $E^{(p)}\mathfrak{A}E^{(p)}$ is a two-sided ideal in $\mathfrak{A}$ and the latter is a direct sum of these ideals.

We have seen that $E^{(p)}\mathfrak{A}E^{(p)}$ is isomorphic to the ring $\mathfrak{A}_p$ of Ω-endomorphisms of $\mathfrak{M}E^{(p)}$. Since the indecomposable parts of $\mathfrak{M}E^{(p)}$ are irreducible and Ω-isomorphic, by the preceding theorem and Schur's lemma, $\mathfrak{A}_p = \Phi^{(p)}_{n_p}$ a matrix ring over a division ring. The division ring $\Phi^{(p)}$ is isomorphic to the ring of Ω-endomorphisms of $\mathfrak{M}_{n_1+\cdots+n_{p-1}+1}$.

THEOREM 6. *The ring of Ω-endomorphisms of a completely reducible group that satisfies the descending (ascending) chain condition is a direct sum of two-sided ideals that are matrix rings over division rings.*

6. Nilpotent endomorphisms. We suppose that $\mathfrak{M}$ is an Ω-group for which both chain conditions hold and that $\mathfrak{B}$ is a set of nilpotent Ω-endomorphisms closed under multiplication. We wish to prove the following

THEOREM 7. *If s is the length of a composition series for $\mathfrak{M}$ and $B_1, \cdots, B_s$ are in $\mathfrak{B}$, then $B_s \cdots B_1 = 0$.*

Let $\mathfrak{N}$ be an Ω-subgroup such that $\mathfrak{N}B_i \leqq \mathfrak{N}$ and suppose that $\mathfrak{N}B_i \cdots B_l \neq 0$. Since $\mathfrak{N} \geqq \mathfrak{N}B_i \geqq \mathfrak{N}B_jB_i \geqq \cdots$ and each $\mathfrak{N}B_s \cdots B$ is an Ω-subgroup, the following is a descending chain of Ω-subgroups:

$$\mathfrak{N} \geqq \Sigma\mathfrak{N}B_i \geqq \Sigma\mathfrak{N}B_iB_j \geqq \cdots.$$

If the equality sign holds between two terms of this chain, it holds for all subsequent terms. Since $\mathfrak{N}$ has length $\leqq s$ and $\mathfrak{N}B_s \cdots B_1 \neq 0$, equality holds between $\Sigma\mathfrak{N}B_{i_1} \cdots B_{i_r} \equiv \mathfrak{N}'$ and $\Sigma\mathfrak{N}B_{i_1} \cdots B_{i_{r+1}}$ with $r < s$. Since $\mathfrak{N}B_s \cdots B_1 \neq 0$, $\mathfrak{N}' \neq 0$, and $\mathfrak{N}' = \Sigma\mathfrak{N}'B_j = \cdots$. There exists an infinite sequence $B_{i_1}, B_{i_2}, \cdots$ such that $\mathfrak{N}'B_{i_p} \cdots B_{i_1} \neq 0$. For suppose that p terms $B_{i_1}, \cdots, B_{i_p}$ have been found such that $\mathfrak{N}'B_{i_p} \cdots B_{i_1} \neq 0$. Then $\mathfrak{N}'B_{i_p} \cdots B_{i_1} = \mathfrak{N}'B_1B_{i_p} \cdots B_{i_1} + \cdots + \mathfrak{N}'B_sB_{i_p} \cdots B_{i_1}$ and since $\mathfrak{N}'B_{i_p} \cdots B_{i_1} \neq 0$, there is an i_{p+1} such that $\mathfrak{N}'B_{i_{p+1}} \cdots B_{i_1} \neq 0$. Let k be one of the indices that occurs infinitely often in the sequence $B_{i_1}, B_{i_2}, \cdots$. By dropping enough terms we may suppose that $i_1 = k$. Thus there exist s endomorphisms $C_1, \cdots, C_s$ in $\mathfrak{B}$, where $C_i = B'_iB_k$ and B'_i is a product of B's, such that $\mathfrak{N}'C_s \cdots C_1 \neq 0$. Since B_k is nilpotent, $\mathfrak{N}'B_k < \mathfrak{N}'$ and since $\Sigma\mathfrak{N}'C_i \leqq \mathfrak{N}'B_k$, $\Sigma\mathfrak{N}'C_i < \mathfrak{N}'$. By the first part of the discussion we can find an Ω-subgroup $\bar{\mathfrak{N}} \neq 0$, $< \mathfrak{N}'$ and therefore $< \mathfrak{N}$, such that $\bar{\mathfrak{N}} = \Sigma\bar{\mathfrak{N}}C_i$. If we repeat this argument, we obtain an $\bar{\bar{\mathfrak{N}}} \neq 0$ and properly contained in $\bar{\mathfrak{N}}$ and endomorphisms D_i in $\mathfrak{B}$ such that $\bar{\bar{\mathfrak{N}}} = \Sigma\bar{\bar{\mathfrak{N}}}D_i$. Thus this process leads to an infinite descending chain of Ω-subgroups and hence the assumption that $\mathfrak{N}B_s \cdots B_1 \neq 0$ is untenable. If we apply this to $\mathfrak{N} = \mathfrak{M}$, we obtain $B_s \cdots B_1 = 0$.

We note as a first consequence of this result that if $\mathfrak{M}$ is indecomposable and satisfies both chain conditions, then the set $\mathfrak{R}$ of Ω-endomorphisms that are not $(1 - 1)$ is a nilpotent ideal: We have $\mathfrak{R}^s = 0$ if s is the length of $\mathfrak{M}$.

7. The radical of the ring of endomorphisms. The assumption that both chain conditions hold for $\mathfrak{M}$ is retained in this section and in the next. Write $\mathfrak{M} = \mathfrak{M}_1 \oplus \cdots \oplus \mathfrak{M}_u$ where the $\mathfrak{M}_i \neq 0$ are indecomposable; $1 = E_1 + \cdots + E_u$

where the E_i are the projections on the $\mathfrak{M}_i$. Any A in $\mathfrak{A}$ may be written in one and only one way as ΣA_{ij} where A_{ij} is in $E_i\mathfrak{A}E_j$. An endomorphism A_{ij} maps each $\mathfrak{M}_k$, $k \neq i$, into 0 and induces an Ω-homomorphism between $\mathfrak{M}_i$ and an Ω-subgroup of $\mathfrak{M}_j$.

Let $\mathfrak{R}$ denote the set of endomorphisms of the form $B = \Sigma B_{ij}$ where no B_{ij} induces an Ω-isomorphism between $\mathfrak{M}_i$ and $\mathfrak{M}_j$. Thus either there are elements $z_i \neq 0$ in $\mathfrak{M}_i$ such that $z_iB_{ij} = 0$ or $\mathfrak{M}_iB_{ij} < \mathfrak{M}_j$. We wish to show that $\mathfrak{R}$ is a nilpotent ideal. If $j \neq k$, $B_{ij}A_{kl} = 0$ and $A_{ij}B_{kl} = 0$ since $E_jE_k = 0$. If $z_iB_{ij} = 0$, then $z_iB_{ij}A_{jl} = 0$. Now suppose that $\mathfrak{M}_j \neq \mathfrak{M}_iB_{ij} \equiv \mathfrak{M}_j'$, but that $B_{ij}A_{jl}$ induces an Ω-isomorphism between $\mathfrak{M}_i$ and $\mathfrak{M}_l$. Then A_{jl} induces an Ω-isomorphism between $\mathfrak{M}_j'$ and $\mathfrak{M}_l$. It follows readily that $\mathfrak{M}_j = \mathfrak{M}_j' \oplus \mathfrak{M}_j''$ where $\mathfrak{M}_j''$ is the subset of $\mathfrak{M}_j$ cf elements sent into 0 by A_{jl}.[1] This contradicts the assumption that $\mathfrak{M}_j$ is indecomposable. Thus we have shown that if $B_{ij} \in \mathfrak{R}$ and A is any Ω-endomorphism, then $B_{ij}A \in \mathfrak{R}$. Furthermore if $A_{ki}B_{ij}$ induces an Ω-isomorphism between $\mathfrak{M}_k$ and $\mathfrak{M}_j$, A_{ki} induces an Ω-isomorphism betwen $\mathfrak{M}_k$ and $\mathfrak{M}_i$, and hence B_{ij} induces an Ω-isomorphism between $\mathfrak{M}_i$ and $\mathfrak{M}_j$ contrary to hypothesis. Thus $AB_{ij} \in \mathfrak{R}$.

Now let B_{ij}, $C_{ij} \in \mathfrak{R}$ and consider $A_{ij} = B_{ij} + C_{ij}$. If A_{ij} induces an Ω-isomorphism $\bar{A}_{ij}$ between $\mathfrak{M}_i$ and $\mathfrak{M}_j$, set $A_{ji} = E_j\bar{A}_{ij}^{-1}E_i$. Then $E_i = B_{ij}A_{ji} + C_{ij}A_{ji}$ and $B_{ij}A_{ji}$, $C_{ij}A_{ji}$ are not $(1 - 1)$ in $\mathfrak{M}_i$. Since $\mathfrak{M}_i$ is indecomposable, this is impossible and so $A_{ij} \in \mathfrak{R}$. If we combine these results, we obtain the result that $\mathfrak{R}$ is a two-sided ideal in $\mathfrak{A}$.

If B is in $\mathfrak{R}$, we decompose $\mathfrak{M}$ as $\mathfrak{M}' \oplus \mathfrak{M}''$ in such a way that B is nilpotent in $\mathfrak{M}'$ and B is an automorphism in $\mathfrak{M}''$ (Fitting's lemma). By the Krull-Schmidt theorem there is an Ω-automorphism U such that $\mathfrak{M}'U = \mathfrak{M}^*$, $\mathfrak{M}''U = \mathfrak{M}^{**}$ where $\mathfrak{M}^* = \mathfrak{M}_1 \oplus \cdots \oplus \mathfrak{M}_t$, $\mathfrak{M}^{**} = \mathfrak{M}_{t+1} \oplus \cdots \oplus \mathfrak{M}_u$, assuming that the order of the $\mathfrak{M}_i$ has been properly chosen. Thus $U^{-1}BU = C$ is in $\mathfrak{R}$ and this endomorphism induces an automorphism $\bar{C}$ in $\mathfrak{M}^{**}$. Hence if $E^{**} = E_{t+1} + \cdots + E_u$, $E^{**}CE^{**}\bar{C}^{-1}E^{**} = E^{**}$ is in $\mathfrak{R}$, and this is impossible since E_{t+1}, E_{t+2}, $\cdots$ are in $E_{t+1}\mathfrak{A}E_{t+1}$, $\cdots$, unless $t = u$. Thus $\mathfrak{M}^{**} = 0 = \mathfrak{M}''$ and B is nilpotent. Since every element of $\mathfrak{R}$ is nilpotent, by the theorem of the preceding section, $\mathfrak{R}$ is nilpotent. Now if $\mathfrak{N}$ is any nilpotent two-sided ideal in $\mathfrak{A}$ and $N = \Sigma N_{ij}$ is in $\mathfrak{N}$, then each $N_{ij} = E_iNE_j$ is in $\mathfrak{N}$. It follows that N_{ij} is in $\mathfrak{R}$, for otherwise, we should find by a suitable multiplication that E_j is in $\mathfrak{N}$. Hence $\mathfrak{N} \leqq \mathfrak{R}$.

THEOREM 8. *Let $\mathfrak{M} = \mathfrak{M}_1 \oplus \cdots \oplus \mathfrak{M}_u$ be a decomposition of the Ω-group $\mathfrak{M}$, satisfying both chain conditions, into indecomposable $\mathfrak{M}_i \neq 0$ and let E_i be the corresponding projections. Then the set of endomorphisms ΣB_{ij}, where $B_{ij} \in E_i\mathfrak{A}E_j$ and B_{ij} is not $(1 - 1)$ between $\mathfrak{M}_i$ and $\mathfrak{M}_j$, forms a nilpotent two-sided ideal $\mathfrak{R}$ in the ring of Ω-endomorphisms $\mathfrak{A}$. $\mathfrak{R}$ contains every nilpotent two-sided ideal of $\mathfrak{A}$.*

The ideal $\mathfrak{R}$ will be called the *radical* of $\mathfrak{A}$.

[1] Cf. the proof of the Krull-Schmidt theorem, Chapter 1.

8. The structure of the ring of endomorphisms of an arbitrary group. We order the components $\mathfrak{M}_i$ so that $\mathfrak{M}_1, \cdots, \mathfrak{M}_{n_1}$ are Ω-isomorphic, $\mathfrak{M}_{n_1+1}, \cdots, \mathfrak{M}_{n_1+n_2}$ are Ω-isomorphic but not Ω-isomorphic to $\mathfrak{M}_1$, etc. Set

$$\mathfrak{M}^{(1)} = \mathfrak{M}_1 \oplus \cdots \oplus \mathfrak{M}_{n_1}, \qquad \mathfrak{M}^{(2)} = \mathfrak{M}_{n_1+1} \oplus \cdots \oplus \mathfrak{M}_{n_1+n_2}, \cdots$$

$$E^{(1)} = E_1 + \cdots + E_{n_1}, \qquad E^{(2)} = E_{n_1+1} + \cdots + E_{n_1+n_2}, \cdots .$$

Then $\mathfrak{M} = \mathfrak{M}^{(1)} \oplus \cdots \oplus \mathfrak{M}^{(t)}$. If i and j are in different ranges $n_1 + \cdots + n_{p-1} + 1, \cdots, n_1 + \cdots + n_p$ and $n_1 + \cdots + n_{q-1} + 1, \cdots, n_1 + \cdots + n_q$, $E_i\mathfrak{A}E_j \leqq$ the radical $\mathfrak{R}$. Hence $E^{(p)}\mathfrak{A}E^{(q)} \leqq \mathfrak{R}$ if $p \neq q$ and $E^{(p)}\mathfrak{A}E^{(p)} + \mathfrak{R}$ is a two-sided ideal in $\mathfrak{A}$, which determines a two-sided ideal $\bar{\mathfrak{A}}_p$ in $\bar{\mathfrak{A}} = \mathfrak{A} - \mathfrak{R}$. Since $E^{(p)}\mathfrak{A}E^{(p)}$ contains $E^{(p)}$, $\bar{\mathfrak{A}}_p \neq 0$. Evidently $\bar{\mathfrak{A}} = \bar{\mathfrak{A}}_1 \oplus \cdots \oplus \bar{\mathfrak{A}}_t$. We have seen that the correspondence between A_p in $E^{(p)}\mathfrak{A}E^{(p)}$ and its effect induced in $\mathfrak{M}^{(p)}$ is an isomorphism between $E^{(p)}\mathfrak{A}E^{(p)}$ and the ring of Ω-endomorphisms of $\mathfrak{M}^{(p)}$. It follows that the radical of $E^{(p)}\mathfrak{A}E^{(p)}$ consists of the elements ΣB_{ij} where $i, j = n_1 + \cdots + n_{p-1} + 1, \cdots, n_1 + \cdots + n_p$ and B_{ij} is not $(1-1)$. Thus the radical of $E^{(p)}\mathfrak{A}E^{(p)}$ is $(E^{(p)}\mathfrak{A}E^{(p)} \wedge \mathfrak{R}) = E^{(p)}\mathfrak{R}E^{(p)}$, and $\bar{\mathfrak{A}}_p \cong E^{(p)}\mathfrak{A}E^{(p)} - E^{(p)}\mathfrak{R}E^{(p)}$.[2]

We suppose now that $\mathfrak{M}$ is *homogeneous* in the sense that all of its indecomposable components $\mathfrak{M}_i$ are Ω-isomorphic. Then we have seen that $\mathfrak{A} = \mathfrak{B}_u$ where $\mathfrak{B}$ is isomorphic to the ring of Ω-endomorphisms of $\mathfrak{M}_i$. We have shown also that $\mathfrak{B} - \mathfrak{S}$ is a division ring if $\mathfrak{S}$ is the radical of $\mathfrak{B}$. If $\mathfrak{R}$ denotes the radical of $\mathfrak{A}$, $(\mathfrak{R} \wedge \mathfrak{B})$ is a nilpotent two-sided ideal in $\mathfrak{B}$ and is therefore contained in $\mathfrak{S}$. On the other hand, if E_{ij} are the matrix units of $\mathfrak{A}$ and $S_{ij} \in \mathfrak{S}$, then the set of elements $\Sigma E_{ij}S_{ij}$ is a nilpotent ideal in $\mathfrak{A}$ and hence is contained in $\mathfrak{R}$. In particular, $\Sigma E_{ii}S \in \mathfrak{R}$ and $(\mathfrak{R} \wedge \mathfrak{B}) = \mathfrak{S}$. If $B = \Sigma E_{ij}B_{ij}$ is any element of $\mathfrak{R}$, $B_{ij} = \Sigma E_{ki}BE_{jk}$ is in $(\mathfrak{R} \wedge \mathfrak{B}) = \mathfrak{S}$. Thus $\mathfrak{R} = \mathfrak{S}_u$ and the difference ring $\bar{\mathfrak{A}} = \mathfrak{A} - \mathfrak{R} \cong (\mathfrak{B} - \mathfrak{S})_u$, a matrix ring over a division ring. Since rings of this form are necessarily simple, we have shown that $\bar{\mathfrak{A}}$ is simple. On the other hand, we have seen that $\bar{\mathfrak{A}} = \bar{\mathfrak{A}}_1 \oplus \cdots \oplus \bar{\mathfrak{A}}_t$ and so if $\bar{\mathfrak{A}}$ is simple, $t = 1$ and $\mathfrak{M}$ is homogeneous. The following implications have therefore been established:

$\mathfrak{M}$ is homogeneous $\rightarrow \mathfrak{A} = \mathfrak{B}_u$, $\mathfrak{B}$ completely primary $\rightarrow \mathfrak{A} - \mathfrak{R}$ is simple $\rightarrow$ $\mathfrak{M}$ is homogeneous. Hence we have

THEOREM 9. *The following conditions are equivalent:*

1. *$\mathfrak{M}$ is homogeneous.*
2. *$\mathfrak{A} = \mathfrak{B}_u$, $\mathfrak{B}$ completely primary.*
3. *$\mathfrak{A} - \mathfrak{R}$ is simple.*

The question of the uniqueness of the representation of $\mathfrak{A}$ as $\mathfrak{B}_u$ is settled in the following

[2] We are using the isomorphism theorem that if $\mathfrak{A} \geqq \mathfrak{B}$ and $\mathfrak{R}$ is an ideal in $\mathfrak{A}$, then $(\mathfrak{B} + \mathfrak{R}) - \mathfrak{R} \cong \mathfrak{B} - (\mathfrak{B} \wedge \mathfrak{R})$.

THEOREM 10. *Suppose that the conditions of the preceding theorem hold. Then if* $\mathfrak{A} = \mathfrak{B}'_{u'}$ *where* $\mathfrak{B}'$ *is completely primary,* $u = u'$ *and* $\mathfrak{B}$ *and* $\mathfrak{B}'$ *are isomorphic.*

Let E'_{ij} be the new set of matrix units. Since $E'_{ii}\mathfrak{A}E'_{ii} \cong \mathfrak{B}'$, the identity E'_{ii} is the only idempotent element in $E'_{ii}\mathfrak{A}E'_{ii}$. Hence E'_{ii} is a primitive idempotent element and the components of the decomposition $\mathfrak{M} = \mathfrak{M}E'_{11} \oplus \cdots \oplus \mathfrak{M}E'_{u'u'}$ are indecomposable. By the Krull-Schmidt theorem, $u = u'$ and there is an Ω-automorphism A such that $A^{-1}E_{ii}A = E'_{i'i'}$ for a suitable permutation $1' \cdots, u'$ of $1, \cdots, u$. Since $\mathfrak{M}$ is homogeneous, there is an Ω-automorphism P such that $\mathfrak{M}_{i'}P = \mathfrak{M}_i$. Hence if $B = PA$,we have $B^{-1}E_{ii}B = E'_{ii}$. Then the endomorphism $B^{-1}E_{ij}B$ induces an Ω-isomorphism between $\mathfrak{M}'_i$ and $\mathfrak{M}'_j$ and so $\sum_1^u B^{-1}E_{j1}BE'_{1j}$ and $C = \Sigma E_{j1}BE'_{1j}$ are Ω-automorphisms. Evidently $E_{ij}C = CE'_{ij}$, $E_{ij} = CE'_{ij}C^{-1}$. Since $\mathfrak{B}$ and $\mathfrak{B}'$ are respectively the sets of endomorphisms commutative with the E_{ij} and the E'_{ij}, we have $\mathfrak{B} = C\mathfrak{B}'C^{-1}$.

9. Direct sums. We consider now the theory of abstract rings. In order to include the case of algebras, we suppose that $\mathfrak{A}$ is a ring and that Φ is a set of endomorphisms of the additive group of $\mathfrak{A}$ which commute with the elements of $\mathfrak{A}_r$ and the elements of $\mathfrak{A}_l$. We begin with some elementary remarks on direct decompositions of $\mathfrak{A}$ into Φ-ideals. The first of these is a special case of the theorem connecting direct decompositions and projections, namely,

LEMMA. *If* $\mathfrak{A}$ *is a* Φ*-ring with an identity and* $\mathfrak{A} = \mathfrak{J}_1 \oplus \cdots \oplus \mathfrak{J}_u$ *is a direct decomposition of* $\mathfrak{A}$ *into left* Φ*-ideals* $\neq 0$, *then* $1 = e_1 + \cdots + e_u$, $e_j^2 = e_j \neq 0$, $e_je_k = 0$ *if* $j \neq k$ *and* $\mathfrak{J}_j = \mathfrak{A}e_j$.

A direct proof is the following. Write $1 = e_1 + \cdots + e_u$, where the $e_j \in \mathfrak{J}_j$. Then any $a = ae_1 + \cdots + ae_u$ and $ae_j \in \mathfrak{J}_j$. If $a = a_j \in \mathfrak{J}_j$, $a_j = a_je_1 + \cdots + a_je_u$. Since the $\mathfrak{J}_j$ are independent, all of the $a_je_k = 0$ with the exception of $a_je_j = a_j$. Hence $\mathfrak{J}_j = \mathfrak{A}e_j$ and $e_j^2 = e_j \neq 0$, $e_je_k = 0$ if $j \neq k$.

Now suppose that $\mathfrak{A} = \mathfrak{A}_1 \oplus \cdots \oplus \mathfrak{A}_t$ is a direct sum of two-sided Φ-ideals. If $j \neq k$, $\mathfrak{A}_j\mathfrak{A}_k \leqq \mathfrak{A}_j \wedge \mathfrak{A}_k = 0$. Hence any Φ-ideal (left, right or two-sided) of $\mathfrak{A}_j$ is a Φ-ideal of $\mathfrak{A}$. On the other hand, let $\mathfrak{J}$ be a left Φ-ideal in $\mathfrak{A}$. Then $\mathfrak{J}_j = \mathfrak{A}_j\mathfrak{J} \leqq \mathfrak{J} \wedge \mathfrak{A}_j$ is a left Φ-ideal in $\mathfrak{A}_j$. Since

$$\mathfrak{J} = 1\mathfrak{J} = \mathfrak{A}\mathfrak{J} = \mathfrak{J}_1 \oplus \cdots \oplus \mathfrak{J}_t, \tag{3}$$

$\mathfrak{J} \wedge \mathfrak{A}_j \leqq \mathfrak{J}_j$ and hence $\mathfrak{J} \wedge \mathfrak{A}_j = \mathfrak{J}_j$. The decomposition (3) shows in particular that $\mathfrak{A}$ satisfies the ascending or descending chain condition for one or two-sided ideals if and only if each $\mathfrak{A}_j$ does.

Suppose that the $\mathfrak{A}_j$ are indecomposable two-sided Φ-ideals $\neq 0$, that is, $\mathfrak{A}_j = \mathfrak{A}'_j \oplus \mathfrak{A}''_j$ occurs only if either $\mathfrak{A}'_j$ or $\mathfrak{A}''_j = 0$. Then the $\mathfrak{A}_j$ are uniquely determined. For if $\mathfrak{A} = \mathfrak{B}_1 \oplus \cdots \oplus \mathfrak{B}_u$ is a decomposition of $\mathfrak{A}$ into indecomposable two-sided Φ-ideals $\neq 0$, $\mathfrak{A}_1 \wedge \mathfrak{B}_j = 0$ for all but one j since $\mathfrak{A}_1 = (\mathfrak{A}_1 \wedge \mathfrak{B}_1) \oplus \cdots \oplus (\mathfrak{A}_1 \wedge \mathfrak{B}_u)$. We may suppose that $j = 1$ and we obtain $\mathfrak{A}_1 = \mathfrak{A}_1 \wedge \mathfrak{B}_1$ and, by symmetry, $\mathfrak{B}_1 = \mathfrak{A}_1 \wedge \mathfrak{B}_1 = \mathfrak{A}_1$. Similarly, if the order is properly chosen, $\mathfrak{A}_2 = \mathfrak{B}_2$, $\cdots$ and $t = u$.

THEOREM 11. *If $\mathfrak{A}$ is a Φ-ring with an identity and $\mathfrak{A} = \mathfrak{A}_1 \oplus \cdots \oplus \mathfrak{A}_t$ is a direct decomposition of $\mathfrak{A}$ into indecomposable two-sided Φ-ideals $\neq 0$, then any decomposition of $\mathfrak{A}$ into indecomposable Φ-ideals $\neq 0$ has the same components as the given decomposition.*

10. The radical. We recall the definition of a *nil ring* as one containing only nilpotent elements and of a *nilpotent ring* as one having the property that a finite power of it is 0. Thus the statement that $\mathfrak{A}$ is nilpotent means that for a suitable integer s, any product $a_1a_2 \cdots a_s = 0$ for any a_i in $\mathfrak{A}$. In particular $a^s = 0$ for all a and so $\mathfrak{A}$ is a nil ring. It is remarkable that the converse of this rather trivial statement holds if $\mathfrak{A}$ satisfies the descending chain condition for Φ-ideals. Before proceeding to the proof we note the following lemmas.

LEMMA 1. *If $\mathfrak{J}_1$ and $\mathfrak{J}_2$ are nilpotent left Φ-ideals of $\mathfrak{A}$, then $\mathfrak{J}_1 + \mathfrak{J}_2$ is nilpotent.*

Let $\mathfrak{J}_1^r = 0$ and $\mathfrak{J}_2^s = 0$. Now $(\mathfrak{J}_1 + \mathfrak{J}_2)^k = \Sigma\mathfrak{J}_{i_1}\mathfrak{J}_{i_2} \cdots \mathfrak{J}_{i_k}$ where $i_j = 1, 2$. If $k = r + s - 1$, each product contains either at least r $\mathfrak{J}_1$'s or at least s $\mathfrak{J}_2$'s. In the first case we replace any $\mathfrak{J}_2\mathfrak{J}_1$ in the product by $\mathfrak{J}_1$. After a finite number of such replacements we obtain $\mathfrak{J}_{i_1} \cdots \mathfrak{J}_{i_k} \leq \mathfrak{J}_1^r\mathfrak{J}_2^t = 0$. Similarly if there are at least s $\mathfrak{J}_2$'s, we have $\mathfrak{J}_{i_1} \cdots \mathfrak{J}_{i_k} \leqq \mathfrak{J}_2^s\mathfrak{J}_1^{t'} = 0$ and so $\mathfrak{J}_{i_1} \cdots \mathfrak{J}_{i_k} = 0$ in all cases and $(\mathfrak{J}_1 + \mathfrak{J}_2)^k = 0$.

LEMMA 2. *If $\mathfrak{J}$ is a nilpotent left Φ-ideal of $\mathfrak{A}$, $\mathfrak{J}$ is contained in a nilpotent two-sided Φ-ideal.*

Since $\mathfrak{A}\mathfrak{J} \leqq \mathfrak{J}$, we have $(\mathfrak{J} + \mathfrak{J}\mathfrak{A})^k \leqq \mathfrak{J}^k + \mathfrak{J}^k\mathfrak{A}$. Hence if $\mathfrak{J}^r = 0$, $(\mathfrak{J} + \mathfrak{J}\mathfrak{A})^r = 0$. Evidently $\mathfrak{J} + \mathfrak{J}\mathfrak{A}$ is a two-sided Φ-ideal.

As a consequence of these lemmas we have

THEOREM 12. *Let $\mathfrak{R}$ be the join of all nilpotent left Φ-ideals of a Φ-ring. Then $\mathfrak{R}$ is a nil two-sided Φ-ideal.*

By the join we mean the smallest subgroup containing all the nilpotent left Φ-ideals. If $b \in \mathfrak{R}$, $b \in \mathfrak{J}_1 + \cdots + \mathfrak{J}_m = \mathfrak{J}$ for suitable nilpotent left Φ-ideals $\mathfrak{J}_j$. By Lemma 1, $\mathfrak{J}$ is nilpotent and hence b is nilpotent. By Lemma 2, $\mathfrak{J} \leqq \mathfrak{S}$ a nilpotent two-sided Φ-ideal. Hence $ba \in \mathfrak{S} \leqq \mathfrak{R}$ for any a and so $\mathfrak{R}$ is a right ideal as well as a left ideal.

We suppose now (and for the remainder of the chapter) that $\mathfrak{A}$ is a Φ-ring satisfying the descending chain condition for left Φ-ideals. Let $\mathfrak{N}$ be a nil left Φ-ideal in $\mathfrak{A}$. Since the product of Φ-ideals is a Φ-ideal and $\mathfrak{N} \geqq \mathfrak{N}^2 \geqq \cdots$, there is an integer k such that $\mathfrak{N}^k = \mathfrak{N}^{k+1}$ and hence $\mathfrak{N}^k = \mathfrak{N}^{k+1} = \mathfrak{N}^{k+2} = \cdots$. We wish to show that $\mathfrak{M} = \mathfrak{N}^k = 0$. Evidently $\mathfrak{M} = \mathfrak{M}^2$ is a nil left Φ-ideal. If $\mathfrak{M} \neq 0$, let $\mathfrak{J}$ be a minimal left Φ-ideal contained in $\mathfrak{M}$ with the property that $\mathfrak{M}\mathfrak{J} \neq 0$. (The existence of such an ideal is assured by the descending chain condition.) Then there is an element b in $\mathfrak{J}$ such that $\mathfrak{M}b \neq 0$. Since $\mathfrak{M}b$ is a left Φ-ideal contained in $\mathfrak{J}$ and $\mathfrak{M}(\mathfrak{M}b) = \mathfrak{M}b$, we have $\mathfrak{M}b = \mathfrak{J}$. It follows that there is an element m in $\mathfrak{M}$ such that $mb = b$ and this is clearly impossible since it implies that $b = mb = m^2b = \cdots = m^rb = 0$ if r is sufficiently large. This contradiction shows that $\mathfrak{M} = 0$ and so we have the following theorem.

THEOREM 13. *If $\mathfrak{A}$ is a Φ-ring satisfying the descending chain condition for left Φ-ideals, then any nil left Φ-ideal of $\mathfrak{A}$ is nilpotent.*[3]

As a consequence of this theorem we see that the ideal $\mathfrak{R}$ defined in Theorem 12 is nilpotent. If $\mathfrak{J}$ is any nil left Φ-ideal of $\mathfrak{A}$, $\mathfrak{J} \leqq \mathfrak{R}$. This is clear since $\mathfrak{J}$ is nilpotent. Furthermore $\mathfrak{R}$ contains every nilpotent right Φ-ideal. For, as before, any such ideal is contained in a nilpotent two-sided Φ-ideal and the latter is contained in $\mathfrak{R}$. We shall call $\mathfrak{R}$ the (left) *radical* of $\mathfrak{A}$. Similarly if $\mathfrak{A}$ satisfies the descending chain condition for right Φ-ideals, $\mathfrak{A}$ has a right radical $\mathfrak{R}'$ that contains all nil right Φ-ideals. If both descending chain conditions hold, $\mathfrak{R} = \mathfrak{R}'$. We prove next

THEOREM 14. *If $\mathfrak{A}$ is a Φ-ring satisfying both chain conditions for left Φ-ideals, then any left Φ-ideal $\mathfrak{J}$ that contains a non-nilpotent element contains an idempotent element $\neq 0$.*

Suppose that $\mathfrak{J} = \mathfrak{J}_1 \oplus \mathfrak{J}_2$ where the $\mathfrak{J}_j$ are left Φ-ideals. If $\mathfrak{J}_1$ and $\mathfrak{J}_2$ are nil ideals, they are nilpotent and hence $\mathfrak{J}$ is nilpotent. Thus at least one of the $\mathfrak{J}_j$ is not a nil ideal and so we may suppose at the start that $\mathfrak{J}$ is indecomposable when regarded as a group relative to the set of endomorphisms Ω, the logical sum of $\mathfrak{A}_l$ and Φ. The mapping $y \to yb \equiv yB$ for y, b in $\mathfrak{J}$ is an Ω-endomorphism. Hence by Fitting's lemma, either B is nilpotent or B is an automorphism. If B is nilpotent, b is a nilpotent element and so by the assumption that $\mathfrak{J}$ is not a nil ideal, there is a b such that B is $(1 - 1)$. Then $\mathfrak{J}B = \mathfrak{J}$ and there is an element e in $\mathfrak{J}$ such that $eB = b$. Then $eb = b$ and $(e^2 - e)b = (e^2 - e)B = 0$, and so $e^2 = e \neq 0$ is an idempotent element in $\mathfrak{J}$.

In a similar manner we may use Schur's lemma to prove

THEOREM 15. *If $\mathfrak{J}$ is an irreducible left Φ-ideal, then either $\mathfrak{J}^2 = 0$ or $\mathfrak{J} = \mathfrak{A}e$ where e is idempotent.*

We consider again the mapping $y \to yb = yB$, y and b in $\mathfrak{J}$. Either $B = 0$ or B is $(1 - 1)$. As before, the second possibility implies that $\mathfrak{J}$ contains an idempotent element e. Then $\mathfrak{J} = \mathfrak{A}e$.

11. The structure of semi-simple rings. We shall call a Φ-ring $\mathfrak{A}$ *semi-simple* if 1) it satisfies the descending chain condition for left Φ-ideals and 2) it has no nilpotent left Φ-ideals. It follows from the preceding section that $\mathfrak{A}$ contains no nil left Φ-ideals and no nilpotent right Φ-ideals $\neq 0$. If $\mathfrak{A}$ is a ring satisfying 1) and $\mathfrak{R}$ is its radical, then $\bar{\mathfrak{A}} = \mathfrak{A} - \mathfrak{R}$ is semi-simple. For if $\bar{\mathfrak{J}}$ is a nilpotent left Φ-ideal of $\bar{\mathfrak{A}}$, $\bar{\mathfrak{J}} = \mathfrak{J} - \mathfrak{R}$ where $\mathfrak{J}$ is a left Φ-ideal of $\mathfrak{A}$ and $\mathfrak{J}^k \leqq \mathfrak{R}$ for a suitable k. Then $\mathfrak{J}^{ks} \leqq \mathfrak{R}^s = 0$ if s is sufficiently large. Hence $\mathfrak{J} \leqq \mathfrak{R}$ and $\bar{\mathfrak{J}} = 0$. The following theorem is fundamental in determining the structure of semi-simple rings.

THEOREM 16. *Any semi-simple Φ-ring has an identity, and its lattice of left Φ-ideals is completely reducible. Conversely, if $\mathfrak{A}$ is a Φ-ring having the properties*

[3] This theorem is due to C. Hopkins. I am indebted to Professor R. Brauer for the present proof.

1) $\mathfrak{A}$ *has an identity and* 2) *the lattice of left* Φ*-ideals is completely reducible and satisfies the descending chain condition, then* $\mathfrak{A}$ *is semi-simple.*

Assume that $\mathfrak{A}$ is semi-simple. We shall show first that any irreducible left Φ-ideal $\mathfrak{J} \neq 0$ has a complement. Since $\mathfrak{J}^2 \neq 0$, $\mathfrak{J} = \mathfrak{A}e$ where e is an idempotent element in $\mathfrak{J}$. Let $\mathfrak{J}'$ be the set of elements b' in $\mathfrak{A}$ such that $b'e = 0$. Then $\mathfrak{J}'$ is a left Φ-ideal and $\mathfrak{J} \wedge \mathfrak{J}' = 0$. Since $a = ae + (a - ae) = b + b'$ where $b \in \mathfrak{J}$, $b' \in \mathfrak{J}'$, $\mathfrak{J}'$ is a complement of $\mathfrak{J}$.

Set $\mathfrak{J} = \mathfrak{J}_1$, $e = e_1$. If $\mathfrak{J}'$ is not minimal, let $\mathfrak{J}_2 = \mathfrak{A}e_2$, where $e_2^2 = e_2$, be an irreducible left Φ-ideal $\neq 0$ contained in $\mathfrak{J}'$. Then $\mathfrak{A} = \mathfrak{J}_2 \oplus \mathfrak{J}_2'$ where $\mathfrak{J}_2'$ is the left Φ-ideal of elements c' such that $c'e_2 = 0$. It follows that $\mathfrak{J}' = \mathfrak{J}_2 \oplus \mathfrak{J}''$, where $\mathfrak{J}'' \equiv \mathfrak{J}_2' \wedge \mathfrak{J}'$ may be characterized as the set of elements b'' such that $b''e_1 = b''e_2 = 0$. Hence $\mathfrak{A} = \mathfrak{J}_1 \oplus \mathfrak{J}_2 \oplus \mathfrak{J}''$ where $\mathfrak{J}'' < \mathfrak{J}'$. If $\mathfrak{J}''$ is not irreducible, we repeat the argument and obtain $\mathfrak{A} = \mathfrak{J}_1 \oplus \mathfrak{J}_2 \oplus \mathfrak{J}_3 \oplus \mathfrak{J}'''$ where $\mathfrak{J}_j = \mathfrak{A}e_j \neq 0$ is irreducible, $e_je_i = 0$ if $i < j$, $e_i^2 = e_i \neq 0$ and $\mathfrak{J}'''$ is a left Φ-ideal $< \mathfrak{J}''$. Continuing in this way, we obtain finally

$$\mathfrak{A} = \mathfrak{J}_1 \oplus \mathfrak{J}_2 \oplus \cdots \oplus \mathfrak{J}_u, \qquad \mathfrak{J}_j = \mathfrak{A}e_j \text{ irreducible},$$

where $e_i^2 = e_i$ and $e_je_i = 0$ if $i < j$. Hence we have proved thé complete reducibility of the lattice.

If we put $v = \sum e_i - \sum_{i<j} e_ie_j + \cdots + (-1)^{u-1}e_1e_2 \cdots e_u$, we may verify that $e_kv = e_k$ for $k = 1, \cdots, u$. Since any $a = \Sigma a_ke_k$, $av = a$ for all a. In particular $v^2 = v$. The set of elements z such that $vz = 0$ is a right Φ-ideal $\mathfrak{B}$. Since $z_1z_2 = (z_1v)z_2 = z_1(vz_2) = 0$, $\mathfrak{B}^2 = 0$ and hence $\mathfrak{B} = 0$. Hence for any a, we have $a - va = 0$, since $v(a - va) = 0$, and so $a = va$. Thus v is a left identity also and we may set $v = 1$.

Conversely, if $\mathfrak{A}$ has an identity and its lattice of left Φ-ideals is completely reducible, any left Φ-ideal $\mathfrak{J} \neq 0$ has the form $\mathfrak{A}e$, $e^2 = e$. For, $\mathfrak{A} = \mathfrak{J} \oplus \mathfrak{J}'$ and hence $1 = e + e'$ where $e \in \mathfrak{J}$, $e' \in \mathfrak{J}'$, $e^2 = e \neq 0$, $e'^2 = e'$ and $ee' = e'e = 0$. Then $\mathfrak{J} = \mathfrak{A}e$. Since $\mathfrak{J}$ contains the idempotent element $e \neq 0$, it can not be nilpotent. If $\mathfrak{A}$ satisfies the descending chain condition, it will follow that $\mathfrak{A}$ is semi-simple. The theorem is therefore proved as is also the

Corollary. *Any left* Φ*-ideal of a semi-simple* Φ*-ring is principal and is generated by an idempotent element.*

If we recall the general lattice theoretic argument of Chapter 3, **4**, we obtain the following dual of Theorem 16.

Theorem 17. *If* $\mathfrak{A}$ *is a semi-simple* Φ*-ring, there exist maximal left* Φ*-ideals* $\mathfrak{M}_1, \cdots, \mathfrak{M}_u$ *in* $\mathfrak{A}$ *such that*

$$0 = \mathfrak{M}_1 \wedge \cdots \wedge \mathfrak{M}_u,$$

$$\mathfrak{M}_i + (\mathfrak{M}_1 \wedge \cdots \wedge \mathfrak{M}_{i-1} \wedge \mathfrak{M}_{i+1} \wedge \cdots \wedge \mathfrak{M}_u) = \mathfrak{A}.$$

For if $\mathfrak{A} = \mathfrak{J}_1 \oplus \cdots \oplus \mathfrak{J}_u$ where the $\mathfrak{J}_j$ are irreducible, then $\mathfrak{M}_i = \mathfrak{J}_1 + \cdots + \mathfrak{J}_{i-1} + \mathfrak{J}_{i+1} + \cdots + \mathfrak{J}_u$ satisfies the conditions of the theorem.

This result leads to an interesting "arithmetic" characterization of the radical, namely

THEOREM 18. *Let $\mathfrak{A}$ be a Φ-ring with an identity for which the descending chain condition for left Φ-ideals holds. Then the radical $\mathfrak{R}$ of $\mathfrak{A}$ is the intersection of all maximal left Φ-ideals of $\mathfrak{A}$.*

Since $\bar{\mathfrak{A}} = \mathfrak{A} - \mathfrak{R}$ is semi-simple, $\bar{0} = \bar{\mathfrak{M}}_1 \wedge \cdots \wedge \bar{\mathfrak{M}}_u$ for suitable maximal left Φ-ideals of $\bar{\mathfrak{A}}$. Hence if $\mathfrak{M}_i$ is the left Φ-ideal of elements mapped into those of $\bar{\mathfrak{M}}_i$, then $\mathfrak{M}_1 \wedge \cdots \wedge \mathfrak{M}_u = \mathfrak{R}$. By the First Isomorphism Theorem, the $\mathfrak{A}_l$-groups[4] $\bar{\mathfrak{A}} - \bar{\mathfrak{M}}_i$ and $\mathfrak{A} - \mathfrak{M}_i$ are isomorphic. Hence $\mathfrak{A} - \mathfrak{M}_i$ is irreducible and $\mathfrak{M}_i$ is a maximal left Φ-ideal. If we denote the intersection of all of the maximal left Φ-ideals by $\mathfrak{S}$, we have, therefore, proved that $\mathfrak{S} \leqq \mathfrak{R}$. On the other hand let $\mathfrak{M}$ be any maximal left Φ-ideal. Then either $\mathfrak{M} + \mathfrak{R} = \mathfrak{M}$ or $\mathfrak{M} + \mathfrak{R} = \mathfrak{A}$. In the latter case $1 = m + r$, m in $\mathfrak{M}$ and r in $\mathfrak{R}$, and so $1 = (1 + r + r^2 + \cdots)(1 - r) = (1 + r + r^2 + \cdots)m \in \mathfrak{M}$. This contradicts the maximality of $\mathfrak{M}$ and proves that $\mathfrak{M} + \mathfrak{R} = \mathfrak{M}$, i.e. $\mathfrak{R} \leqq \mathfrak{M}$. Thus $\mathfrak{R} \leqq \mathfrak{S}$ and the theorem is proved.

We proceed now to the fundamental structure theorem for semi-simple Φ-rings. We base the proof on two facts: 1) $\mathfrak{A}$ is isomorphic to $\mathfrak{A}_r$ and 2) $\mathfrak{A}_r$ is the complete set of $\mathfrak{A}_l$-endomorphisms. Both of these are consequences of the fact that $\mathfrak{A}$ has an identity. Now we have seen that $\mathfrak{A} = \mathfrak{J}_1 \oplus \cdots \oplus \mathfrak{J}_u$ where the $\mathfrak{J}_j$ are irreducible $\mathfrak{A}_l$-groups. Hence by the general theory of **5**, $\mathfrak{A}_r$ is a direct sum of two-sided ideals that are matrix rings over division rings. Thus it follows that $\mathfrak{A} = P^{(1)}_{n_1} \oplus \cdots \oplus P^{(t)}_{n_t}$, $P^{(i)}$ a division ring. Now the two-sided ideals $\mathfrak{A}_i = P^{(i)}_{n_i}$ are Φ-ideals since the elements of Φ are multiplications by elements of the center $\mathfrak{C}$ of $\mathfrak{A}$. If 1_i is the identity of $\mathfrak{A}_i$, the endomorphism induced by α in $\mathfrak{A}_i$ is also the multiplication by $1_i\alpha$, an element of the center $\mathfrak{C}_i$ of $\mathfrak{A}_i$. Since $P^{(i)} \geqq \mathfrak{C}_i$, $P^{(i)}$ is a Φ-subring of $\mathfrak{A}$. This proves the first part of the structure theorem:

THEOREM 19. *Any semi-simple Φ-ring is a direct sum of two-sided ideals that are matrix rings over Φ-division rings and conversely.*

To prove the converse it suffices, by the considerations of **9**, to prove that a matrix ring over a division ring is semi-simple. We saw in Chapter 2 that a ring of this type is a direct sum of irreducible left ideals. These are Φ-ideals. Hence $\mathfrak{A}$ is semi-simple by the preceding theorem.

From the theory of matrix rings we obtain also the

COROLLARY. *A semi-simple Φ-ring satisfies both chain conditions for left (right) Φ-ideals.*

This corollary shows, in particular, that the conditions imposed on the left ideals in the definition of a semi-simple ring hold also for right ideals. We can also begin with the conditions on the right ideals. We then obtain the result that $\mathfrak{A}_l$ is a direct sum of matrix rings over division rings. Hence $\mathfrak{A}$ is anti-

[4] We need not mention Φ since $\mathfrak{A}_l \geqq \Phi$.

isomorphic to a ring having this structure, and since a ring anti-isomorphic to a matrix ring over a division ring is a matrix ring over a division ring, we see that $\mathfrak{A}$ is semi-simple. Any theorem that holds for left (right) ideals in a semi-simple ring has a dual for right (left) ideals. For example, by the above corollary, any right ideal in $\mathfrak{A}$ is principal.

If $\mathfrak{A}$ is a simple Φ-ring satisfying the descending chain condition for left Φ-ideals and $\mathfrak{A}$ is not semi-simple, then $\mathfrak{A}$ is nilpotent. Since $\mathfrak{A}^2 < \mathfrak{A}$ is a two-sided Φ-ideal, we have $\mathfrak{A}^2 = 0$. Hence any element $b \neq 0$ generates a two-sided ideal and therefore b generates the whole of $\mathfrak{A}$. Thus $\mathfrak{A} = \{b\}$, the set of elements $\Sigma b\alpha_1\alpha_2 \cdots \alpha_m$, α_i in Φ or $\alpha_i = \pm 1$, for any $b \neq 0$, and $b^2 = 0$. A ring of this type is called a *zero-Φ-ring*. Hence if a simple $\mathfrak{A}$ satisfies the descending chain condition and is not a zero Φ-ring, it is semi-simple. The above theorem then applies and we obtain

THEOREM 20. *A simple Φ-ring satisfying the descending chain condition for left ideals is either a zero Φ-ring or a matrix ring P_n over a Φ-division ring P, and conversely. If $\mathfrak{A} = P_n = \Psi_m$ where Ψ is a division ring, then $n = m$ and P and Ψ are isomorphic.*

The direct part of the theorm is an immediate consequence of the theorem on semi-simple rings. The converse and the uniqueness of n and of P in the sense of isomorphism were proved in Chapter 2.

It is interesting to note that the corresponding statement for rings satisfying the ascending chain condition for left ideals does not hold. For let $P = P_0(\xi)$ be the field of rational functions in one indeterminate over a field P_0 of characteristic 0 and let $\mathfrak{A} = P[t,']$ the ring of differential polynomials over P, i.e. polynomials in t where $\alpha t = t\alpha + \alpha'$, α' the ordinary derivative of α. $\mathfrak{A}$ is a principal ideal domain and hence satisfies the ascending chain condition. It is easy to show that there are no proper two-sided ideals in $\mathfrak{A}$ so that $\mathfrak{A}$ is simple. However, since $\mathfrak{A}$ is not a division ring, it does not have the form Ψ_n, Ψ a division ring.

If we use **9** and the above theorems, we obtain

THEOREM 21. *The lattice of two-sided Φ-ideals of a semi-simple Φ-ring $\mathfrak{A}$ is completely reducible. If $\mathfrak{A} = \mathfrak{A}_1 \oplus \cdots \oplus \mathfrak{A}_t = \mathfrak{B}_1 \oplus \cdots \oplus \mathfrak{B}_s$ are decompositions of $\mathfrak{A}$ into irreducible two-sided Φ-ideals, then $s = t$ and $\mathfrak{A}_i = \mathfrak{B}_i$ for a suitable ordering of the $\mathfrak{B}_i$.*

Suppose that $\mathfrak{A} = \mathfrak{A}_1 \oplus \cdots \oplus \mathfrak{A}_t$ where the $\mathfrak{A}_i$ are irreducible two-sided Φ-ideals. If $\mathfrak{B}$ is a two-sided Φ-ideal of $\mathfrak{A}$, we have seen that $\mathfrak{B} = \mathfrak{B}_1 \oplus \cdots \oplus \mathfrak{B}_t$ where $\mathfrak{B}_i = \mathfrak{B} \wedge \mathfrak{A}_i$ is a two-sided Φ-ideal in $\mathfrak{A}$. Hence either $\mathfrak{B}_i = \mathfrak{A}_i$ or $= 0$. Thus $\mathfrak{B} = \mathfrak{A}_{i_1} \oplus \cdots \oplus \mathfrak{A}_{i_r}$ and there are exactly 2^t distinct two-sided Φ-ideals in $\mathfrak{A}$.

We consider now the connection between the decomposition of $\mathfrak{A}$ into left Φ-ideals and into two-sided Φ-ideals. If $\mathfrak{J}$ is an irreducible left Φ-ideal then we have seen that $\mathfrak{J}$ is contained in one of the $\mathfrak{A}_i$, say $\mathfrak{A}_1$. If $\mathfrak{J}'$ is a second left Φ-ideal and $\mathfrak{J}' \leqq \mathfrak{A}_2$, $\mathfrak{J}$ and $\mathfrak{J}'$ are not $\mathfrak{A}_l$-isomorphic since for 1_1, the identity of $\mathfrak{A}_1$, we have $1_1\mathfrak{J} = \mathfrak{J}$ while $1_1\mathfrak{J}' = 0$. Thus if $\mathfrak{B}$ is the join of all the irreducible left Φ-ideals $\mathfrak{A}_l$-isomorphic to $\mathfrak{J}$, $\mathfrak{B} \leqq \mathfrak{A}_1$. We assert that $\mathfrak{B}$ is a two-sided ideal. For if $b \in \mathfrak{B}$, $b = y_1 + \cdots + y_r$ where $y_i \in \mathfrak{J}_i$, an irreducible Φ-ideal isomorphic to $\mathfrak{J}$. If a is arbitrary, $\mathfrak{J}_i a$ is either 0 or is $\mathfrak{A}_l$-isomorphic to $\mathfrak{J}$. Hence $y_i a \in \mathfrak{B}$ and $ba \in \mathfrak{B}$, and so $\mathfrak{B}$ is a right ideal as well as a left ideal. Since $\mathfrak{A}_1$ is irreducible, $\mathfrak{B} = \mathfrak{A}_1$. Hence we have

THEOREM 22. *If $\mathfrak{A}$ is semi-simple and $\mathfrak{A} = \mathfrak{A}_1 \oplus \cdots \oplus \mathfrak{A}_t$ where the $\mathfrak{A}_i$ are irreducible two-sided Φ-ideals, then each $\mathfrak{A}_i$ is the join of the set of irreducible left (right) Φ-ideals that are $\mathfrak{A}_l$- ($\mathfrak{A}_r$-)isomorphic to a fixed irreducible left (right) Φ-ideal.*

COROLLARY. *The number t of irreducible two-sided components is the same as the number of classes of non-isomorphic (relative to $\mathfrak{A}_l$ or $\mathfrak{A}_r$) irreducible left (right) ideals.*

If $\mathfrak{J}$ is a left Φ-ideal of $\mathfrak{A}$, we define $\mathfrak{Z}_r(\mathfrak{J})$ to be the set of elements b in $\mathfrak{A}$ such that $\mathfrak{J}b = 0$. $\mathfrak{Z}_r(\mathfrak{J})$ is a right Φ-ideal. Similarly, if $\mathfrak{J}'$ is a right Φ-ideal, we can obtain a left Φ-ideal $\mathfrak{Z}_l(\mathfrak{J}')$ as the set of elements c such that $c\mathfrak{J}' = 0$. If $\mathfrak{A}$ is semi-simple, we may suppose that $\mathfrak{J} = \mathfrak{A}e$ and we may write $\mathfrak{A} = \mathfrak{A}e \oplus \mathfrak{A}e'$, where $e^2 = e$, $e'^2 = e'$, $ee' = e'e = 0$. Then we obtain also $\mathfrak{A} = e\mathfrak{A} \oplus e'\mathfrak{A}$. We assert that $\mathfrak{Z}_r(\mathfrak{A}e) = e'\mathfrak{A}$. For, $e'\mathfrak{A} \leqq \mathfrak{Z}_r(\mathfrak{A}e)$ and if $b \in \mathfrak{Z}_r(\mathfrak{A}e)$, we have $eb = 0$ and hence $b = (e + e')b = e'b \in e'\mathfrak{A}$. By symmetry, $\mathfrak{Z}_l(e'\mathfrak{A}) = \mathfrak{A}e$. Thus $\mathfrak{Z}_l(\mathfrak{Z}_r(\mathfrak{J})) = \mathfrak{J}$ and similarly, $\mathfrak{Z}_r(\mathfrak{Z}_l(\mathfrak{J}')) = \mathfrak{J}'$. The correspondences $\mathfrak{J} \rightarrow \mathfrak{Z}_r(\mathfrak{J})$ and $\mathfrak{J}' \rightarrow \mathfrak{Z}_l(\mathfrak{J}')$ are inverses of each other and are $(1 - 1)$ between the lattice of left Φ-ideals and the lattice of right Φ-ideals. Evidently if $\mathfrak{J}_1 \leqq \mathfrak{J}_2$, then $\mathfrak{Z}_r(\mathfrak{J}_1) \geqq \mathfrak{Z}_r(\mathfrak{J}_2)$. This result may be expressed as follows.

THEOREM 23. *If $\mathfrak{A}$ is semi-simple, then the correspondence $\mathfrak{J} \rightarrow \mathfrak{Z}_r(\mathfrak{J})$ is an anti-isomorphism between the lattice of left Φ-ideals and the lattice of right Φ-ideals of $\mathfrak{A}$.*

Let $\mathfrak{C}$ be the center of the semi-simple ring $\mathfrak{A} = \mathfrak{A}_1 \oplus \cdots \oplus \mathfrak{A}_t$. We have seen that $\mathfrak{C}$ is a Φ-subring. If $c \in \mathfrak{C}$, $c = c_1 + \cdots + c_t$ where $c_i \in \mathfrak{A}_i$. Since $a = \Sigma a_i$, a_i in $\mathfrak{A}_i$ and $ac = ca$, we have $a_i c_i = c_i a_i$. In addition $a_j c_i = c_i a_j = 0$ if $j \neq i$. Hence the $c_i \in \mathfrak{C}$ and $\mathfrak{C} = \mathfrak{C}_1 \oplus \cdots \oplus \mathfrak{C}_t$ where $\mathfrak{C}_i = \mathfrak{C} \wedge \mathfrak{A}_i$ is a Φ-ring. $\mathfrak{C}_i$ is the center of $\mathfrak{A}_i$. For if $d_i \in \mathfrak{A}_i$ and $d_i a_i = a_i d_i$, then $d_i \in \mathfrak{C}$ and hence $\in \mathfrak{C}_i$. Since $\mathfrak{A}_i = \mathrm{P}^{(i)}_{n_i}$, $\mathrm{P}^{(i)}$ a division ring, its center is contained in $\mathrm{P}^{(i)}$ and is therefore a field.

THEOREM 24. *If $\mathfrak{A}$ is a semi-simple Φ-ring and $\mathfrak{A} = \mathfrak{A}_1 \oplus \cdots \oplus \mathfrak{A}_t$ where the $\mathfrak{A}_i$ are simple two-sided ideals, then the center $\mathfrak{C}$ is a Φ-ring and $\mathfrak{C} = \mathfrak{C}_1 \oplus \cdots \oplus \mathfrak{C}_t$, where $\mathfrak{C}_i \equiv \mathfrak{C} \wedge \mathfrak{A}_i$ is a field.*

12. Representation of semi-simple rings. We suppose first that $\mathfrak{A}$ is any ring and that $\mathfrak{M}$ is an $\mathfrak{A}$-module. If $\mathfrak{J}$ is a right ideal of $\mathfrak{A}$, $x\mathfrak{J}$ the set of elements of xb, x fixed and b variable in $\mathfrak{J}$, is an $\mathfrak{A}$-sub-module of $\mathfrak{M}$. The association $b \rightarrow xb$ is an $\mathfrak{A}$-homomorphism between $\mathfrak{J}$ and $x\mathfrak{J}$. Hence if $\mathfrak{J}$ is irreducible, either $\mathfrak{J}$ is $\mathfrak{A}$-isomorphic to $x\mathfrak{J}$ or $x\mathfrak{J} = 0$. If $\mathfrak{M}$ is irreducible, $x\mathfrak{J} = 0$ or $\mathfrak{M}$, for any x and any right ideal $\mathfrak{J}$. Now if $\mathfrak{A}$ is semi-simple, $\mathfrak{A} = \mathfrak{J}_1 \oplus \cdots \oplus \mathfrak{J}_u$ where the $\mathfrak{J}_j$ are irreducible right Φ-ideals. Then if $\mathfrak{M}$ is irreducible and $\mathfrak{M}\mathfrak{A} \neq 0$, there exists an x in $\mathfrak{M}$ such that $x\mathfrak{A} \neq 0$ and there exists an $\mathfrak{J}_j$ such that $x\mathfrak{J}_j \neq 0$. It follows that $\mathfrak{M} = x\mathfrak{J}_j$ is $\mathfrak{A}$-isomorphic to $\mathfrak{J}_j$.

We assume now that $\mathfrak{A}$ is semi-simple and that $x1 = x$ for all x in $\mathfrak{M}$. Then each $x = x1 \in x\mathfrak{A} = x\mathfrak{J}_1 + \cdots + x\mathfrak{J}_u$. Since the $x\mathfrak{J}_j$ are either irreducible or 0, $\mathfrak{M}$ is the join of its irreducible submodules. If $\mathfrak{M}$ satisfies the ascending chain condition or, what is equivalent, if $\mathfrak{M}$ is finitely generated, $\mathfrak{M} = \mathfrak{M}_1 + \cdots + \mathfrak{M}_t$ for suitable irreducible $\mathfrak{M}_i \neq 0$. It follows that $\mathfrak{M} = \mathfrak{M}_{i_1} \oplus \cdots \oplus \mathfrak{M}_{i_t}$. For if $\mathfrak{M} \neq \mathfrak{M}_1 \equiv \mathfrak{M}_{i_1}$, there is a least $j = i_2$ such that $\mathfrak{M}_j \nleqq \mathfrak{M}_{i_1}$. Then $\mathfrak{M}_{i_1} \wedge \mathfrak{M}_{i_2} = 0$ and $\mathfrak{M}' = \mathfrak{M}_1 + \cdots + \mathfrak{M}_{i_2} = \mathfrak{M}_{i_1} \oplus \mathfrak{M}_{i_2}$. If $\mathfrak{M}' \neq \mathfrak{M}$, let i_3 be the least index such that $\mathfrak{M}_{i_3} \nleqq \mathfrak{M}'$. Then $\mathfrak{M}'' = \mathfrak{M}_1 + \cdots + \mathfrak{M}_{i_3} = \mathfrak{M}_{i_1} \oplus \mathfrak{M}_{i_2} \oplus \mathfrak{M}_{i_3}$. This process evidently leads to the required decomposition. Thus we have shown that the $\mathfrak{A}$-module $\mathfrak{M}$ is completely reducible and satisfies the descending chain condition.

On the other hand, suppose that the descending chain condition holds, and let $\mathfrak{M}_1 < \mathfrak{M}_2 < \cdots$ be an ascending chain of $\mathfrak{A}$-submodules. If x_i is an element of $\mathfrak{M}_{i+1}$ not in $\mathfrak{M}_i$, then $x_i\mathfrak{A} \leqq \mathfrak{M}_{i+1}$. Since $x_i \in x_i\mathfrak{A} = x_i\mathfrak{J}_1 + \cdots + x_i\mathfrak{J}_u$, at least one of the irreducible $x_i\mathfrak{J}_j \neq 0$ is in $\mathfrak{M}_{i+1}$ but not in $\mathfrak{M}_i$. Choose one of these $x_i\mathfrak{J}_j$ and denote it as $\mathfrak{N}_i$. Consider the chain $(\mathfrak{N}_1 + \mathfrak{N}_2 + \cdots) \geqq (\mathfrak{N}_2 + \mathfrak{N}_3 + \cdots) \geqq \cdots$, where $(\mathfrak{N}_k + \mathfrak{N}_{k+1} + \cdots)$ denotes the join of all the $\mathfrak{N}_i$ with $i \geqq k$. We assert that $(\mathfrak{N}_1 + \mathfrak{N}_2 + \cdots) > (\mathfrak{N}_2 + \mathfrak{N}_3 + \cdots)$. For if $(\mathfrak{N}_1 + \mathfrak{N}_2 + \cdots) = (\mathfrak{N}_2 + \mathfrak{N}_3 + \cdots)$ for any y_1 in $\mathfrak{N}_1$, we have $y_1 = y_2 + \cdots + y_m$ where $y_i \in \mathfrak{N}_i$, and we may suppose that $y_m \neq 0$. Thus $y_m = y_1 - y_2 - \cdots - y_{m-1} \in (\mathfrak{N}_1 + \cdots + \mathfrak{N}_{m-1}) \leqq \mathfrak{M}_m$. Hence $y_m\mathfrak{A} \leqq \mathfrak{M}_m$, and this is impossible since $y_m\mathfrak{A}$ is an $\mathfrak{A}$-module $\neq 0$ in $\mathfrak{N}_m$ and therefore $y_m\mathfrak{A} = \mathfrak{N}_m$. Thus we have the inequalities $(\mathfrak{N}_1 + \mathfrak{N}_2 + \cdots) > (\mathfrak{N}_2 + \mathfrak{N}_3 + \cdots) > \cdots$. By the descending chain condition, this chain is finite in length and hence the original ascending chain $\mathfrak{M}_1 < \mathfrak{M}_2 < \cdots$ is finite.

THEOREM 25. *Let $\mathfrak{A}$ be a semi-simple Φ-ring and let $\mathfrak{M}$ be an $\mathfrak{A}$-module such that $x1 = x$ for all x in $\mathfrak{M}$. Then if $\mathfrak{M}$ satisfies either chain condition for $\mathfrak{A}$-submodules, it is completely reducible and satisfies the other chain condition. Any irreducible $\mathfrak{M}$ is $\mathfrak{A}$-isomorphic to an irreducible right ideal of $\mathfrak{A}$. The number of non-isomorphic irreducible $\mathfrak{A}$-modules is the same as the number of irreducible two-sided ideals $\mathfrak{A}_i$ in the decomposition $\mathfrak{A} = \mathfrak{A}_1 \oplus \cdots \oplus \mathfrak{A}_t$.*

A partial converse of this theorem holds. In order to prove it, we require the following general remarks: Suppose that $\mathfrak{M}$ is an $\mathfrak{A}$-module and that $\mathfrak{B}$ is a two-sided ideal in $\mathfrak{A}$ annihilating $\mathfrak{M}$ in the sense that $xb = 0$ for all x in $\mathfrak{M}$ and all b in $\mathfrak{B}$. If we denote the coset $a + \mathfrak{B}$ by $\bar{a}$, then it is clear that the function $x\bar{a} \equiv xa$ is a single valued function of x and $\bar{a}$ in $\bar{\mathfrak{A}} = \mathfrak{A} - \mathfrak{B}$. It follows that $\mathfrak{M}$ is an $\bar{\mathfrak{A}}$-module relative to this product. Evidently $\mathfrak{A}$-submodules of $\mathfrak{M}$ are $\bar{\mathfrak{A}}$-submodules, and conversely, and so $\mathfrak{A}$-reducibility, $\mathfrak{A}$-decomposability, etc. are equivalent to $\bar{\mathfrak{A}}$-reducibility, $\bar{\mathfrak{A}}$-decomposability, etc. It is also readily seen that if $\mathfrak{M}$ and $\mathfrak{N}$ are two $\mathfrak{A}$-modules annihilated by $\mathfrak{B}$, then $\mathfrak{A}$-homomorphisms and $\mathfrak{A}$-isomorphisms between them are at the same time $\bar{\mathfrak{A}}$-homomorphisms and $\bar{\mathfrak{A}}$-isomorphisms, and conversely.

Now let $\mathfrak{A}$ be a Φ-ring satisfying the descending chain condition for left Φ-ideals and let $\mathfrak{R}$ be the radical of $\mathfrak{A}$. If $\mathfrak{M}$ is an irreducible $\mathfrak{A}$-module, then

$x\mathfrak{R}$ is a submodule for any x, and so either $x\mathfrak{R} = 0$ or $x\mathfrak{R} = \mathfrak{M}$. If $x\mathfrak{R} = \mathfrak{M}$, $\mathfrak{M} = x\mathfrak{R} = x\mathfrak{R}^2 = \cdots = 0$. Hence $\mathfrak{R}$ annihilates $\mathfrak{M}$ and $\mathfrak{M}$ is an $\bar{\mathfrak{A}}$-module $\bar{\mathfrak{A}} = \mathfrak{A} - \mathfrak{R}$. If $\bar{\mathfrak{A}} = \bar{\mathfrak{A}}_1 \oplus \cdots \oplus \bar{\mathfrak{A}}_t$, either $\mathfrak{M}\mathfrak{A} = 0$ or $\mathfrak{M}$ is isomorphic to a right ideal contained in one of the $\bar{\mathfrak{A}}_i$. It follows that $\mathfrak{M}$ is an $\bar{\mathfrak{A}}_i$-module. Similarly if $\mathfrak{M}$ is a join of irreducible submodules, $x\mathfrak{R} = 0$ for all x, and $\mathfrak{M}$ is an $\bar{\mathfrak{A}}$-module.

THEOREM 26. *Let $\mathfrak{A}$ be a Φ-ring satisfying the descending chain condition for left Φ-ideals and let $\mathfrak{M}$ be an $\mathfrak{A}$-module such that $\mathfrak{M}\mathfrak{A} \neq 0$. If $\mathfrak{M}$ is irreducible, then $\mathfrak{M}$ is an $\bar{\mathfrak{A}}_i$-module, where $\bar{\mathfrak{A}}_i$ is one of the irreducible two-sided ideals of $\bar{\mathfrak{A}} = \mathfrak{A} - \mathfrak{R}$. If $\mathfrak{M}$ is a join of irreducible $\mathfrak{A}$-modules, then $\mathfrak{M}$ is an $\bar{\mathfrak{A}}$-module.*

A more striking form of this result is the following

THEOREM 27. *Let $\mathfrak{A} \neq 0$ be a Φ-ring of endomorphisms in $\mathfrak{M}$ and suppose that $\mathfrak{A}$ satisfies the descending chain condition for left Φ-ideals. Then if $\mathfrak{M}$ is irreducible, $\mathfrak{A}$ is simple, and if $\mathfrak{M}$ is a join of irreducible $\mathfrak{A}$-groups, then $\mathfrak{A}$ is semi-simple.*

For in this case, $\mathfrak{M}$ is an $\mathfrak{A}$-module and the representation of $\mathfrak{A}$ by itself is clearly $(1 - 1)$.

Now suppose that $\mathfrak{A}$ is any ring of endomorphisms in $\mathfrak{M}$ including the identity endomorphism and let $\mathfrak{A} = \mathfrak{A}_1 \oplus \cdots \oplus \mathfrak{A}_t$ where the $\mathfrak{A}_i$ are two-sided ideals. Then $\mathfrak{M} = \mathfrak{M}\mathfrak{A}_1 + \cdots + \mathfrak{M}\mathfrak{A}_t$ where $\mathfrak{M}\mathfrak{A}_i$ denotes the smallest submodule containing all the elements xa_i, a_i in $\mathfrak{A}_i$. If $1 = 1_1 + \cdots + 1_t$ where $1_i \in \mathfrak{A}_i$, 1_i is the identity of $\mathfrak{A}_i$ since $\mathfrak{A}_i = \mathfrak{A}1_i = 1_i\mathfrak{A}$. Hence if $x_i \in \mathfrak{M}\mathfrak{A}_i$, $x_i1_i = x_i$ and since $1_i1_j = 0$, $x_i1_j = 0$ if $i \neq j$. If $x_1 + \cdots + x_t = 0$ with x_i in $\mathfrak{A}_i$, $(x_1 + \cdots + x_t)1_i = x_i1_i = x_i = 0$. Thus $\mathfrak{M} = \mathfrak{M}\mathfrak{A}_1 \oplus \cdots \oplus \mathfrak{M}\mathfrak{A}_t$.

Again let $\mathfrak{A}$ be semi-simple and suppose that $\mathfrak{M}$ satisfies the chain conditions. Then $\mathfrak{M}\mathfrak{A}_i$ is a join of irreducible submodules isomorphic to the right ideals of $\mathfrak{A}_i$. As we have seen in **5**, the ring of $\mathfrak{A}$-endomorphisms of $\mathfrak{M}$ has the form $\mathfrak{B} = \mathfrak{B}_1 \oplus \cdots \oplus \mathfrak{B}_t$ where $\mathfrak{B}_i$ is obtained from the $\mathfrak{A}$-endomorphisms b_i of $\mathfrak{M}_i \equiv \mathfrak{M}\mathfrak{A}_i$ by extending the definition of these endomorphisms so that $\mathfrak{M}_ib_j = 0$ for $j \neq i$. Thus $\mathfrak{M}_i = \mathfrak{M}\mathfrak{B}_i$. Since the ring of endomorphisms induced by $\mathfrak{A}$ in $\mathfrak{M}_i$ is a matrix ring over a division ring if $\mathfrak{A}_i$ has this structure, it follows readily from our result on matrix rings[5] that if $\mathfrak{A}_i = \mathrm{P}^{(i)}_{n_i}$, then $\mathfrak{B}_i = \bar{\mathrm{P}}^{(i)}_{m_i}$, $\bar{\mathrm{P}}^{(i)}$ anti-isomorphic to $\mathrm{P}^{(i)}$, and $\mathfrak{A}$ is the complete set of $\mathfrak{B}$-endomorphisms of $\mathfrak{M}$.

THEOREM 28. *Let $\mathfrak{A}$ be a semi-simple Φ-ring of endomorphisms including the identity in $\mathfrak{M}$ and suppose that $\mathfrak{M}$ satisfies either one of the chain conditions for $\mathfrak{A}$-groups. If $\mathfrak{A} = \mathrm{P}^{(1)}_{n_1} \oplus \cdots \oplus \mathrm{P}^{(t)}_{n_t}$, where $\mathrm{P}^{(i)}_{n_i}$ is a two-sided ideal and $\mathrm{P}^{(i)}$ is a division ring, then the ring of $\mathfrak{A}$-endomorphisms $\mathfrak{B}$ has the form $\mathfrak{B} = \bar{\mathrm{P}}^{(1)}_{n_1} \oplus \cdots \oplus \bar{\mathrm{P}}^{(t)}_{n_t}$, where $\bar{\mathrm{P}}^{(i)}_{n_i}$ is a two-sided ideal and $\bar{\mathrm{P}}^{(i)}$ is anti-isomorphic to $\mathrm{P}^{(i)}$. $\mathfrak{A}$ is the ring of $\mathfrak{B}$-endomorphisms of $\mathfrak{M}$.*

13. Rings satisfying the descending chain condition. The results of the preceding section yield the interesting

[5] See Chapter 2, **6**.

THEOREM 29. *If $\mathfrak{A}$ is a Φ-ring with an identity satisfying the descending chain condition for right (left) Φ-ideals, $\mathfrak{A}$ satisfies the ascending chain condition for right (left) Φ-ideals also.*

Let $\mathfrak{R}$ be the radical of $\mathfrak{A}$ and $\mathfrak{R}^{s+1} = 0$, $\mathfrak{R}^s \neq 0$. Then $\mathfrak{A} > \mathfrak{R} > \cdots > \mathfrak{R}^s > 0$ is a descending chain of $\mathfrak{A}$-modules ($\mathfrak{A}_r$-groups). The difference modules $\mathfrak{A} - \mathfrak{R}$, $\mathfrak{R} - \mathfrak{R}^2$, $\cdots$ are mapped into 0 by the elements of $\mathfrak{R}$ and hence may be regarded as ($\mathfrak{A} - \mathfrak{R}$)-modules. Since $\mathfrak{A}$ satisfies the descending chain condition, the difference modules $\mathfrak{A} - \mathfrak{R}$, $\mathfrak{R} - \mathfrak{R}^2$, $\cdots$ satisfy this condition also. Since $\bar{\mathfrak{A}} = \mathfrak{A} - \mathfrak{R}$ is semi-simple and its identity is the identity endomorphism in $\mathfrak{A} - \mathfrak{R}$, $\mathfrak{R} - \mathfrak{R}^2$, $\cdots$ these modules satisfy the ascending chain condition and hence have composition series. For example, $\mathfrak{A} - \mathfrak{R} = \bar{\mathfrak{J}}_1 > \bar{\mathfrak{J}}_2 > \cdots > \bar{\mathfrak{J}}_m = 0$ where $\bar{\mathfrak{J}}_j - \bar{\mathfrak{J}}_{j+1}$ is $\mathfrak{A}$-irreducible. Hence $\mathfrak{A} = \mathfrak{J}_1 > \mathfrak{J}_2 > \cdots > \mathfrak{J}_m = \mathfrak{R}$ where $\mathfrak{J}_j$ is the right Φ-ideal mapped into 0 in the homomorphism between the groups $\mathfrak{A}$ and $\mathfrak{A} - \mathfrak{R}$. By the First Isomorphism Theorem, $\mathfrak{J}_j - \mathfrak{J}_{j+1}$ is $\mathfrak{A}$-isomorphic to $\bar{\mathfrak{J}}_j - \bar{\mathfrak{J}}_{j+1}$ and hence is $\mathfrak{A}$-irreducible. Similarly, we obtain $\mathfrak{R} = \mathfrak{J}_m > \cdots > \mathfrak{J}_{m+p} = \mathfrak{R}^2$ where the $\mathfrak{J}_k$ are right Φ-ideals and $\mathfrak{J}_k - \mathfrak{J}_{k+1}$ are irreducible. Thus $\mathfrak{A}$ has a composition series $\mathfrak{J}_1 > \cdots > \mathfrak{J}_m > \mathfrak{J}_{m+1} > \cdots > 0$ and so both chain conditions hold for right Φ-ideals of $\mathfrak{A}$.

This theorem enables us to apply directly our results on rings of endomorphisms to abstract rings. Thus, as a consequence of **6**, we obtain the following

THEOREM 30. *If $\mathfrak{A}$ is a Φ-ring with an identity satisfying the descending chain condition for left (right) Φ-ideals, then any nil subring $\mathfrak{B}$ of $\mathfrak{A}$ is nilpotent.*

Consider $\bar{\mathfrak{B}}$ the ring of right multiplications in $\mathfrak{A}$ corresponding to the elements of $\mathfrak{B}$. The elements of $\bar{\mathfrak{B}}$ are $\mathfrak{A}_l$-endomorphisms. Hence if s is the length of a composition series for the $\mathfrak{A}_l$-group $\mathfrak{A}$, $\bar{b}_1 \cdots \bar{b}_s = 0$ for any $\bar{b}_i$ in $\bar{\mathfrak{B}}$. Thus $xb_1 \cdots b_s = 0$ for any x in $\mathfrak{A}$ and b_i in $\mathfrak{B}$. Hence $\mathfrak{B}^{s+1} = 0$.

We note also the following theorem which is an immediate consequence of **8.**

THEOREM 31. *If $\mathfrak{A}$ is a Φ-ring with an identity satisfying the descending chain condition for left (right) Φ-ideals, then the following conditions on $\mathfrak{A}$ are equivalent:*

1. *$\mathfrak{A}$ is a direct sum of $\mathfrak{A}_l(\mathfrak{A}_r$-)isomorphic indecomposable left (right) Φ-ideals.*
2. *$\mathfrak{A} - \mathfrak{R}$ is simple, $\mathfrak{R}$ the radical of $\mathfrak{A}$.*
3. *$\mathfrak{A} = \mathfrak{B}_u$ where $\mathfrak{B}$ is a completely primary ring.*

If any one of these conditions holds, $\mathfrak{A}$ is a *primary ring.*

14. The regular representations. Let $\mathfrak{A}$ be an arbitrary Φ-ring with an identity satisfying the descending chain conditions for one-sided ideals. Suppose that $\mathfrak{A} = e_1\mathfrak{A} \oplus \cdots \oplus e_u\mathfrak{A}$ is a decomposition of $\mathfrak{A}$ into right ideals $\neq 0$ where

$$(4) \qquad 1 = \Sigma e_i, \qquad e_i^2 = e_i \neq 0 \qquad e_ie_j = 0 \quad \text{if} \quad i \neq j.$$

If $\mathfrak{R}$ is a nilpotent two-sided ideal of $\mathfrak{A}$ and $\bar{e}_i = e_i + \mathfrak{R}$, then

$$(5) \qquad \bar{1} = \Sigma \bar{e}_i, \qquad \bar{e}_i^2 = \bar{e}_i \neq 0, \qquad \bar{e}_i\bar{e}_j = 0 \quad \text{if} \quad i \neq j$$

in $\bar{\mathfrak{A}} = \mathfrak{A} - \mathfrak{N}$, and so $\bar{\mathfrak{A}} = \bar{e}_1\bar{\mathfrak{A}} \oplus \cdots \oplus \bar{e}_u\bar{\mathfrak{A}}$ is a direct decomposition of $\bar{\mathfrak{A}}$ into right ideals. Since $\bar{e}_i \neq 0$, $\bar{e}_i\bar{\mathfrak{A}} \neq 0$. We wish to show that any decomposition of $\bar{\mathfrak{A}}$ into right ideals may be obtained in this way. For this purpose we require the

LEMMA. *If $\bar{e}_1, \cdots, \bar{e}_v$ are idempotent elements $\neq 0$ in $\bar{\mathfrak{A}}$ such that $\bar{e}_i\bar{e}_j = 0$ for $i \neq j$, then it is possible to choose elements e_i in the cosets $\bar{e}_i$ such that $e_i^2 = e_i$ and $e_ie_j = 0$.*

Suppose that $e_1, \cdots, e_m$ have already been determined so that these conditions hold for $i, j = 1, \cdots, m$. Let u be any element in $\bar{e}_{m+1}$ and set $v = u - eu - ue + eue$ where $e = \sum_1^m e_i$. Then $e_iv = ve_i = 0$ for $i = 1, \cdots, m$ and $v \equiv u \pmod{\mathfrak{N}}$. Hence $v^2 = v + z$ where z is nilpotent, say $z^s = 0$. Evidently $zv = vz$. We now try to determine an element $w = f(z)v + g(z)$ so that $w^2 = w$ and $f(z)$ and $g(z)$ are polynomials in z with integer coefficients. This leads to the consideration of the equations

$$(6) \qquad f^2 + 2fg = f, \qquad g^2 + f^2z = g.$$

We shall first solve these equations for power series in an indeterminate t. By elimination, we obtain $f(t) = (1 + 4t)^{-\frac{1}{2}}$ and $g(t) = \frac{1}{2}(1 - f(t))$. Now consider the expansion of $f(t) = (1 + 4t)^{-\frac{1}{2}}$. It is readily seen that $f(t) = 1 + \sum(-1)^n \binom{2n-1}{n} 2t^n$ and so the coefficients of $f(t)$ and of $g(t)$ are all integers. The formal identities

$$(f(t))^2 + 2f(t)g(t) = f(t), \qquad (g(t))^2 + (f(t))^2t = g(t)$$

are satisfied. It follows that if $f_s(t)$ and $g_s(t)$ are the sth partial sums of $f(t)$ and $g(t)$, respectively, then $f = f_s(z)$ and $g = g_s(z)$ satisfy (6). Hence $w = fv + g$ is idempotent. Since $e_iv = ve_i = 0$, $we_i = e_iw = 0$. The formula for w shows also that $w \equiv v \pmod{\mathfrak{N}}$ and hence $w \equiv u \pmod{\mathfrak{N}}$. Thus w may serve as the element e_{m+1} and the lemma follows by induction.

Remark. The above proof is also valid for an algebra $\mathfrak{A}$ without an identity. For we may adjoin an identity to $\mathfrak{A}$ and then construct the idempotent element e_{m+1} in the manner indicated from an element u chosen in $\mathfrak{A}$. It is readily seen that $e_{m+1} \in \mathfrak{A}$.

If $\Sigma\bar{e}_i = \bar{1}$ in the lemma, then $\Sigma\bar{e}_i = \bar{1}$ and $\Sigma e_i = 1 + y$, y in $\mathfrak{N}$. Since Σe_i is idempotent, $(1 + y)^2 = (1 + y)$. Hence $y^2 + y = 0$. Thus $y = -y^2 = y^3 = \cdots = 0$ and so $\Sigma e_i = 1$. It follows that if $\bar{\mathfrak{A}} = \bar{e}_1\bar{\mathfrak{A}} \oplus \cdots \oplus \bar{e}_v\bar{\mathfrak{A}}$ is a decomposition of $\bar{\mathfrak{A}}$ into right ideals $\neq 0$, where the $\bar{e}_i$ satisfy (5), then $\mathfrak{A} = e_1\mathfrak{A} \oplus \cdots \oplus e_v\mathfrak{A}$, where the $e_i \in \bar{e}_i$ and satisfy (4). The lemma together with the Krull-Schmidt Theorem shows also that the idempotent elements e_i are primitive if and only if the $\bar{e}_i$ are. Thus $e_i\mathfrak{A}$ is indecomposable if and only if $\bar{e}_i\bar{\mathfrak{A}}$ is. If $\mathfrak{N}$ is the radical of $\mathfrak{A}$, $\bar{\mathfrak{A}}$ is semi-simple and hence $\bar{e}_i\bar{\mathfrak{A}}$ is indecomposable if and only if it is irreducible. We have therefore proved

THEOREM 32. *Let $\mathfrak{A}$ be a Φ-ring with an identity satisfying the descending chain conditions for one-sided ideals and let $\mathfrak{N}$ be its radical. If $\mathfrak{A} = e_1\mathfrak{A} \oplus \cdots \oplus e_u\mathfrak{A}$*

where the e_i *satisfy* (4), *then* $\bar{\mathfrak{A}} \equiv \mathfrak{A} - \mathfrak{R} = \bar{e}_1\bar{\mathfrak{A}} \oplus \cdots \oplus \bar{e}_u\bar{\mathfrak{A}}$. *The ideal* $e_i\mathfrak{A}$ *is indecomposable if and only if* $\bar{e}_i\bar{\mathfrak{A}}$ *is irreducible.*

Consider $\mathfrak{M} = e_i\mathfrak{A} \oplus e_j\mathfrak{A} = e\mathfrak{A}$, $e = e_i + e_j$. Since $\mathfrak{A} = \mathfrak{M} \oplus (1 - e)\mathfrak{A}$, the projections E and E' determined by this decomposition are the left multiplications by the elements e and $e' = 1 - e$, respectively. Since $\mathfrak{A}_l$ is the ring of $\mathfrak{A}_r$-endomorphisms of $\mathfrak{A}$, by **4**, $E\mathfrak{A}_lE$ is the ring of $\mathfrak{A}_r$-endomorphisms of $\mathfrak{M}$. Thus the ring of $\mathfrak{A}_r$-endomorphisms of $\mathfrak{M}$ is anti-isomorphic to $e\mathfrak{A}e$. Hence by **8**, $e_i\mathfrak{A}$ and $e_j\mathfrak{A}$ are $\mathfrak{A}_r$-isomorphic if and only if $e\mathfrak{A}e$ is primary. Now $e\mathfrak{A}e - (\mathfrak{R} \wedge e\mathfrak{A}e) \cong (e\mathfrak{A}e + \mathfrak{R}) - \mathfrak{R} = \bar{e}\bar{\mathfrak{A}}\bar{e}$ and $\bar{e}\bar{\mathfrak{A}}\bar{e}$ is anti-isomorphic to the ring of $\bar{\mathfrak{A}}_r$-endomorphisms of $\bar{\mathfrak{M}} = \bar{e}\bar{\mathfrak{A}}$. Since $\bar{\mathfrak{M}}$ is completely reducible, $\bar{e}\bar{\mathfrak{A}}\bar{e}$ is semisimple. It follows that $\mathfrak{R} \wedge e\mathfrak{A}e = e\mathfrak{R}e$, which is evidently contained in the radical of $e\mathfrak{A}e$, coincides with this radical. Hence $e\mathfrak{A}e$ is primary if and only if $\bar{e}\bar{\mathfrak{A}}\bar{e}$ is simple and we have proved the following

THEOREM 33. *Let* $\mathfrak{A}$, $\bar{\mathfrak{A}}$, *etc. be as in the preceding theorem with the* e_i *primitive. Then* $e_i\mathfrak{A}$ *and* $e_j\mathfrak{A}$ *are* $\mathfrak{A}$*-isomorphic if and only if* $\bar{e}_i\bar{\mathfrak{A}}$ *and* $\bar{e}_j\bar{\mathfrak{A}}$ *are* $\bar{\mathfrak{A}}$*-isomorphic.*

Since $e_i\mathfrak{R} = e_i\mathfrak{A} \wedge \mathfrak{R}$, the $\mathfrak{A}$-module $e_i\mathfrak{A} - e_i\mathfrak{R}$ is isomorphic to $(e_i\mathfrak{A} + \mathfrak{R}) - \mathfrak{R}$. The latter is essentially the $\bar{\mathfrak{A}}$-module $\bar{e}_i\bar{\mathfrak{A}}$. Hence $e_i\mathfrak{A} - e_i\mathfrak{R}$ is irreducible and so $e_i\mathfrak{R}$ is a maximal submodule of $e_i\mathfrak{A}$ and $\bar{e}_i\bar{\mathfrak{A}}$ is a first composition factor of $e_i\mathfrak{A}$. On the other hand, if $\mathfrak{M}$ is any maximal submodule of $e_i\mathfrak{A}$, $e_i\mathfrak{A} - \mathfrak{M}$ is irreducible and therefore this module is annihilated by $\mathfrak{R}$. It follows that $e_i\mathfrak{R} \leqq \mathfrak{M}$ and hence $e_i\mathfrak{R} = \mathfrak{M}$. Thus $e_i\mathfrak{R}$ is the only maximal submodule of the $\mathfrak{A}$-module $e_i\mathfrak{A}$. This is the first part of

THEOREM 34. *If* $\mathfrak{I}$ *is an indecomposable right ideal that occurs in a direct decomposition of* $\mathfrak{A}$, *then* $\mathfrak{I}$ *contains only one maximal right ideal of* $\mathfrak{A}$. *If* $\mathfrak{I}$ *and* $\mathfrak{I}'$ *are indecomposable right ideals that occur in direct decompositions of* $\mathfrak{A}$, *then a necessary and sufficient condition that* $\mathfrak{I}$ *and* $\mathfrak{I}'$ *be* $\mathfrak{A}$*-isomorphic is that their first composition factors be* $\mathfrak{A}$*-isomorphic.*

The second part of the theorem follows from the Krull-Schmidt Theorem and Theorem 33. For, by the former, $\mathfrak{I}$ is $\mathfrak{A}$-isomorphic to one of the right ideals $\mathfrak{I}_1$ that occur in a decomposition of $\mathfrak{A}$ that includes $\mathfrak{I}'$. By Theorem 33, $\mathfrak{I}_1$ and $\mathfrak{I}'$ are isomorphic if and only if their first composition factors are isomorphic. Hence this holds also for $\mathfrak{I}$ and $\mathfrak{I}'$.

In order to obtain a connection between the decomposition of $\mathfrak{A}$ into indecomposable right ideals and into indecomposable two-sided ideals, we require the following

LEMMA. *Let* $\mathfrak{N}$ *and* $\mathfrak{N}'$ *be* $\mathfrak{A}$*-modules with composition series. Suppose that* $\mathfrak{N}$ *has only one maximal submodule* $\mathfrak{M}$ *and that* $\mathfrak{N}$ *is* $\mathfrak{A}$*-homomorphic to a submodule* $\mathfrak{N}^*$ *of* $\mathfrak{N}'$. *Then any composition series for* $\mathfrak{N}'$ *includes a factor* $\mathfrak{A}$*-isomorphic to* $\mathfrak{N} - \mathfrak{M}$.

We note first that if $\mathfrak{Z}$ is a proper submodule of $\mathfrak{N}$, then $\mathfrak{Z} \leqq \mathfrak{M}$. For $\mathfrak{N} - \mathfrak{Z}$ contains a maximal submodule $\bar{\mathfrak{M}}_1$ and the corresponding submodule $\mathfrak{M}_1$ of $\mathfrak{N}$ is maximal and contains $\mathfrak{Z}$. Since $\mathfrak{M}$ is the only maximal submodule of $\mathfrak{N}$, $\mathfrak{M} = \mathfrak{M}_1$. Now let $\mathfrak{Z}$ be the submodule of elements mapped into 0 by the homomorphism between $\mathfrak{N}$ and $\mathfrak{N}^*$. Then $\mathfrak{N}^*$ is $\mathfrak{A}$-isomorphic to $\mathfrak{N} - \mathfrak{Z}$. Since

$(\mathfrak{N} - \mathfrak{Z}) > (\mathfrak{M} - \mathfrak{Z}) \geqq 0$, $\mathfrak{N} - \mathfrak{Z}$ has the composition factor $(\mathfrak{N} - \mathfrak{Z}) - (\mathfrak{M} - \mathfrak{Z})$, $\mathfrak{A}$-isomorphic to $\mathfrak{N} - \mathfrak{M}$. Hence $\mathfrak{N}^*$, and consequently $\mathfrak{N}'$, have a composition factor $\mathfrak{A}$-isomorphic to $\mathfrak{N} - \mathfrak{M}$.

We shall confine our attention to the indecomposable ideals of $\mathfrak{A}$ that occur in direct decompositions of $\mathfrak{A}$. These are necessarily of the form $e\mathfrak{A}$ where e is a primitive idempotent element. Conversely, any ideal of this form is indecomposable and belongs to a decomposition of $\mathfrak{A}$. We shall say that two such ideals $e\mathfrak{A}$ and $e'\mathfrak{A}$ belong to the *same block* if there is a sequence of indecomposable ideals $e\mathfrak{A} = e_1\mathfrak{A}, e_2\mathfrak{A}, \cdots, e_h\mathfrak{A} = e'\mathfrak{A}$ such that $e_i^2 = e_i$ and each $e_i\mathfrak{A}$ has a composition factor $\mathfrak{A}$-isomorphic to one of the composition factors of $e_{i+1}\mathfrak{A}$. This relation between $e\mathfrak{A}$ and $e'\mathfrak{A}$ is evidently an equivalence. The sequence $\{e_i\mathfrak{A}\}$ is a sequence *connecting* $e\mathfrak{A}$ and $e'\mathfrak{A}$. With these definitions, we have the following

THEOREM 35. *Let $\mathfrak{A} = \mathfrak{A}_1 \oplus \cdots \oplus \mathfrak{A}_t$ be the decomposition of $\mathfrak{A}$ into indecomposable two-sided ideals. Then any two indecomposable ideals $e\mathfrak{A}$ and $e'\mathfrak{A}$ belong to the same block if and only if they are contained in the same component $\mathfrak{A}_i$. Hence $\mathfrak{A}_i$ is a join of a set of indecomposable ideals $e\mathfrak{A}$ belonging to the same block.*

We have seen in **9** that any indecomposable right ideal is contained in one of the $\mathfrak{A}_i$. If $e\mathfrak{A}$ and $e'\mathfrak{A}$ belong to the same block, they are contained in the same two-sided component. For suppose that $e\mathfrak{A}$ and $e'\mathfrak{A}$ are in different components say, $\mathfrak{A}_1$ and $\mathfrak{A}_2$ respectively. If 1_1 is the identity of $\mathfrak{A}_1$, then $(e\mathfrak{A})1_1 = e\mathfrak{A}$ and $(e'\mathfrak{A})1_1 = 0$ and so no composition factor of $e\mathfrak{A}$ is $\mathfrak{A}$-isomorphic to one of $e'\mathfrak{A}$. Thus if $\{e_i\mathfrak{A}\}$ is a sequence connecting $e\mathfrak{A}$ and $e'\mathfrak{A}$, each pair $e_i\mathfrak{A}$, $e_{i+1}\mathfrak{A}$ is contained in the same component and this holds also for $e\mathfrak{A}$ and $e'\mathfrak{A}$. If $e\mathfrak{A}$ and $e'\mathfrak{A}$ are in different blocks, then $e\mathfrak{A}e'\mathfrak{A} = 0$. For otherwise, there is an element $b = eae' \neq 0$ and so the left multiplication determined by b is an $\mathfrak{A}$-homorphism $\neq 0$ between $e'\mathfrak{A}$ and a submodule of $e\mathfrak{A}$. Hence by the lemma, both $e'\mathfrak{A}$ and $e\mathfrak{A}$ have composition factors isomorphic to $\bar{e}'\bar{\mathfrak{A}}$, contrary to assumption. Now let $\mathfrak{A} = e_1\mathfrak{A} \oplus \cdots \oplus e_u\mathfrak{A}$ be a decomposition of $\mathfrak{A}$ into indecomposable right ideals $\neq 0$. We suppose that the e_i satisfy (4) and that $e_1\mathfrak{A}, \cdots, e_{n_1}\mathfrak{A}$ belong to the same block, $e_{n_1+1}\mathfrak{A}, \cdots, e_{n_1+n_2}\mathfrak{A}$, belong to the same block but not to the same block as $e_1\mathfrak{A}$, etc. Set $\mathfrak{B}_1 = e_1\mathfrak{A} \oplus \cdots \oplus e_{n_1}\mathfrak{A}$, $\mathfrak{B}_2 = e_{n_1+1}\mathfrak{A} \oplus \cdots \oplus e_{n_1+n_2}\mathfrak{A}, \cdots$. Then $e_j\mathfrak{A}e_i\mathfrak{A} = 0$ if $i \leqq n_1$ and $j > n_1$. Hence $\mathfrak{A}(e_i\mathfrak{A}) \leqq \mathfrak{B}_1(e_i\mathfrak{A}) \leqq \mathfrak{B}_1$ and $\mathfrak{B}_1$ is a two-sided ideal. Similarly, each $\mathfrak{B}$ is a two-sided ideal and since $\mathfrak{B}$ is a join of ideals that belong to the same block, it is contained in one of the $\mathfrak{A}$'s. Hence we may suppose that $\mathfrak{B}_1 = \mathfrak{A}_1, \cdots, \mathfrak{B}_t = \mathfrak{A}_t$. Now suppose again that $e\mathfrak{A}$ and $e'\mathfrak{A}$ are in different blocks. We may suppose that $e\mathfrak{A} = e_1\mathfrak{A} \leqq \mathfrak{A}_1$. Then we have seen that $e_i\mathfrak{A}e'\mathfrak{A} = 0$ for $i \leqq n_1$. Hence there is a $j > n_1$ such that $e_j\mathfrak{A}e'\mathfrak{A} \neq 0$ and so $e_j\mathfrak{A}$ and $e'\mathfrak{A}$ are in the same component. Consequently $e\mathfrak{A}$ and $e'\mathfrak{A}$ are in different components.

The above results are, of course, also valid for left ideals. The following theorem gives a connection between decompositions into right ideals and into left ideals.

THEOREM 36. *If $\mathfrak{A} = e_1\mathfrak{A} \oplus \cdots \oplus e_u\mathfrak{A}$ where the e_i satisfy* (4), *then $\mathfrak{A} = \mathfrak{A}e_1 \oplus \cdots \oplus \mathfrak{A}e_u$. The ideal $\mathfrak{A}e_i$ is indecomposable if and only if $e_i\mathfrak{A}$ is in-*

decomposable. If $e_i\mathfrak{A}$ and $e_j\mathfrak{A}$ are indecomposable, then these ideals are $\mathfrak{A}$-isomorphic if and only if $\mathfrak{A}e_i$ and $\mathfrak{A}e_j$ are $\mathfrak{A}$-isomorphic.

The first part of this theorem is evident, since the condition for indecomposability in either case is that e_i be primitive. To prove the second part, we suppose first that $\mathfrak{A}$ is semi-simple. Then the condition that $e_i\mathfrak{A}$ and $e_j\mathfrak{A}$ be isomorphic is that they be contained in the same irreducible two-sided ideal $\mathfrak{B}$ of $\mathfrak{A}$. Since $e_i\mathfrak{A} \leqq \mathfrak{B}$ if and only if $\mathfrak{A}e_i \leqq \mathfrak{B}$, the theorem is true in this case. The theorem in the general case then follows directly from Theorem 33.

It may be noted that we have succeeded in obtaining extensions of all of the main theorems on the structure of semi-simple rings with the exception of the theorem establishing an anti-isomorphism between the lattice of left ideals and the lattice of right ideals. The class of rings for which this theorem holds has been the subject of a very interesting investigation by Nakayama. We refer the reader to his papers [**10**], [**14**] for this discussion.

15. Principal ideal rings. In this, and in the next, section we shall indicate, following Asano, that the main results of Chapter 3 are valid for principal ideal rings satisfying the descending chain conditions for one-sided ideals. These results will play an important role in the multiplicative ideal theory that will be considered in Chapter 6.

By a *principal ideal ring* we mean here a ring with an identity in which every left ideal is a principal left ideal and every right ideal is a principal right ideal. For the sake of simplicity we assume that the set of endomorphisms Φ is vacuous. We prove first the following

THEOREM 37. *If $\mathfrak{A}$ is a ring with an identity satisfying the descending chain conditions for one-sided ideals, and every two-sided ideal of $\mathfrak{A}$ is a principal right ideal and a principal left ideal, then $\mathfrak{A}$ is a direct sum of two-sided ideals that are primary rings having these properties.*

Let $\mathfrak{A}_1$ be a minimal non-nilpotent two-sided ideal in $\mathfrak{A}$. Then $\mathfrak{A}_1 = \mathfrak{A}c = c'\mathfrak{A}$ for suitable c and c', and $\mathfrak{A}c^2 = (c')^2\mathfrak{A}$ is a two-sided ideal of $\mathfrak{A}$ contained in $\mathfrak{A}_1$. Since $(\mathfrak{A}c)^2 = \mathfrak{A}(c\mathfrak{A})c \leqq \mathfrak{A}(\mathfrak{A}c)c = \mathfrak{A}c^2$, $\mathfrak{A}c^2$ is not nilpotent. Hence by the minimality of $\mathfrak{A}_1$, $\mathfrak{A}c^2 = \mathfrak{A}c = c'\mathfrak{A} = (c')^2\mathfrak{A}$. Since the ascending chain condition holds for $\mathfrak{A}_1$ regarded as an $\mathfrak{A}_l$-group, the $\mathfrak{A}_l$-endomorphism $x \to xc$ is $(1-1)$ in $\mathfrak{A}_1$. Hence the only element z in $\mathfrak{A}_1$ such that $zc = 0$ is $z = 0$. Thus if $\mathfrak{A}^*$ denotes the set of elements a^* in $\mathfrak{A}$ such that $a^*c = 0$, $\mathfrak{A}^*$ is a left ideal and $\mathfrak{A}^* \wedge \mathfrak{A}_1 = 0$. If x is any element of $\mathfrak{A}$, $xc = yc^2$ for a suitable y. Hence $x = (x - yc) + yc \in \mathfrak{A}^* + \mathfrak{A}_1$ and so $\mathfrak{A} = \mathfrak{A}_1 \oplus \mathfrak{A}^*$. Similarly, $\mathfrak{A} = \mathfrak{A}_1 \oplus \mathfrak{A}''$ where $\mathfrak{A}''$ is the right ideal of elements a'' such that $c'a'' = 0$. If a'' is arbitrary in $\mathfrak{A}''$, $a'' = a_1 + a^*$, a_1 in $\mathfrak{A}_1$, a^* in $\mathfrak{A}^*$. Then $0 = c'a'' = c'a_1 + c'a^* \in \mathfrak{A}_1 \oplus \mathfrak{A}^*$ and $c'a_1 = 0$. Hence $a_1 = 0$ and $a'' = a^* \in \mathfrak{A}^*$. Similarly if $a^* \in \mathfrak{A}^*$, $a^* \in \mathfrak{A}''$ and so $\mathfrak{A}^* = \mathfrak{A}''$ is a two-sided ideal. It follows that $\mathfrak{A}_1$ is primary. For otherwise, we should have a non-nilpotent two-sided ideal $\mathfrak{B}_1 \neq \mathfrak{A}_1$ in $\mathfrak{A}_1$ and this would contradict the minimality of $\mathfrak{A}_1$. In order to complete the proof we require the

LEMMA. *Let $\mathfrak{A}$ be a ring with an identity and $\mathfrak{A} = \mathfrak{A}_1 \oplus \cdots \oplus \mathfrak{A}_t$ where the $\mathfrak{A}_i$ are two-sided ideals. A necessary and sufficient condition that every right (left, two-sided) ideal of $\mathfrak{A}$ be a principal right (left, left and right) ideal is that this holds for each $\mathfrak{A}_i$.*

If $\mathfrak{J}_i$ is a right ideal of $\mathfrak{A}_i$, it is one of $\mathfrak{A}$ since $\mathfrak{A}_i\mathfrak{A}_j = 0$ if $i \neq j$. Hence $\mathfrak{J}_i = c_i\mathfrak{A} = c_i\mathfrak{A}_i$ since c_i is in $\mathfrak{A}_i$. On the other hand, any right ideal $\mathfrak{J}$ of $\mathfrak{A}$ has the form $\mathfrak{J} = \mathfrak{J}_1 \oplus \cdots \oplus \mathfrak{J}_t$ where $\mathfrak{J}_i = \mathfrak{A}_i \wedge \mathfrak{J}$ is a right ideal in $\mathfrak{A}_i$. If $\mathfrak{J}_i = c_i\mathfrak{A}_i$, c_i in $\mathfrak{J}_i$, then $\mathfrak{J} = c_1\mathfrak{A}_1 + \cdots + c_t\mathfrak{A}_t = c\mathfrak{A}$ where $c = c_1 + \cdots + c_t$ is in $\mathfrak{J}$.

This lemma implies that the rings $\mathfrak{A}_1$ and $\mathfrak{A}^*$ determined above satisfy the same conditions as $\mathfrak{A}$. If $\mathfrak{A}^*$ is not primary, we may repeat this process and write $\mathfrak{A}^* = \mathfrak{A}_2 \oplus \mathfrak{A}''$ where $\mathfrak{A}_2$, $\mathfrak{A}''$ are two-sided ideals of $\mathfrak{A}'$ and hence of $\mathfrak{A}$, and $\mathfrak{A}_2$ is primary. After a finite number of repetitions of this process we obtain $\mathfrak{A} = \mathfrak{A}_1 \oplus \cdots \oplus \mathfrak{A}_t$ where the $\mathfrak{A}_i$ are primary and satisfy the conditions of the theorem.

THEOREM 38. *Let $\mathfrak{A} = \mathfrak{B}_u$ where $\mathfrak{B}$ is completely primary and $\mathfrak{A}$ satisfies the chain conditions for one-sided ideals. Suppose that the radical $\mathfrak{R}$ of $\mathfrak{A}$ is a principal right ideal and a principal left ideal. Then $\mathfrak{R} = w\mathfrak{A} = \mathfrak{A}w$ for any w which belongs to $(\mathfrak{R} \wedge \mathfrak{B})$ but not to $(\mathfrak{R}^2 \wedge \mathfrak{B})$.*

We have seen that if $\mathfrak{R} \wedge \mathfrak{B} = \mathfrak{S}$ and e_{ij} are matrix units such that $\mathfrak{A} = \Sigma e_{ij}\mathfrak{B}$ and $e_{ij}b = be_{ij}$, b in $\mathfrak{B}$, then $\mathfrak{R} = \Sigma e_{ij}\mathfrak{S}$. Then $\mathfrak{R}^k = \Sigma e_{ij}\mathfrak{S}^k$ and $\mathfrak{S}^k = \mathfrak{R}^k \wedge \mathfrak{B}$. We note next that if u and $\bar{v}$ are elements of $\mathfrak{A}$ such that $u\bar{v} \equiv 1(\mathfrak{R})$, then $u\bar{v} = 1 - r$, r in $\mathfrak{R}$ and hence $u\bar{v}(1 + r + r^2 + \cdots + r^{s-1}) = uv = 1$ if $r^s = 0$. Evidently $v \equiv \bar{v}\ (\mathfrak{R})$. Also, since $\mathfrak{A} - \mathfrak{R}$ is a matrix ring over a division ring, $\bar{v}u \equiv 1(\mathfrak{R})$. Hence there is a v' such that $v'u = 1$, $v' \equiv \bar{v}\ (\mathfrak{R})$. It follows that $v' = v$, and u is a unit with v as its inverse.

After these preliminaries we may begin the proof of the theorem. Let $w \in \mathfrak{S}$, $\notin \mathfrak{S}^2$. Then $w = zu$ if $\mathfrak{R} = z\mathfrak{A}$. We consider u modulo $\mathfrak{R}$. Since $\mathfrak{A} - \mathfrak{R}$ is a matrix ring over a division ring, there exist elements v_1 and v_2 which are units modulo $\mathfrak{R}$ such that $u \equiv v_1e_sv_2(\mathfrak{R})$, $e_s = \sum_1^s e_{ii}$. We may suppose that v_1 and v_2 are units in $\mathfrak{A}$. Thus $u = v_1e_sv_2 + r$, r in $\mathfrak{R}$, and $w = z(v_1e_sv_2 + r)$, $wv_2^{-1} = (zv_1)e_s + zrv_2^{-1}$. Hence $wv_2^{-1} \equiv (zv_1)e_s\ (\mathfrak{R}^2)$. If we write $wv_2^{-1} = \Sigma e_{ij}w_{ij}$, w_{ij} in $\mathfrak{B}$, this shows that the w_{ij}, with $j > s$, are in $\mathfrak{S}^2$. If $v_2^{-1} = \Sigma e_{ij}v_{ij}$, v_{ij} in $\mathfrak{B}$, then each v_{ij}, $j > s$, is in $\mathfrak{S}$. Otherwise v_{ij} is a unit and since $w_{ij} = wv_{ij}$, we should have w in $\mathfrak{S}^2$ contrary to assumption. We have therefore proved that $v_{ij} = 0\ (\mathfrak{S})$ for $j > s$. Since v is a unit modulo $\mathfrak{R}$, this is impossible unless $s = n$, i.e. unless $u \equiv v_1v_2\ (\mathfrak{R})$. Since v_1 and v_2 are units, it follows that u is a unit and $\mathfrak{R} = z\mathfrak{A} = w\mathfrak{A}$. Similarly, $\mathfrak{R} = \mathfrak{A}w$.

THEOREM 39. *Under the assumptions of the preceding theorem, the ideals $\mathfrak{S}^k$, $k = 0, 1, \cdots$ ($\mathfrak{S}^0 = \mathfrak{B}$), are the only right (left) ideals of $\mathfrak{B}$. $\mathfrak{S}^k$ is a principal right (left) ideal. The only two-sided ideals of $\mathfrak{A}$ are $\mathfrak{R}^k$.*

Let b be any element $\neq 0$ of $\mathfrak{B}$. If $b \notin \mathfrak{S}$, b is a unit. Now suppose that $b \in \mathfrak{S}^k$, $\notin \mathfrak{S}^{k+1}$, $k > 0$. Then $b \in \mathfrak{R}^k$, $\notin \mathfrak{R}^{k+1}$ and hence $b = w^ku$ where $w \in \mathfrak{S}$,

$\notin \mathfrak{S}^2$. If we write $u = \Sigma e_{ij}u_{ij}$, we obtain $b = w^k u_{ii}$, $w^k u_{ij} = 0$. Hence we may replace u by $u_1 = u_{11}$ and obtain $b = w^k u_1$ with u_1 in $\mathfrak{B}$. Then u_1 is a unit. By a parallel argument we may show that $b = \bar{u}_1 w^k$ with $\bar{u}_1$ a unit in $\mathfrak{B}$. If now b is any element of $\mathfrak{S}^k$, b is in $\mathfrak{S}^l$ but not in $\mathfrak{S}^{l+1}$, $l \geqq k$, and so $b = w^l u_2 = w^k c$, c in $\mathfrak{B}$, and likewise $b = \bar{c}w^k$, $\bar{c}$ in $\mathfrak{B}$. Hence we have proved that $\mathfrak{S}^k = w^k\mathfrak{B} = \mathfrak{B}w^k$. Now let $\mathfrak{J}$ be a right ideal of $\mathfrak{B}$. Suppose that $\mathfrak{J} \leqq \mathfrak{S}^k$ but $\nleqq \mathfrak{S}^{k+1}$ and let b be an element of $\mathfrak{J}$ not in $\mathfrak{S}^{k+1}$. Then $b = w^k u_1$, u_1 a unit. Hence w^k is in $\mathfrak{J}$ and $\mathfrak{S}^k = \mathfrak{J}$. Since any two-sided ideal $\mathfrak{B}_1$ in $\mathfrak{B}$ has the form $\Sigma e_{ij}\mathfrak{J}$, where $\mathfrak{J}$ is a two-sided ideal in $\mathfrak{B}$, $\mathfrak{B}_1$ must be one of the ideals $\Sigma e_{ij}\mathfrak{S}^k = \mathfrak{R}^k$.

We prove next

THEOREM 40. *If $\mathfrak{B}$ is a principal ideal ring, then so is the matrix ring $\mathfrak{A} = \mathfrak{B}_u$.*

Let $\mathfrak{F}$ denote the free $\mathfrak{B}$-module with u generators. The elements of $\mathfrak{F}$ are the u-tuples $(b_1, \cdots, b_u)$, b_i in $\mathfrak{B}$. With any right ideal $\mathfrak{J}$ in $\mathfrak{A}$ we associate the set $\mathfrak{F}(\mathfrak{J})$ of elements in $\mathfrak{F}$ consisting of the columns of the matrices in $\mathfrak{J}$. Evidently $\mathfrak{F}(\mathfrak{J})$ is a submodule of $\mathfrak{F}$ and hence by the argument on p. 43, $\mathfrak{F}(\mathfrak{J})$ has a set of $m \leqq u$ generators. Let these be $(b_{1j}, b_{2j}, \cdots, b_{uj})$, $j = 1, \cdots, m$, and let b be the matrix (b_{ij}) where $b_{ij} = 0$ if $j > m$. We wish to show that $\mathfrak{J} = b\mathfrak{A}$. To prove this, we note that if $c = \Sigma e_{ij}c_{ij}$ is an element in $\mathfrak{J}$, then so is ce_{pq}, and this matrix has as its q-th column the p-th column of c and all other columns are 0. Since the columns of b occur in matrices of $\mathfrak{J}$, the matrices $\Sigma e_{ij}b_{ij} \in \mathfrak{J}$ for $j = 1, \cdots, u$ and hence $b = \Sigma e_{ij}b_{ij}$ is in $\mathfrak{J}$. The u^2 matrices be_{pq} contain the columns of b in all possible positions, and since these columns form a basis for $\mathfrak{F}(\mathfrak{J})$, any element of $\mathfrak{J}$ has the form bv for a suitable v in $\mathfrak{A}$. A similar argument holds for left ideals.

Now let $\mathfrak{A}$ be any primary ring with an identity satisfying the descending chain condition for one-sided ideals. Suppose that the radical of $\mathfrak{A}$ is a principal right ideal and a principal left ideal. Then by Theorem 39, $\mathfrak{A} = \mathfrak{B}_u$ where $\mathfrak{B}$ is a completely primary principal ideal ring. Hence by Theorem 40, $\mathfrak{A}$ itself is a principal ideal ring.

THEOREM 41. *If $\mathfrak{A}$ is a primary ring satisfying the descending chain conditions for one-sided ideals and the radical $\mathfrak{R}$ of $\mathfrak{A}$ is a principal left ideal and a principal right ideal, then $\mathfrak{A}$ is a principal ideal ring.*

By this theorem and Theorem 37, we obtain

THEOREM 42. *If $\mathfrak{A}$ is a ring with an identity satisfying the descending chain conditions for one-sided ideals, and if every two-sided ideal of $\mathfrak{A}$ is a principal left ideal and a principal right ideal, then $\mathfrak{A}$ is a principal ideal ring.*

16. $\mathfrak{A}$-modules, $\mathfrak{A}$ a principal ideal ring. We wish to determine the structure of finitely generated $\mathfrak{A}$-modules, $\mathfrak{A}$ of the type of **15**. As usual, we suppose that $x1 = x$ for all x in the module $\mathfrak{M}$ and 1, the identity of $\mathfrak{A}$. Since $\mathfrak{A} = \mathfrak{A}_1 \oplus \cdots \oplus \mathfrak{A}_t$, $\mathfrak{M} = \mathfrak{M}\mathfrak{A}_1 \oplus \cdots \oplus \mathfrak{M}\mathfrak{A}_t$. Since $\mathfrak{A}_i\mathfrak{A}_j = 0$ if $i \neq j$, any $\mathfrak{A}_i$-submodule of $\mathfrak{M}\mathfrak{A}_i$ is an $\mathfrak{A}$-submodule and $\mathfrak{A}_i$-isomorphism of submodules of $\mathfrak{M}\mathfrak{A}_i$ implies $\mathfrak{A}$-isomorphism. Hence we may assume that $t = 1$, that is, that $\mathfrak{A}$ is primary. We consider first the case where $\mathfrak{A} = \mathfrak{B}$ is completely primary.

Then if $\mathfrak{S}$ is its radical and $w \in \mathfrak{S}$ but $\notin \mathfrak{S}^2$, we have seen that every element of $\mathfrak{B}$ has the form $uw^k = w^k u'$ where u and u' are units. Hence if (a_{ij}) is a matrix in $\mathfrak{B}_m$, we may use elementary transformations (p. 42) to reduce it to diagonal form. Thus there exist units u_1 and u_2 in $\mathfrak{B}_m$ such that

$$u_1(a_{ij})u_2 = \begin{pmatrix} w^{k_1} & & & & & & \\ & w^{k_2} & & & & & \\ & & \ddots & & & & \\ & & & w^{k_s} & & & \\ & & & & 0 & & \\ & & & & & \ddots & \\ & & & & & & 0 \end{pmatrix} \tag{7}$$

where $0 \leqq k_1 \leqq k_2 \leqq \cdots < l$ if $\mathfrak{S}^l = 0$, $\mathfrak{S}^{l-1} \neq 0$. As we saw in Chapter 3, this implies that any finitely generated $\mathfrak{B}$-module is a direct sum of cyclic $\mathfrak{B}$-modules isomorphic to the $\mathfrak{B}$-modules $\mathfrak{B} - w^k\mathfrak{B} = \mathfrak{B} - \mathfrak{S}^k$. Since the mapping $x \rightarrow w^{l-k}x$ is a $\mathfrak{B}$-homomorphism between $\mathfrak{B}$ and $\mathfrak{S}^{l-k}$ sending the elements of $\mathfrak{S}^k$ and only these into 0, $\mathfrak{S}^{l-k}$ and $\mathfrak{B} - \mathfrak{S}^k$ are isomorphic. Now $\mathfrak{S}^{l-k}$ is indecomposable, for otherwise, $\mathfrak{S}^{l-k} = \mathfrak{J}_1 \oplus \mathfrak{J}_2$ where the $\mathfrak{J}_j$ are right ideals $\neq 0$ of $\mathfrak{B}$, and this is impossible since each $\mathfrak{J}_j$ is a power of $\mathfrak{S}$. Thus the cyclic modules $\mathfrak{B} - w^k\mathfrak{B}$ are indecomposable.

Now let $\mathfrak{A} = \mathfrak{B}_u = \Sigma e_{ij}\mathfrak{B}$ where $\mathfrak{B}$ is a completely primary principal ideal ring. If $\mathfrak{M}$ is an $\mathfrak{A}$-module, $\mathfrak{M}$ is a $\mathfrak{B}$-module and if $x_1, \cdots, x_n$ generate $\mathfrak{M}$ relative to $\mathfrak{A}$, the elements $x_i e_{jk}$ generate $\mathfrak{M}$ relative to $\mathfrak{B}$. Thus $\mathfrak{M}$ is finitely generated relative to $\mathfrak{B}$. The sets $\mathfrak{M}e_{ii}$ are $\mathfrak{B}$-modules since $e_{ii}b = be_{ii}$ for all b in $\mathfrak{B}$ and are $\mathfrak{B}$-isomorphic since, as is readily verified, the correspondence $x \rightarrow xe_{1i}$ is a $\mathfrak{B}$-isomorphism between $\mathfrak{M}e_{11}$ and $\mathfrak{M}e_{ii}$. Evidently $\mathfrak{M} = \mathfrak{M}e_{11} \oplus \cdots \oplus \mathfrak{M}e_{uu}$. We fix our attention now on $\mathfrak{M}e_{11}$. Since $\mathfrak{B}$ is a principal ideal ring, $\mathfrak{M}e_{11}$ is a finitely generated $\mathfrak{B}$-module. Hence, by our assumption, we may write $\mathfrak{M}e_{11} = \mathfrak{M}^{(1)} \oplus \cdots \oplus \mathfrak{M}^{(g)}$ where $\mathfrak{M}^{(j)}$ is a cyclic $\mathfrak{B}$-module generated by y_j. By the isomorphism noted, we have $\mathfrak{M}e_{ii} = \mathfrak{M}^{(1)}e_{1i} \oplus \cdots \oplus \mathfrak{M}^{(g)}e_{1i}$ where $\mathfrak{M}^{(j)}e_{1i}$ is cyclic and has the generator $y_j e_{1i}$. It follows that the elements y_j are generators of $\mathfrak{M}$ relative to $\mathfrak{A}$. Now suppose that $y_1a_1 + \cdots + y_ra_r = 0$ for $a_j = \Sigma e_{ik}b_{ik}^{(j)}$. Since $y_je_{11} = y_j$, we have

$$0 = (y_1e_{11}a_1 + \cdots + y_ge_{11}a_g)e_{pp} = (y_1b_{1p}^{(1)} + \cdots + y_gb_{1p}^{(g)})e_{1p}, \qquad p = 1, 2, \cdots.$$

This implies that $b_{1p}^{(j)} = 0$ and hence $y_ga_g = 0$. Thus $\mathfrak{M} = (y_1) \oplus \cdots \oplus (y_g)$ where (y_j) denotes the cyclic $\mathfrak{A}$-module generated by y_j. This proves

THEOREM 43. *If $\mathfrak{A}$ is a principal ideal ring satisfying the descending chain conditions for one-sided ideals, then any finitely generated $\mathfrak{A}$-module is a direct sum of cyclic $\mathfrak{A}$-modules.*

Our argument shows also that if $\mathfrak{M}$ is an indecomposable $\mathfrak{A}$-module, $\mathfrak{M}e_{11}$ is an indecomposable $\mathfrak{B}$-module. For if $\mathfrak{M}e_{11} = \mathfrak{M}' \oplus \mathfrak{M}''$, we may decompose $\mathfrak{M}'$ and $\mathfrak{M}''$ into cyclic modules $\mathfrak{M}^{(j)}$, $j = 1, 2, \cdots, g$, $g \geqq 2$, and then obtain a decomposition of $\mathfrak{M}$ into g $\mathfrak{A}$-modules. On the other hand, suppose that $\mathfrak{M}e_{11}$

is an indecomposable $\mathfrak{B}$-module. If $\mathfrak{M} = \mathfrak{M}' \oplus \mathfrak{M}''$, we may write $\mathfrak{M}' = \mathfrak{M}'e_{11} \oplus \cdots \oplus \mathfrak{M}'e_{uu}$, $\mathfrak{M}'' = \mathfrak{M}''e_{11} \oplus \cdots \oplus \mathfrak{M}''e_{uu}$ and obtain a direct decomposition of $\mathfrak{M}$ into $2u$ components. Since we have $\mathfrak{M} = \mathfrak{M}e_{11} \oplus \cdots \oplus \mathfrak{M}e_{uu}$ where the $\mathfrak{M}e_{ii}$ are indecomposable, we obtain a contradiction to the Krull-Schmidt Theorem.

If $\mathfrak{M} = (y_1)$ is an indecomposable $\mathfrak{B}_u$-module, we may suppose that $y_1e_{11} = y_1$ and that y_1 generates an indecomposable $\mathfrak{B}$-module. It follows, on considering the set of elements a of $\mathfrak{A} = \mathfrak{B}_u$ such that $y_1a = 0$, that (y_1) is $\mathfrak{A}$-isomorphic to $\mathfrak{A} - \mathfrak{J}$ where $\mathfrak{J}$ is the right ideal generated by e_{ij} with $i > 1$ and by $e_{ij}w^k$. If we use the $\mathfrak{A}$-homomorphism $x \to e_{11}w^{l-k}x$, we may prove that $\mathfrak{A} - \mathfrak{J}$, and hence (y_1), is isomorphic to the ideal $e_{11}\mathfrak{R}^{l-k}$. Now, as in Chapter 3, we define the *bound* of an $\mathfrak{A}$-module $\mathfrak{M}$ to be the two-sided ideal of elements d of $\mathfrak{A}$ such that $xd = 0$ for all x in $\mathfrak{M}$. In the present case we allow the bound to be 0. It is readily seen that the bound of (y_1) (or $e_{11}\mathfrak{R}^{l-k}$) is $\mathfrak{R}^k$. Hence if $\mathfrak{A}$ is primary, then a necessary and sufficient condition that two indecomposable $\mathfrak{A}$-modules be isomorphic is that they have the same bounds. If we use the decomposition $\mathfrak{M} = \mathfrak{M}\mathfrak{A}_1 \oplus \cdots \oplus \mathfrak{M}\mathfrak{A}_t$, we see that this result is also valid for arbitrary principal ideal rings $\mathfrak{A}$ and finally, by the Krull-Schmidt Theorem, we obtain the following general criterion:

THEOREM 44. *If $\mathfrak{A}$ is a principal ideal ring, then a necessary and sufficient condition for $\mathfrak{A}$-isomorphism of any two finitely generated $\mathfrak{A}$-modules is that the totality of bounds of the indecomposable components that occur in a decomposition of one of the modules coincide with the totality of bounds occurring in a decomposition of the second.*

The main results of Chapter 3 may now be proved for the rings considered here. We mention, for example, the following theorem that will be required later.

THEOREM 45. *If $\mathfrak{M}$ is an indecomposable $\mathfrak{A}$-module and $\mathfrak{Q}$ is its bound, then $\bar{\mathfrak{A}} = \mathfrak{A} - \mathfrak{Q}$ is a primary ring. If e is the exponent of the radical of $\bar{\mathfrak{A}}$, $\mathfrak{M}$ has length e. An indecomposable $\mathfrak{A}$-module has only one composition series.*

The proof is left to the reader.

We remark that if $\mathfrak{o}$ is an arbitrary principal ideal domain and $\mathfrak{J}$ is a two-sided ideal $\neq 0$ in $\mathfrak{o}$, then $\mathfrak{o} - \mathfrak{J}$ is a principal ideal ring satisfying both chain conditions for one-sided ideals. Hence if $\mathfrak{M}$ is a bounded $\mathfrak{o}$-module in the sense of Chapter 3 and $\mathfrak{J}$ is the bound of $\mathfrak{M}$, then $\mathfrak{M}$ is an $(\mathfrak{o} - \mathfrak{J})$-module, and so the results on bounded $\mathfrak{o}$-modules are consequences of the present theory. The treatment of Chapter 3 is, however, of a more elementary character.

17. Projective and affine representations of a group. In the remainder of this chapter we shall consider some applications of the theory developed thus far. We begin with the problem of representation of groups.

It is a classical result that any projective space $\mathfrak{P}$ of dimensionality $(n - 1) \geqq 3$ may be regarded as the system of one dimensional subspaces $\{x\alpha\}$, x fixed, in a suitable n-dimensional vector space $\mathfrak{M}$ over a division ring Φ.

The k-dimensional subspaces of $\mathfrak{M}$ correspond to $(k - 1)$-dimensional subspaces of $\mathfrak{P}$. It is well known that the collineations, or, projective transformations of $\mathfrak{P}$, i.e. the $(1 - 1)$ transformations that preserve incidence, are induced by non-singular semi-linear transformations in $\mathfrak{M}$ over Φ. Two semi-linear transformations T_1 and T_2 have the same effect in $\mathfrak{P}$ if and only if $T_1 = T_2\mu$, μ in Φ. The complete projective group is therefore isomorphic to $\mathfrak{S}/\Phi^*$, where $\mathfrak{S}$ is the group of non-singular semi-linear transformations and Φ^*, the set of mappings $x \to x\mu$, $\mu \neq 0$ in Φ.[6] We recall also that the collineations which are generated by perspectivities are the ones whose corresponding semi-linear transformations induce inner automorphisms in Φ. We shall call these collineations *special.*

Consider the following problem: Given a group $\mathfrak{g} = (1, s, t, \cdots)$ and a projective space $\mathfrak{P}$, to determine the homomorphisms between $\mathfrak{g}$ and groups of collineations in $\mathfrak{P}$. Such homomorphisms are called *projective representations* of $\mathfrak{g}$. Two representations $s \to c_s$, $s \to d_s$ are *equivalent* (*strictly equivalent*) if there is a collineation (special collineation) $u \to u'$ such that $(uc_s)' = u'd_s$, or, if $u \to u'$ is denoted as f, then $d_s = f^{-1}c_sf$.

If we transfer this to $\mathfrak{M}$, we obtain the following formulation: A projective representation of $\mathfrak{g}$ corresponds to an association $s \to T_s$ where T_s is a non-singular semi-linear transformation in $\mathfrak{M}$ such that

$$T_sT_t = T_{st}\rho_{s,t}, \qquad \rho_{s,t} \text{ in } \Phi.$$

If $\bar{s}$ denotes the automorphism in Φ determined by T_s, then

$$\xi^{\bar{s}\bar{t}} = \rho_{s,t}^{-1}\xi^{\overline{st}}\rho_{s,t} \tag{8}$$

for all ξ in Φ. Thus if we call the factor group of the group of automorphisms of Φ relative to the invariant subgroup of inner automorphisms the *group of outer automorphisms* of Φ, we see that the correspondence $s \to \bar{s}$ determines a homomorphism between $\mathfrak{g}$ and a subgroup of the group of outer automorphisms of Φ. It follows that the subset $\mathfrak{h}$ of elements h of $\mathfrak{g}$ for which T_h is a special collineation is an invariant subgroup of $\mathfrak{g}$. The set $\rho = \{\rho_{s,t}\}$ will be called the *factor set* of the representation. The associative law imposes the condition

$$\rho_{s,tu}\rho_{t,u} = \rho_{st,u}\rho_{s,t}^{\bar{u}}. \tag{9}$$

The projective representations $s \to T_s$, $s \to U_s$ are equivalent if there is a non-singular semi-linear transformation A with automorphism $\bar{a}$ and elements μ_s such that

$$U_s = A^{-1}T_sA\mu_s = A^{-1}(T_s\mu_s^{\bar{a}^{-1}})A.$$

We have strict equivalence if A may be taken to be linear. If s' is the automorphism associated with U_s and σ is the factor set of this representation, we have as necessary conditions for equivalence

$$s' = \bar{a}^{-1}\bar{s}\bar{a}\mu_s, \qquad \sigma_{s,t} = \mu_{st}^{-1}\rho_{s,t}^{\bar{a}}\mu_s^{\bar{a}^{-1}\bar{t}\bar{a}}\mu_t, \tag{10}$$

[6] Since the group operation is denoted as multiplication, we use the terms: factor group, power, etc.

where in the first equation μ_s denotes the inner automorphism $\xi \to \mu_s^{-1}\xi\mu_s$. Necessary conditions for strict equivalence are

$$(11) \qquad s' = \bar{s}\mu_s, \qquad \sigma_{s,t} = \mu_{st}^{-1}\rho_{s,t}\mu_s^{\bar{t}}\mu_t.$$

If Φ is commutative, the only inner automorphism is the identity mapping. Hence the correspondence $s \to \bar{s}$ is a homomorphism between $\mathfrak{g}$ and a subgroup of the group of automorphisms of Φ. Strictly equivalent representations have the same automorphisms in Φ.

An important class of projective representations consists of those representations for which the factor sets $\rho_{s,t} = 1$. In this case we have $T_{st} = T_sT_t$ and $s \to \bar{s}$ is a homomorphism. We call a representation of this type an *affine representation*, and we are usually concerned with equivalence of pairs of these representations defined by the condition that $U_s = A^{-1}T_sA$ where A is a linear transformation (i.e. $\mu_s = 1$). Finally, we may impose the further condition that $\bar{s}$ is the identity automorphism for all s. Then T_s is linear. If, in addition, Φ is commutative, we obtain the classical case for which there is a very extensive literature.

From now on we suppose that $\mathfrak{g}$ is finite. Let r be its order and p the characteristic of Φ. We wish to prove the following

THEOREM 46. *If $p \nmid r$, any projective representation of $\mathfrak{g}$ is completely reducible.*

By complete reducibility we mean complete reducibility of $\mathfrak{M}$ relative to the set of endomorphisms $\{\Phi, T_1, T_s, \cdots\}$. Let $\mathfrak{N}$ be a subspace of $\mathfrak{M}$ invariant under the transformations T_s and let $\mathfrak{N}^*$ be any complementary space, i.e. $\mathfrak{M} = \mathfrak{N} \oplus \mathfrak{N}^*$. We wish to show that we may choose $\mathfrak{N}^*$ so that it, too, is invariant relative to the T_s. If $x \in \mathfrak{M}$, we may write $x = y + y^*$ where $y \in \mathfrak{N}$, $y^* \in \mathfrak{N}^*$. The mapping $x \to y = xD$ determined by the decomposition is then an idempotent linear transformation such that $\mathfrak{M}D = \mathfrak{N}$. Now any linear transformation which maps $\mathfrak{M}$ into $\mathfrak{N}$ and acts as the identity in $\mathfrak{N}$ is idempotent. Hence if $D_1, \cdots, D_m$ have this property and $p \nmid m$, then $\frac{1}{m}(D_1 + \cdots + D_m)$ has the property. Thus $E = \frac{1}{r}(\sum_{s \in \mathfrak{g}} T_s^{-1}DT_s)$ has the property since $T_s^{-1}DT_s$ is linear, $\mathfrak{M}T_s^{-1}DT_s \leqq \mathfrak{N}T_s = \mathfrak{N}$ and $yT_s^{-1}DT_s = y$ for all y in $\mathfrak{N}$. Now

$$T_t^{-1}ET_t = \frac{1}{r}\sum_s T_t^{-1}T_s^{-1}DT_sT_t = \frac{1}{r}\sum_s \rho_{s,t}^{-1}T_{st}^{-1}DT_{st}\rho_{s,t} = E$$

since $T_{st}^{-1}DT_{st}$ commutes with $\rho_{s,t}$ and st ranges over $\mathfrak{g}$ when s does. Thus E, and hence $1 - E$, commutes with all the T_s. Then $\mathfrak{M} = \mathfrak{M}E \oplus \mathfrak{M}(1 - E) \equiv \mathfrak{N} \oplus \mathfrak{N}'$ and $\mathfrak{N}'$ is invariant relative to all T_s.

18. Crossed products. The preceding theorem may be strengthened by replacing the hypothesis $p \nmid r$ by the weaker one, $p \nmid q$, q the order of the invariant subgroup $\mathfrak{h}$ of elements h such that $\bar{h}$ is inner. In order to prove this, and for other purposes, we introduce a certain ring $\mathfrak{A}$ determined by $\mathfrak{g}$, Φ, the correspondence $s \rightarrow \bar{s} \equiv s^H$ and the factor set ρ. We need not suppose that the $\bar{s}$ and the ρ are obtained from a projective representation but merely that they satisfy (8) and (9) and that $\rho_{s,t} \neq 0$. The elements of $\mathfrak{A} \equiv \Phi(\mathfrak{g}, H, \rho)$ are the expressions $\sum_{s \in \mathfrak{g}} t_s \xi_s$ where the ξ_s vary in Φ. We consider $\Sigma t_s \xi_s \equiv \Sigma t_s \eta_s$ if and only if $\xi_s = \eta_s$ for all s and we define

$$\Sigma t_s \xi_s + \Sigma t_s \eta_s \equiv \Sigma t_s(\xi_s + \eta_s),$$

$$(\Sigma t_s \xi_s)(\Sigma t_t \eta_t) \equiv \sum_{s,t} t_{st} \rho_{s,t} \xi_s^{\bar{t}} \eta_t .$$

It is readily verified that $\mathfrak{A}$ is a ring. We shall call it the *crossed product* of Φ and $\mathfrak{g}$ with correspondence H and factor set ρ.

The conditions on ρ imply in particular that $\rho_{s,1}\rho_{1,1} = \rho_{s,1}\rho_{s,1}^{\bar{1}}$. Hence $\rho_{s,1}^{\bar{1}} = \rho_{1,1}$, $\rho_{1,1}^{\bar{1}} = \rho_{1,1}$ and $\rho_{s,1} = \rho_{1,1}$. Similarly, $\rho_{1,s} = \rho_{1,1}^{\bar{s}}$. We note also that $\xi^{\bar{1}} = \rho_{1,1}^{-1} \xi \rho_{1,1}$. It follows that the element $t_1 \rho_{1,1}^{-1}$ is an identity 1 for $\mathfrak{A}$ and the elements 1ξ form a division subring of $\mathfrak{A}$ that may be identified with Φ. We may set $t_s \equiv t_s 1$, 1, the identity of Φ. Then any element of $\mathfrak{A}$ has the form $\Sigma t_s \xi_s$ where $t_s \xi_s$ now indicates the product of t_s and ξ_s in Φ. The ring $\mathfrak{A}$ is a vector space over Φ relative to the endomorphisms $x \rightarrow x\xi$. Since the expression of an element in the form $\Sigma t_s \xi_s$ is unique, $(\mathfrak{A} : \Phi) = r$. We note that

$$\xi t_s = t_s \xi^{\bar{s}}, \qquad t_s t_t = t_{st} \rho_{s,t} .$$

Similarly, the endomorphisms $x \rightarrow \xi x \equiv x\xi'$ form a division ring Φ' anti-isomorphic to Φ, and $(\mathfrak{A} : \Phi') = r$ also. Since right (left) ideals of $\mathfrak{A}$ are subspaces over Φ (Φ'), $\mathfrak{A}$ satisfies both chain conditions for right (left) ideals.

If $s \rightarrow T_s$ is a projective representation of $\mathfrak{g}$ in $\mathfrak{M}$ over Φ with the correspondence $s \rightarrow \bar{s}$ and the factor set ρ, the correspondence $\Sigma t_s \xi_s \rightarrow \Sigma T_s \xi_s$ is a representation of $\mathfrak{A}$ by endomorphisms in $\mathfrak{M}$ in which 1 is mapped into the identity endomorphism. If two $\mathfrak{A}$-modules thus determined are $\mathfrak{A}$-isomorphic, then the projective representations are strictly equivalent. Conversely if we have a representation of $\mathfrak{A}$ by endomorphisms in $\mathfrak{M}$ such that $1 \rightarrow 1$, $\mathfrak{M}$ may be regarded as a vector space relative to the endomorphisms of $\Phi \leqq \mathfrak{A}$. If $(\mathfrak{M} : \Phi)$ is finite and T_s is the correspondent of t_s, we obtain a projective representation $s \rightarrow T_s$ of $\mathfrak{g}$ having the same correspondence and factor set as $\mathfrak{A}$. Thus the theory of representations of the ring $\mathfrak{A}$ is closely connected with that of the projective representations of $\mathfrak{g}$ that have the same correspondence and factor set.

Let $\mathfrak{B}$ denote the subring of $\mathfrak{A}$ of elements $\sum_{h \in \mathfrak{h}} t_h \xi_h$. Then $\mathfrak{B}$ is the crossed product of Φ and $\mathfrak{h}$ with the factor set $\rho_{h,k}$ and the correspondence $h \rightarrow \bar{h}$. Let $1, u, \cdots, w$ be representatives of the cosets of $\mathfrak{g}/\mathfrak{h}$. Then if $s \in \mathfrak{g}$, $s = uh$, h in $\mathfrak{h}$. Hence $t_s = t_u t_h \rho_{u,h}^{-1} = t_u b$ where b is in $\mathfrak{B}$. It follows that the elements of $\mathfrak{A}$ may be written in the form $\Sigma t_u b_u$, b_u in $\mathfrak{B}$ and the summation extending over the representatives $1, u, \cdots$. This representation is unique. For if $\Sigma t_u b_u = 0$, we set $b_u = \Sigma t_h \xi_{h,u}$ and obtain $\Sigma t_{uh} \rho_{u,h} \xi_{h,u} = 0$. Since the r elements uh are distinct, $\rho_{u,h}\xi_{h,u} = 0$, $\xi_{h,u} = 0$ and hence $b_u = 0$. Now $t_s^{-1} = (\rho_{1,1}\rho_{s^{-1},s})^{-1}t_{s^{-1}}$ and so

$$t_s^{-1}t_h t_s = (\rho_{1,1}\rho_{s^{-1},s})^{-1}t_{s^{-1}}t_h t_s = (\rho_{1,1}\rho_{s^{-1},s})^{-1}t_{s^{-1}hs}\rho_{s^{-1}h,s}\rho_{s^{-1},h}^{\bar{s}}$$
$$= t_{s^{-1}hs}(\rho_{1,1}^{\overline{s^{-1}hs}}\rho_{s^{-1},s}^{\overline{s^{-1}hs}})^{-1}\rho_{s^{-1}h,s}\rho_{s^{-1},h}^{\bar{s}}$$

is in $\mathfrak{B}$. Hence the mapping $b \to t_s^{-1}bt_s \equiv b^{s'}$ is an automorphism in $\mathfrak{B}$. By means of this automorphism we may write $(t_u b_u)(t_v b_v) = t_{uv}\rho_{u,v}b_u^{v''}b_v$.

We may now prove the following theorem.

THEOREM 47. *A necessary and sufficient condition that $\mathfrak{A}$ be semi-simple is that $\mathfrak{B}$ be semi-simple.*

Let $\mathfrak{S}$ be the radical of $\mathfrak{B}$. Since the automorphism $b \to t_s^{-1}bt_s$ maps a nilpotent ideal into a nilpotent ideal, $t_s^{-1}\mathfrak{S}t_s \leq \mathfrak{S}$. It follows that the totality $\mathfrak{N}$ of elements $\Sigma t_u s_u$, s_u in $\mathfrak{S}$, is a two-sided ideal in $\mathfrak{A}$. Since $\mathfrak{N} \geq \mathfrak{S}$, it contains $\mathfrak{A}\mathfrak{S}\mathfrak{A}$. On the other hand, by definition, $\mathfrak{N} \leq \mathfrak{A}\mathfrak{S}$. Hence $\mathfrak{N} = \mathfrak{A}\mathfrak{S} = \mathfrak{A}\mathfrak{S}\mathfrak{A}$. Then $\mathfrak{N}^k = \mathfrak{A}\mathfrak{S}^k$ and so $\mathfrak{N}$ is nilpotent. Hence if $\mathfrak{S} \neq 0$, the radical $\mathfrak{N}$ of $\mathfrak{A}$ is $\neq 0$. Now suppose that $\mathfrak{S} = 0$. We wish to show that if $\mathfrak{J}$ is any two-sided ideal $\neq 0$ in $\mathfrak{A}$, then $(\mathfrak{B} \wedge \mathfrak{J}) \neq 0$. For this purpose let $z = t_u b_u + \cdots$ be an element $\neq 0$ in $\mathfrak{J}$ for which the least number of coefficients b_u are $\neq 0$. If $b \in \mathfrak{B}$, $zb = t_u b_u b + \cdots$ and $bz = t_u b^{u'} b_u + \cdots$ are in $\mathfrak{J}$. We fix our attention on a particular u for which $b_u \neq 0$. Since $b^{u'}$ ranges over $\mathfrak{B}$, the coefficients c_u of the elements $t_u c_u + \cdots$ of $\mathfrak{J}$ that have the same form[7] as z form a two-sided ideal $\mathfrak{J}_u \neq 0$. Since $\mathfrak{B}$ is semi-simple, $\mathfrak{J}_u$ has an identity e_u and e_u is in the center of $\mathfrak{B}$. Now we may suppose that $z = t_u e_u + \cdots$. We assert that $z = t_u e_u$. For suppose that $z = t_u e_u + t_v b_v + \cdots$ with $b_v \neq 0$. For any ξ in Φ, $\xi z - z\xi^{\bar{u}} = t_v(\xi^{\bar{v}}b_v - b_v\xi^{\bar{u}}) + \cdots$ is in $\mathfrak{B}$ and has fewer non-zero terms than z and is $\neq 0$ unless $\xi^{\bar{v}}b_v = b_v\xi^{\bar{u}}$. Now if $b_v = \Sigma t_h\beta_h$,

$$\xi^{\bar{v}}b_v - b_v\xi^{\bar{u}} = \Sigma t_h(\xi^{\bar{v}\bar{h}}\beta_h - \beta_h\xi^{\bar{u}}).$$

Since $b_v \neq 0$, there is a $\beta_h \neq 0$ and so $\xi^{\bar{u}} = \beta_h^{-1}\xi^{\bar{v}\bar{h}}\beta_h$. This holds for all ξ and it implies that u and v differ by an inner automorphism contrary to the assumption that u and v are in different cosets of $\mathfrak{h}$ in $\mathfrak{g}$. Hence $z = t_u e_u$ and $t_u^{-1}z = e_u$ is an element $\neq 0$ in $(\mathfrak{B} \wedge \mathfrak{J})$. Now if $\mathfrak{J} = \mathfrak{N}$, we obtain $\mathfrak{N} = 0$ since $(\mathfrak{B} \wedge \mathfrak{N})$ is a nilpotent ideal in $\mathfrak{B}$.

We have seen that if $p \nmid q$, the order of $\mathfrak{h}$, any representation of $\mathfrak{B}$ such that $1 \to 1$ is completely reducible. If we apply this to the regular representation (by $\mathfrak{B}_r$) we see that the lattice of right ideals of $\mathfrak{B}$ is completely reducible. Hence $\mathfrak{B}$ is semi-simple and we have proved the following theorems.

THEOREM 48. *The crossed product $\mathfrak{A}$ is semi-simple if $p \nmid q$, q, the order of $\mathfrak{h}$.*

THEOREM 49. *A projective representation $s \to T_s$ of a finite group is completely reducible if $p \nmid q$, q, the order of the subgroup of elements h such that T_h is a special collineation.*

The proof of the main theorem implies also

[7] i.e. $c_v = 0$ if $b_v = 0$.

THEOREM 50. *If* $\mathfrak{h} = (1)$, *then* $\mathfrak{A}$ *is simple.*

For in this case $\mathfrak{B} = \Phi$ so that $\mathfrak{S} = 0$. Then if $\mathfrak{J}$ is a two-sided ideal $\neq 0$ in $\mathfrak{A}$, $(\Phi \wedge \mathfrak{J})$ is an ideal $\neq 0$ in Φ and hence $(\Phi \wedge \mathfrak{J}) = \Phi$. Thus $\mathfrak{J}$ contains 1 and $\mathfrak{J} = \mathfrak{A}$.

Suppose that $\mathfrak{h} = (1)$. Then if $(\Sigma t_s\beta_s)\xi = \xi(\Sigma t_s\beta_s)$ for all ξ, $\xi^{\bar{s}}\beta_s = \beta_s\xi$. Hence if $\beta_s \neq 0$, $\xi^{\bar{s}} = \beta_s\xi\beta_s^{-1}$. Thus $s = 1$ and we have proved that the only elements of $\mathfrak{A}$ which commute with all ξ in Φ are in Φ. Consequently the center of $\mathfrak{A} \leqq \Gamma$, the center of Φ. If $\gamma \in \Gamma$, $\gamma t_s = t_s\gamma$ for all s implies that $\gamma \in \Gamma_0$ the subfield of Γ of elements invariant under all $\bar{s}$. It follows that Γ_0 is the center of $\mathfrak{A}$. If $\Phi = \Gamma$ so that Φ is commutative, it is well known that $(\Gamma: \Gamma_0) = r$.[8] Hence $(\mathfrak{A}: \Gamma_0) = r^2$.

Suppose now that $\rho_{s,t} = 1$ in addition to $\mathfrak{h} = (1)$. Since $\mathfrak{A}$ is simple, $\mathfrak{A} = \mathfrak{J}_1 \oplus \cdots \oplus \mathfrak{J}_u$ where the $\mathfrak{J}_j$ are $\mathfrak{A}$-isomorphic irreducible right ideals. An $\mathfrak{A}$-isomorphism between right ideals is in particular a $(1-1)$ linear transformation between them regarded as subspaces of the vector space $\mathfrak{A}$ over Φ. Hence $(\mathfrak{J}_j : \Phi) = (\mathfrak{J}_k : \Phi) = m$ and $r = (\mathfrak{A}: \Phi) = um$. Now let $\mathfrak{J}$ be the right ideal of multiples $e\alpha$ of $e = \Sigma t_s$. Since $et_s = t_s$, $e\alpha = e\alpha$ for a suitable α in Φ. Hence $(\mathfrak{J}: \Phi) = 1$ so that $\mathfrak{J}$ is irreducible. Then $(\mathfrak{J}_j : \Phi) = (\mathfrak{J}: \Phi) = 1$ and $r = u$. It follows that $\mathfrak{A} = \Psi_r$ where Ψ is a division ring. Ψ contains the center Γ_0 of $\mathfrak{A}$ and since $(\mathfrak{A}: \Psi) = r^2$, it follows that when $\Phi = \Gamma$, $\Psi = \Gamma_0$.

THEOREM 51. *If* $\mathfrak{A} = \Phi(\mathfrak{g}, H, 1)$ *and* $\mathfrak{h} = 1$, $\mathfrak{A} = \Psi_r$ *where* Ψ *is a division ring and* r *is the order of* $\mathfrak{g}$. *If, in addition,* $\Phi = \Gamma$ *is commutative, then* $\mathfrak{A} = \Gamma_{0r}$, Γ_0 *the center of* $\mathfrak{A}$.

19. Galois theory of division rings. Let Φ be an arbitrary division ring and $\mathfrak{G} = (1, S, \cdots, U)$ a finite group of r outer automorphisms acting in Φ. The subset of invariant elements $(\alpha^S = \alpha)$ is a division subring Φ_0 of Φ. We denote the set of left (right) multiplications in Φ corresponding to the element of Φ_0 by Φ_0' (Φ_0) and the set of endomorphisms $\Sigma S\xi_S$ $(\Sigma S\xi_S')$, where $\xi_S(\xi_S')$ is a right (left) multiplication by ξ_S in Φ, by $(\Phi, \mathfrak{G})$ $((\Phi', \mathfrak{G}))$. We note that $\xi S = S\xi^S$. Hence if $\mathfrak{A}$ is the crossed product of Φ and an abstract group $\mathfrak{g}$ isomorphic to $\mathfrak{G}$ defined by the isomorphism $s \rightarrow S$ and the factor set $\rho_{s,t} = 1$, the correspondence $\Sigma t_s\xi_s \rightarrow \Sigma S\xi_s$ is a representation of $\mathfrak{A}$ by endomorphisms acting in Φ. Since $\mathfrak{A}$ is simple, the representation is $(1-1)$. Hence by the last theorem, the ring $(\Phi, \mathfrak{G}) = \Psi_r$ where Ψ is a division ring.

If $\alpha \neq 0$ is in Φ and β is arbitrary, there is a ξ in Φ such that $\alpha\xi = \beta$. It follows that $(\Phi, \mathfrak{G}) = \Psi_r$ is an irreducible set of endomorphisms and hence by the $\mathfrak{A}$-isomorphism of any two irreducible $\mathfrak{A}$-groups, there exist r elements $\alpha_1, \cdots, \alpha_r$ in Φ such that every element of Φ may be represented in one and only one way as $\alpha_1\psi_1 + \cdots + \alpha_r\psi_r$, ψ_i in Ψ. To prove this again directly, let E_{ij} be the matrix basis in Ψ_r and choose E_{pp} and α so that $\alpha E_{pp} \neq 0$. Then it follows readily that the elements $\alpha_1 = \alpha E_{p1}, \cdots, \alpha_r = \alpha E_{pr}$ are independent over Ψ. Since any $\beta = (\alpha E_{pp})\Sigma E_{ij}\psi_{ij}$ for suitable ψ_{ij}, we have $\beta = \alpha_1\psi_{p1} + \cdots + \alpha_r\psi_{pr}$.

[8] This will be proved in the next section.

Let Ψ' be the ring of linear transformations ψ' in Φ over Ψ defined by the equations $\alpha_i\psi' \equiv \alpha_i\psi$. We have seen in Chapter 2 that Φ is an r-dimensional space over Ψ', that Ψ' is the complete set of endomorphisms commutative with those of Ψ_r and that Ψ_r is the complete set of linear transformations of Φ over Ψ'.

We recall now that $\Psi_r = (\Phi, \mathfrak{G})$. If A is an endomorphism commutative with all the endomorphisms of Φ, it is a left multiplication, say, $\xi \rightarrow \alpha\xi \equiv \xi\alpha'$. If in addition it commutes with all the elements of $\mathfrak{G}$, $(\alpha\xi)^S = \alpha^S\xi^S = \alpha\xi^S$ and $\alpha' \in \Phi_0'$. Thus $\Psi' = \Phi_0'$. In a similar fashion we may treat Φ_0 and $(\Phi', \mathfrak{G})$ and obtain the following

THEOREM 52. *Let Φ be an arbitrary division ring, $\mathfrak{G}$ a finite group of r outer automorphisms acting in Φ and Φ_0 the division subring of invariant elements. Then the dimensionality of Φ over Φ_0' (Φ over Φ_0) is r and $(\Phi, \mathfrak{G})$ $((\Phi', \mathfrak{G}))$ is the complete set of linear transformations of Φ over Φ_0' (Φ over Φ_0).*

Suppose that V is any automorphism in Φ leaving the elements of Φ_0 unaltered. Then V is a linear transformation of Φ over Φ_0' and hence $V = \Sigma S\xi_S$. For every endomorphism η we have $\eta V - V\eta^V = 0$. Hence $\Sigma S(\eta^S\xi_S - \xi_S\eta^V) = 0$ and if $\xi_S \neq 0$, $\eta^V = \xi_S^{-1}\eta^S\xi_S$. Since no $S \neq 1$ is inner, this holds for just one S and so $V = S\xi$. Since V is an automorphism, $\xi = 1$ and $V = S \in \mathfrak{G}$. In particular if γ is any element of Φ commutative with all the elements of Φ_0, then the inner automorphism $\eta \rightarrow \gamma^{-1}\eta\gamma$ is in $\mathfrak{G}$ and hence is the identity mapping. Thus γ is in the center of Φ.

If $\mathfrak{H}$ is a subgroup of $\mathfrak{G}$, we denote the division subring of elements invariant under the transformations of $\mathfrak{H}$ by $\Phi(\mathfrak{H})$, and if Σ is any division subring between Φ_0 and Φ, we denote the subgroup of $\mathfrak{G}$ leaving the elements of Σ invariant by $\mathfrak{G}(\Sigma)$. Note that $\Phi_0 \leqq \Phi(\mathfrak{H})$, $\Phi(\mathfrak{G}) = \Phi_0$, $\Phi(1) = \Phi$. The following is the fundamental theorem of the Galois theory.

THEOREM 53. *The correspondences $\mathfrak{H} \rightarrow \Phi(\mathfrak{H})$ and $\Sigma \rightarrow \mathfrak{G}(\Sigma)$ are inverses of each other. Each one is $(1 - 1)$ between the subgroups of $\mathfrak{G}$ and the division rings Σ between Φ_0 and Φ. The dimensionality $(\Phi\colon \Sigma) = (\Phi\colon \Sigma') =$ order of $\mathfrak{G}(\Sigma)$ and $(\Sigma\colon \Phi_0) = (\Sigma\colon \Phi_0') =$ index of $\mathfrak{G}(\Sigma)$.*

Let $\mathfrak{H}$ be a subgroup of $\mathfrak{G}$ and $\Phi(\mathfrak{H})$ the set of invariant elements. If S is an automorphism of $\mathfrak{G}$ leaving the elements of $\Phi(\mathfrak{H})$ invariant, we have seen that S is in $\mathfrak{H}$. Thus $\mathfrak{G}(\Phi(\mathfrak{H})) = \mathfrak{H}$. Now suppose that Σ is given where $\Phi_0 \leqq \Sigma \leqq \Phi$ and let Λ be the set of linear transformations of Φ over Σ'. Then $\Lambda \leqq (\Phi, \mathfrak{G})$. If $\Sigma S\xi_S \in \Lambda$ and μ' is any element of Σ'.

$$\Sigma S\mu'^S\xi_S = \mu'\Sigma S\xi_S = (\Sigma S\xi_S)\mu' = \Sigma S\mu'\xi_S$$

where μ'^S denotes the left multiplication corresponding to μ^S. Hence $\Sigma S(\mu'^S - \mu')\xi_S = 0$. If $S \in \mathfrak{G}(\Sigma) \equiv \mathfrak{H}$, $\mu'^S = \mu'$. Now suppose that $S \notin \mathfrak{H}$. Then we assert that $\xi_S = 0$. For let $\xi_S \neq 0$. Since $S \notin \mathfrak{H}$, there exists a μ such that $\mu^S \neq \mu$. On the other hand

$$S(\mu'^S - \mu')\xi_S + T(\mu'^T - \mu')\xi_T + \cdots = 0.$$

Clearly this relation cannot reduce to $S(\mu'^S - \mu')\xi_S = 0$ and so we may suppose that $\xi_T \neq 0$ and $\mu'^T - \mu' \neq 0$ so that $T \notin \mathfrak{H}$ either. Then by multiplying on the left by the endomorphism η and on the right by $\xi_S^{-1}\eta^S\xi_S$ and subtracting we obtain

$$T(\mu'^T - \mu')(\eta^T\xi_T - \xi_T\xi_S^{-1}\eta^S\xi_S) + \cdots = 0.$$

Since TS^{-1} is not inner, we may choose an η so that $(\eta^T\xi_T - \xi_T\xi_S^{-1}\eta^S\xi_S) \neq 0$. If we continue this process, we obtain finally a single term $U(\mu'^U - \mu')\zeta_U = 0$ with U not in $\mathfrak{H}$ and $\zeta_U \neq 0$. Since this has been excluded, we have proved that $\xi_S = 0$ for all S not in $\mathfrak{H}$. Hence Λ consists of the transformations $\sum_{S \in \mathfrak{H}} S\xi_S$. Now Σ' is the complete set of endomorphisms commutative with those of Λ. On the other hand, the form of the elements of Λ shows that these transformations are precisely the ζ' such that $\zeta \in \Phi(\mathfrak{H})$. Hence $\Phi(\mathfrak{G}(\Sigma)) = \Sigma$. The dimensionality relations follow from Theorem 1.

If $\Sigma = \Phi(\mathfrak{H})$, $\Sigma^S = \Phi(S^{-1}\mathfrak{H}S)$. Hence $\mathfrak{H}$ is invariant if and only if Σ is transformed into itself by all the elements of $\mathfrak{G}$. If $1, S, \cdots$ are representatives of the cosets of $\mathfrak{H}$, the transformations induced in Σ by these elements are distinct and depend only on the cosets. Their totality is a group $\bar{\mathfrak{G}} \cong \mathfrak{G}/\mathfrak{H}$. The elements S not in $\mathfrak{H}$ induce outer automorphisms in Σ. For if S is inner in Σ, there is an inner automorphism A in Φ such that SA leaves the elements of Σ invariant. Then $SA = H \in \mathfrak{G}(\Sigma) = \mathfrak{H}$ and $S^{-1}H = A$ is inner contrary to assumption.

If Φ is commutative and ξ is any element of this field, let $\xi, \cdots, \xi^T$ be its distinct conjugates. The coefficients of

$$(t - \xi) \cdots (t - \xi^T)$$

are invariant under $\mathfrak{G}$ and therefore belong to Φ_0. Thus every element of Φ satisfies a separable equation with coefficients in Φ_0. Since $\xi, \cdots, \xi^T$ are in Φ, it follows that Φ is separable and normal over Φ_0. To complete the Galois theory for (finite extension) fields along these lines it would be necessary to prove the converse theorem that if Φ is finite, separable and normal over Φ_0 then the elements of Φ_0 are the only ones left invariant by the automorphisms of the Galois group of Φ over Φ_0.

20. Finite groups of semi-linear transformations. We consider a projective representation of a finite group such that $\rho = 1$ and the group $\mathfrak{h} = (1)$. Thus the ring $\Phi(\mathfrak{g}, H, 1) = \Psi_r$, the semi-linear transformations T_s form a group and the automorphisms $\bar{s}$ in Φ associated with the T_s are distinct and outer. Let $\mathfrak{M} = \mathfrak{M}_1 \oplus \cdots \oplus \mathfrak{M}_m$ be a decomposition of the vector space into irreducible Ψ_r-modules. In each $\mathfrak{M}_i$ we may choose a vector x_i such that $y_i = x_i(\Sigma t_s) = x_i(\Sigma T_s) \neq 0$. Then $y_iT_s = y_i$ and since each element in $\mathfrak{M}_i$ has the form $y_i(\Sigma T_s\xi_s) = y_i\xi$, $\mathfrak{M}_i$ is one-dimensional over Φ. Hence $y_1, \cdots, y_m$ is a basis for $\mathfrak{M}$ over Φ.

Let $\mathfrak{M}_0$ be the set of vectors in $\mathfrak{M}$ invariant under all of the T_s. $\mathfrak{M}_0$ is a vector space over Φ_0 the division subring of elements of Φ invariant under the automorphisms $\bar{s}$. If $y = \Sigma y_i \xi_i \in \mathfrak{M}_0$, $\xi_i^{\bar{s}} = \xi_i \in \Phi_0$ and hence $y_1, \cdots, y_m$ is also a basis for $\mathfrak{M}_0$ over Φ_0.

THEOREM 54. *Let $\mathfrak{M}$ be an m-dimensional vector space over a division ring Φ and $T_1 = 1, T_s, \cdots, T_u$ a finite group of semi-linear transformations whose induced automorphisms $\bar{1}, \bar{s}, \cdots, \bar{u}$ are distinct and outer. If $\mathfrak{M}_0$ is the set of vectors invariant under all T_s and Φ_0 the division subring invariant under the $\bar{s}$, then $\mathfrak{M}_0$ is a vector space over Φ_0 of m-dimensions and the extension $\mathfrak{M}_0\Phi = \mathfrak{M}$.*

If we use the correspondence between semi-linear transformations and matrices, we may state this theorem also in the following way:

THEOREM 55. *If $\mathfrak{S}$ is a finite group of outer automorphisms $\bar{1}, \bar{s}, \cdots, \bar{u}$ in a division ring Φ and τ_s are matrices with elements in Φ such that $\tau_1 = 1$ and $\tau_t \tau_s^{\bar{t}} = \tau_{st}$, then there exists a non-singular matrix α such that $\tau_s = \alpha^{-1}\alpha^{\bar{s}}$ for all $\bar{s}$.*

CHAPTER 5

ALGEBRAS OVER A FIELD

1. The direct product of algebras. In the preceding chapter we have been concerned mainly with absolute properties of rings. The role of the set of endomorphisms Φ has been a rather minor one, its sole function having been to weaken the assumption that the set of ideals of the ring satisfies the chain conditions. The results which we obtained apply in particular to algebras. On the other hand, a considerable part of the theory of algebras is concerned with "relative" properties—that depend essentially on the field Φ over which the algebras are defined. This phase of the theory is the subject of the present chapter. We consider first the theory of simple algebras and later we take up again the study of an arbitrary algebra.

The discussion in Chapter 4 has been concerned to a large extent with additive decompositions of a ring, as a direct sum of ideals. In the theory of simple algebras a type of multiplicative decomposition, the direct product, is of fundamental importance. Let $\mathfrak{A}$ be an algebra over Φ and suppose that $(\mathfrak{A}:\Phi) = n < \infty$.[1] We say that $\mathfrak{A}$ is the *direct product* $\mathfrak{A}_1 \times \mathfrak{A}_2$ of the subalgebras $\mathfrak{A}_1$ and $\mathfrak{A}_2$ if the following conditions obtain:

1. The elements of $\mathfrak{A}_1$ commute with those of $\mathfrak{A}_2$.
2. $\mathfrak{A} = \mathfrak{A}_1\mathfrak{A}_2 = \mathfrak{A}_2\mathfrak{A}_1$.
3. $(\mathfrak{A}:\Phi) = (\mathfrak{A}_1:\Phi)(\mathfrak{A}_2:\Phi)$.

Evidently $\mathfrak{A}_1$ and $\mathfrak{A}_2$ are interchangeable in these conditions so that if $\mathfrak{A} = \mathfrak{A}_1 \times \mathfrak{A}_2$, $\mathfrak{A} = \mathfrak{A}_2 \times \mathfrak{A}_1$. It is clear from 3. that this concept depends essentially on the field Φ. We remark that if Σ is a proper subfield of Φ and $\mathfrak{A} = \mathfrak{A}_1 \times \mathfrak{A}_2$ when these are regarded as algebras over Φ, then $\mathfrak{A} \neq \mathfrak{A}_1 \times \mathfrak{A}_2$ when these are regarded over Σ. For then $(\mathfrak{A}:\Sigma)(\Phi:\Sigma) = (\mathfrak{A}_1:\Sigma)(\mathfrak{A}_2:\Sigma)$.

Let $y_1, \cdots, y_{n_1}$ be a basis for $\mathfrak{A}_1$ over Φ with the multiplication table $y_i y_{i'} = \Sigma y_p \gamma^{(1)}_{pii'}$ and $z_1, \cdots, z_{n_2}$, one for $\mathfrak{A}_2$ over Φ with the multiplication table $z_j z_{j'} = \Sigma z_q \gamma^{(2)}_{qjj'}$, $\gamma^{(1)}$ and $\gamma^{(2)}$ in Φ. Then every b in $\mathfrak{A}_1$ has the form $\Sigma y_i \varphi_i$ and every c in $\mathfrak{A}_2$ has the form $\Sigma z_j \varphi_j$. By 2. every a in $\mathfrak{A}$ is a sum $\Sigma a_k^{(1)} a_k^{(2)}$ where $a_k^{(i)} \in \mathfrak{A}_i$. Hence $a = \Sigma y_i z_j \varphi_{ij}$. By 3. the elements $x_{ij} = y_i z_j$, $i = 1, \cdots, n_1$; $j = 1, \cdots, n_2$, are linearly independent and hence form a basis for $\mathfrak{A}$ over Φ. The multiplication table $x_{ij} x_{i'j'} = \Sigma x_{pq} \gamma^{(1)}_{pii'} \gamma^{(2)}_{qjj'}$ of this basis is determined by that of the bases y_i and z_j of $\mathfrak{A}_1$ and $\mathfrak{A}_2$. Hence if $\mathfrak{B}$ is a second algebra over Φ, $\mathfrak{B} = \mathfrak{B}_1 \times \mathfrak{B}_2$, and $a_1 \to a_1^S$, $a_2 \to a_2^S$ are isomorphisms of $\mathfrak{A}_1$ and $\mathfrak{B}_1$ over Φ and of $\mathfrak{A}_2$ and $\mathfrak{B}_2$ over Φ, respectively, then $\Sigma x_{ij} \varphi_{ij} \to \Sigma x_{ij}^S \varphi_{ij}$, where $x_{ij} = y_i z_j$, $x_{ij}^S = y_i^S z_j^S$, is an isomorphism between $\mathfrak{A}$ and $\mathfrak{B}$ over Φ. In this sense the algebra $\mathfrak{A} = \mathfrak{A}_1 \times \mathfrak{A}_2$ is determined by its components $\mathfrak{A}_1$ and $\mathfrak{A}_2$. More gen-

[1] We assume throughout this chapter that our algebras have finite dimensionalities. Some of the results are valid under less stringent conditions but, for the sake of simplicity, we shall not indicate these extensions of the theory.

erally, if $\mathfrak{B}$ is an algebra containing two subalgebras $\mathfrak{B}_1$ and $\mathfrak{B}_2$ such that $b_1b_2 = b_2b_1$ for b_i in $\mathfrak{B}_i$ and if $a_i \to a_i^S$ is a homomorphism between $\mathfrak{A}_i$ and $\mathfrak{B}_i$, then $\Sigma x_{ij}\varphi_{ij} \to \Sigma x_{ij}^S\varphi_{ij}$, $x_{ij}^S = y_i^S z_j^S$, is a homomorphism between $\mathfrak{A}_1 \times \mathfrak{A}_2$ and $\mathfrak{B}_1\mathfrak{B}_2 = \mathfrak{B}_2\mathfrak{B}_1$.

If $a = \Sigma y_i z_j \varphi_{ij}$, $a = y_1 a_1^{(2)} + \cdots + y_{n_1} a_{n_1}^{(2)}$, where $a_i^{(2)} \in \mathfrak{A}_2$ and since the elements $y_i z_j$ are linearly independent, $a = 0$ implies that every $a_i^{(2)} = 0$. Now if $y_1, \cdots, y_r$ is an arbitrary set of linearly independent elements of $\mathfrak{A}_1$, we may add to it $y_{r+1}, \cdots, y_{n_1}$ to obtain a basis for $\mathfrak{A}_1$ over Φ. Similarly if $z_1, \cdots, z_s$ are linearly independent in $\mathfrak{A}_2$, we may add to these elements and obtain a basis for $\mathfrak{A}_2$. It follows that the elements $y_i z_j$, $i = 1, \cdots, r; j = 1, \cdots, s$, are linearly independent. As a special case of this we see that if $\mathfrak{B}_i$ is a subalgebra of $\mathfrak{A}_i$, then $\mathfrak{B}_1\mathfrak{B}_2 = \mathfrak{B}_2\mathfrak{B}_1 = \mathfrak{B}_1 \times \mathfrak{B}_2$. If $\mathfrak{A}_1 = \mathfrak{A}_{11} \times \mathfrak{A}_{12}$, $\mathfrak{A} = (\mathfrak{A}_{11} \times \mathfrak{A}_{12}) \times \mathfrak{A}_2 = \mathfrak{A}_{11} \times (\mathfrak{A}_{12} \times \mathfrak{A}_2)$, Thus the associative law holds for direct multiplication. We note also that the intersection $\mathfrak{A}_1 \wedge \mathfrak{A}_2$ is at most one dimensional. For if a, b are elements of $\mathfrak{A}_1 \wedge \mathfrak{A}_2$, a^2, ab, ba and b^2 are linearly dependent since $ab = ba$. If $\mathfrak{A}_1$ and $\mathfrak{A}_2$ have identities 1_1 and 1_2, respectively, then $1 = 1_1 1_2$ is the identity element of $\mathfrak{A}$. Now $1_1 = 1_1(1_1 1_2) = 1_1 1_2 = 1$ and similarly $1_2 = 1$. Hence $\mathfrak{A}_1 \wedge \mathfrak{A}_2$ consists of the multiples 1α, α in Φ.

Now let $\mathfrak{A}_1$ and $\mathfrak{A}_2$ be arbitrary algebras with identities. Suppose that $y_1 = 1_1, y_2, \cdots, y_{n_1}$ and $z_1 = 1_2, z_2, \cdots, z_{n_2}$ are bases for these algebras and that $y_i y_{i'} = \Sigma y_p \gamma_{pii'}^{(1)}$, $z_j z_{j'} = \Sigma z_q \gamma_{qjj'}^{(2)}$ are the multiplication tables. We define an algebra $\mathfrak{A}$ by using the basis x_{ij}, $i = 1, \cdots, n_1$; $j = 1, \cdots, n_2$, subjected to the multiplication table $x_{ij}x_{i'j'} = \Sigma x_{pq}\gamma_{pii'}^{(1)}\gamma_{qjj'}^{(2)}$. It is readily verified that the subset of elements $\Sigma x_{i1}\varphi_i$, φ_i in Φ, is a subalgebra $\bar{\mathfrak{A}}_1$ of $\mathfrak{A}$ isomorphic to $\mathfrak{A}_1$ and that the subset of elements $\Sigma x_{1j}\varphi_j$ is a subalgebra $\bar{\mathfrak{A}}_2$ isomorphic to $\mathfrak{A}_2$. From the multiplication table we obtain $x_{i1}x_{1j} = x_{ij} = x_{1j}x_{i1}$ and $(x_{i1}x_{1j})(x_{i'1}x_{1j'}) = (x_{i1}x_{i'1})(x_{1j}x_{1j'})$. The latter relation and the associative laws in $\bar{\mathfrak{A}}_1$ and $\bar{\mathfrak{A}}_2$ imply the associative law in $\mathfrak{A}$. Evidently $\mathfrak{A} = \bar{\mathfrak{A}}_1 \times \bar{\mathfrak{A}}_2$. We have, therefore, constructed an algebra $\mathfrak{A}$ that is a direct product of algebras isomorphic to the given algebras $\mathfrak{A}_1$ and $\mathfrak{A}_2$. As we saw above, $\mathfrak{A}$ is the only algebra (in the sense of isomorphism) having this property. We shall identify the algebra $\mathfrak{A}_i$ with $\bar{\mathfrak{A}}_i$ and shall call $\mathfrak{A}$ the direct product ($\mathfrak{A} = \mathfrak{A}_1 \times \mathfrak{A}_2$) of $\mathfrak{A}_1$ and $\mathfrak{A}_2$. The restriction that the $\mathfrak{A}_i$ have identities is not essential in this discussion. For we may adjoin an identity 1_i to $\mathfrak{A}_i$ obtaining an algebra $\mathfrak{B}_i$. We then form $\mathfrak{B}_1 \times \mathfrak{B}_2$ and take the subalgebra $\mathfrak{A}_1 \times \mathfrak{A}_2$ as the direct product of $\mathfrak{A}_1$ and $\mathfrak{A}_2$.

2. Extension of the field. An algebra that is closely related to the direct product is obtained as follows. Let $\mathfrak{A}$ be an algebra with the basis $x_1, \cdots, x_n$ over Φ and let $\mathfrak{B}$ be an algebra over Φ containing an identity. We consider the set of expressions $x_1b_1 + \cdots + x_nb_n$ where the $b_i \in \mathfrak{B}$. Two such expressions Σx_ib_i and $\Sigma x_ib_i'$ are regarded as equal if and only if $b_i = b_i'$. We define

$$\Sigma x_ib_i + \Sigma x_ib_i' = \Sigma x_i(b_i + b_i'),$$

$$(\Sigma x_ib_i)(\Sigma x_jb_j') = \Sigma x_k \Sigma b_ib_j'\gamma_{kij},$$

if $x_i x_j = \Sigma x_k \gamma_{kij}$, γ in Φ. It is readily seen that the system thus defined is a ring. It is independent of the choice of the basis x_i in the sense that the rings determined by different bases are isomorphic. Hence we may denote this ring as $\mathfrak{A}_\mathfrak{B}$.

Since $\mathfrak{B}$ contains an identity, it contains a subfield 1Φ of elements 1α isomorphic to Φ. The ring $\mathfrak{A}_\mathfrak{B}$ contains the subset of elements $\Sigma x_i(1\alpha_i)$ which forms a subring isomorphic to $\mathfrak{A}$. We identify this subring with $\mathfrak{A}$. Now the definition $\Sigma(x_i b_i)b = \Sigma x_i(b_i b)$ turns $\mathfrak{A}_\mathfrak{B}$ into a $\mathfrak{B}$-module. From this definition we obtain

$$u1 = u, \qquad (uv)b = u(vb), \qquad (ub)x = (ux)b$$

for all u, u in $\mathfrak{A}_\mathfrak{B}$, all x in $\mathfrak{A}$ and all b in $\mathfrak{B}$. Thus the module operation commutes with all the left multiplications and with the right multiplications by the elements of $\mathfrak{A}$. Since $\mathfrak{A}_\mathfrak{B}$ is a $\mathfrak{B}$-module and $\mathfrak{B} \geq 1\Phi$, $\mathfrak{A}_\mathfrak{B}$ is a Φ-module. (We set $u\alpha = u(1\alpha)$.) If $\alpha \epsilon \Phi$ and u and v are arbitrary, then $(uv)\alpha = u(v\alpha) = (u\alpha)v$. Hence $\mathfrak{A}_\mathfrak{B}$ is an algebra over Φ. If $y_1, \cdots, y_m$ is a basis for $\mathfrak{B}$ over Φ, the mn elements $x_i y_j$ form a basis for $\mathfrak{A}_\mathfrak{B}$ over Φ.

These properties characterize $\mathfrak{A}_\mathfrak{B}$. For suppose that $\mathfrak{K}$ is an algebra such that

1. $\mathfrak{K}$ contains $\mathfrak{A}$.
2. $\mathfrak{K}$ is a $\mathfrak{B}$-module, $\mathfrak{B}$ an algebra with an identity element 1, and $u\alpha = u(1\alpha)$ for all u in $\mathfrak{K}$ and all α in Φ. $\mathfrak{K}$ is generated by $\mathfrak{A}$ in the sense that the smallest $\mathfrak{B}$-submodule of $\mathfrak{K}$ containing $\mathfrak{A}$ is $\mathfrak{K}$ itself.
3. $(uv)b = u(vb)$, $(ub)x = (ux)b$ for all u, v in $\mathfrak{K}$, all x in $\mathfrak{A}$ and all b in $\mathfrak{B}$.
4. $(\mathfrak{K}:\Phi) = (\mathfrak{A}:\Phi)(\mathfrak{B}:\Phi)$.

Then if $x_1, \cdots, x_n$ is a basis for $\mathfrak{A}$ over Φ, the elements of $\mathfrak{K}$ may be represented in one and only one way in the form $\Sigma x_i b_i$, b_i in $\mathfrak{B}$. If $x_i x_j = \Sigma x_k \gamma_{kij}$, then $(\Sigma x_i b_i)(\Sigma x_j b_j') = \Sigma(x_i b_i)(x_j b_j') = \Sigma((x_i b_i)x_j)b_j' = \Sigma((x_i x_j)b_i)b_j' = \Sigma x_k \gamma_{kij} b_i b_j' = \Sigma x_k b_i b_j' \gamma_{kij}$. Hence $\mathfrak{K}$ is isomorphic to $\mathfrak{A}_\mathfrak{B}$.

If $\mathfrak{A}$ has an identity 1, $(\Sigma x_i b_i)1 = \Sigma(x_i b_i)1 = \Sigma(x_i 1)b_i = \Sigma x_i b_i$ and similarly, $1(\Sigma x_i b_i) = \Sigma x_i b_i$. Hence 1 is the identity of $\mathfrak{A}_\mathfrak{B}$. The set of elements $1b$ forms a subalgebra isomorphic to $\mathfrak{B}$. We note that $u(1b) = ub$ and that $(1b)x = xb = x(1b)$ if $u \epsilon \mathfrak{A}_\mathfrak{B}$ and $x \epsilon \mathfrak{A}$. Hence if we identify the algebra of elements $1b$ with $\mathfrak{B}$, we may write $\mathfrak{A}_\mathfrak{B} = \mathfrak{A} \times \mathfrak{B}$.

If $\mathfrak{A}_1$ is a subalgebra of $\mathfrak{A}$, we may suppose that $x_1, \cdots, x_r$ is a basis for $\mathfrak{A}_1$ where $x_1, \cdots, x_n$ is one for $\mathfrak{A}$. The elements $\sum_1^r x_i b_i$ form an algebra and this set is the smallest $\mathfrak{B}$-module containing $\mathfrak{A}_1$. It is clear that this algebra is isomorphic to $\mathfrak{A}_{1\mathfrak{B}}$ and it may therefore be denoted as $\mathfrak{A}_{1\mathfrak{B}}$. If $\mathfrak{A}_1$ is an ideal (nilpotent ideal) in $\mathfrak{A}$, $\mathfrak{A}_{1\mathfrak{B}}$ is an ideal (nilpotent ideal) in $\mathfrak{A}_\mathfrak{B}$. Hence if $\mathfrak{A}_\mathfrak{B}$ is simple (semi-simple), $\mathfrak{A}$ is simple (semi-simple).

We suppose now that $\mathfrak{B} = P$ is a field.[2] Then $(uv)\rho = u(v\rho) = (u\rho)v$ for all

---

[2] It should be observed that in defining $\mathfrak{A}_\mathfrak{B}$, no use has been made of the assumption that $\mathfrak{B}$ is an algebra with a finite basis. The abstract characterization in the general case is given by 1., 2., 3. and 4'.: If $x_1, \cdots, x_r$ are linearly independent in $\mathfrak{A}$ and $y_1, \cdots, y_s$ are linearly independent in $\mathfrak{B}$, then the rs elements $x_i y_j$ are linearly independent in $\mathfrak{A}_\mathfrak{B}$. The extensions $\mathfrak{A}_P$, P an infinite field, have many important applications.

u, v in $\mathfrak{A}_P$ and all ρ in P. Hence we may regard $\mathfrak{A}_P$ as an algebra over P. Unless otherwise stated, this is, in fact, what we shall do. Evidently $(\mathfrak{A}_P:P) = (\mathfrak{A}:\Phi)$. The following rules may be noted:

$$(\mathfrak{A}_1 \oplus \mathfrak{A}_2)_P = \mathfrak{A}_{1P} \oplus \mathfrak{A}_{2P},$$
$$(\mathfrak{A}_1 \times \mathfrak{A}_2)_P = \mathfrak{A}_{1P} \times \mathfrak{A}_{2P},$$
$$(\mathfrak{A}_P)_\Sigma = \mathfrak{A}_\Sigma,$$

if Σ is a field containing P.

3. Representation by matrices and representation spaces. A second important tool in our study of algebras is the theory of representations of an algebra $\mathfrak{A}$ by matrices. In the usual theory we are interested in the representations of an algebra by matrices with elements in the field Φ. For the investigation of simple algebras we shall require a generalization, in which the elements of the matrices are taken from a simple algebra $\mathfrak{B}$ unrelated to $\mathfrak{A}$. However, before considering this more general case, it will be well to discuss the simpler one.

As in the case of representations by endomorphisms, there are two types of representations by matrices. First, we define an (ordinary) *representation* of an algebra $\mathfrak{A}$ over Φ *by matrices* as a homomorphism $a \to A$ between $\mathfrak{A}$ and a subalgebra of a matrix algebra Φ_N: If $a \to A$ and $b \to B$, then

$$a + b \to A + B, \qquad a\alpha \to A\alpha, \qquad ab \to AB.$$

Similarly, we define an *anti-representation by matrices* as an anti-homomorphism between $\mathfrak{A}$ and a subalgebra of a matrix algebra. Now suppose that $\mathfrak{R}$ is a commutative group that satisfies the following conditions:

1. $\mathfrak{R}$ is a Φ-module such that $x1 = x$ for all x in $\mathfrak{R}$ and 1 the identity of Φ, and $(\mathfrak{R}:\Phi) = N$.
2. $\mathfrak{R}$ is a left $\mathfrak{A}$-module.
3. $(a\alpha)x = (ax)\alpha = a(x\alpha)$ for all a in $\mathfrak{A}$, all α in Φ and all x in $\mathfrak{R}$.

Then $\mathfrak{R}$ is a vector space over Φ of N dimensions, and the endomorphisms corresponding to the elements a are linear transformations. Since $\mathfrak{R}$ is a left $\mathfrak{A}$-module, the correspondence between a and the transformation a is an anti-homomorphism between the ring $\mathfrak{A}$ and a ring of linear transformations. By 3. the linear transformation corresponding to $a\alpha$ is the product of the linear transformation a with the scalar multiplication α. Hence the correspondence is an anti-homomorphism between the algebra $\mathfrak{A}$ and a subalgebra of the algebra of linear transformations. We recall that the correspondence between the linear transformations of a vector space and the matrices that they determine relative to a fixed basis is an algebra anti-isomorphism. It follows that if $x_1, \cdots, x_N$ is such a basis and $ax_i = \Sigma x_j\alpha_{ji}$, then the correspondence between a and the matrix $A = (\alpha_{ij})$ is a representation of $\mathfrak{A}$ by matrices in Φ_N. We may also reverse the steps of this argument and thus associate with any representation of $\mathfrak{A}$ by matrices a group $\mathfrak{R}$ satisfying 1., 2. and 3. We shall call such a group a *representation space of* $\mathfrak{A}$. A similar discussion holds for anti-representations. The modules in this case satisfy 1. and

2′. $\mathfrak{R}$ is an $\mathfrak{A}$-module.

3′. $x(a\alpha) = (xa)\alpha = (x\alpha)a$, a in $\mathfrak{A}$, α in Φ, x in $\mathfrak{R}$.

$\mathfrak{R}$ will be called an *anti-representation space of* $\mathfrak{A}$. We shall restrict our attention now to ordinary representations, since the modifications necessary to treat anti-representations will be obvious.

We recall that if $y_1, \cdots, y_N$ is a second basis for the representation space $\mathfrak{R}$ and $y_i = \Sigma x_j \mu_{ji}$, then the matrix of a relative to this basis is $M^{-1}AM$ where $M = (\mu_{ij})$. The representation $a \to M^{-1}AM$ is said to be *similar* to the representation $a \to A$. Thus a representation space determines a class of similar representations by matrices. We shall call the representation spaces $\mathfrak{R}_1$ and $\mathfrak{R}_2$ *isomorphic* if there is a $(1 - 1)$ correspondence between them which is at the same time a Φ-isomorphism and an $\mathfrak{A}$-isomorphism. If U is such an isomorphism, and $x_1, \cdots, x_N$ is a basis for $\mathfrak{R}_1$ over Φ, then $z_1 = x_1U, \cdots, z_N = x_NU$ is a basis for $\mathfrak{R}_2$ over Φ. Moreover, if $ax_i = \Sigma x_j \alpha_{ji}$, then also $az_i = \Sigma z_j \alpha_{ji}$. Thus isomorphic representation spaces determine the same similarity class of representations by matrices. The converse is also true.

We shall call a representation *reducible, decomposable, completely reducible* according as the group $\mathfrak{R}$ relative to the endomorphisms of Φ and of $\mathfrak{A}$ is reducible, decomposable, completely reducible. It is clear from the discussion in **8** of Chapter 2 that a representation is reducible if and only if it is similar to a representation of the form

$$\begin{pmatrix} A_1 & * \\ 0 & A_2 \end{pmatrix}. \tag{1}$$

The representation $a \to A_1$ corresponds to the proper subspace $\mathfrak{S}$ which is invariant relative to the endomorphisms a. The condition that $\mathfrak{R}$ be a direct sum, $\mathfrak{R} = \mathfrak{R}_1 \oplus \mathfrak{R}_2$ where the $\mathfrak{R}_i$ are invariant subspaces $\neq 0$ is that the representation determined by $\mathfrak{R}$ be similar to one of the form

$$\begin{pmatrix} A_1 & 0 \\ 0 & A_2 \end{pmatrix}. \tag{2}$$

Here $a \to A_i$ is the representation determined by the representation space $\mathfrak{R}_i$. We recall also that if $\mathfrak{R} = \mathfrak{R}_s > \mathfrak{R}_{s-1} > \cdots > \mathfrak{R}_1 > 0$ is a chain of subspaces invariant relative to the transformations a, then our representation is similar to

$$\begin{bmatrix} A_1 & & & * \\ & A_2 & & \\ & & \ddots & \\ 0 & & & A_s \end{bmatrix} \tag{3}$$

where the representations $a \to A_i$ are associated with the spaces $\mathfrak{R}_i - \mathfrak{R}_{i-1}$. The chain of subspaces is a composition series if and only if the representations $a \to A_i$ are irreducible. The condition for complete reducibility is that the representation be similar to one of the form (3) in which the blocks * above the "diagonal" are 0 and in which the representations $a \to A_i$ are irreducible.

Our discussion takes on a much simpler form if the algebra $\mathfrak{A}$ has an identity 1 and 1 is mapped into the identity matrix. This, of course, means that $1x = x$ for all x in $\mathfrak{R}$. Then $(1\alpha)x = x\alpha$. Thus in this case it suffices to regard $\mathfrak{R}$ as a left $\mathfrak{A}$-module. On the other hand, if $\mathfrak{R}$ is any left $\mathfrak{A}$-module in which $1x = x$ for all x, then $\mathfrak{R}$ is a left Φ-module relative to the composition $\alpha x \equiv (1\alpha)x$. Since Φ is commutative, $\mathfrak{R}$ may also be regarded as a Φ-module by setting $x\alpha \equiv \alpha x$. Now if $(\mathfrak{R}:\Phi)$ is finite, $\mathfrak{R}$ is a representation space. We remark that the condition $(\mathfrak{R}:\Phi)$ finite is equivalent to the requirement that $\mathfrak{R}$ be finitely generated relative to $\mathfrak{A}$. For, if $y_1, \cdots, y_r$ are generators of $\mathfrak{R}$ relative to $\mathfrak{A}$ and if $a_1, \cdots, a_n$ is a basis for $\mathfrak{A}$ over Φ, then the nr elements $a_i y_j$ generate $\mathfrak{R}$ relative to Φ. Hence $(\mathfrak{R}:\Phi)$ is finite.

If $\mathfrak{A}$ has an identity 1 but 1 is not mapped into the identity transformation, we write $\mathfrak{R} = \mathfrak{S} \oplus \mathfrak{Z}$ where $\mathfrak{S}$ is the totality of elements $1x$ and $\mathfrak{Z}$ is the totality of elements $x - 1x$ annihilated by 1. If we choose a basis $y_1, \cdots, y_N$ of $\mathfrak{R}$ so that $y_1, \cdots, y_r$ is a basis for $\mathfrak{S}$ and $y_{r+1}, \cdots, y_N$ is a basis for $\mathfrak{Z}$, then the matrix of a in $\mathfrak{A}$ relative to this basis is

$$\begin{pmatrix} A & 0 \\ 0 & 0 \end{pmatrix}.$$

In the representation $a \rightarrow A$ associated with $\mathfrak{S}$ we have $1 \rightarrow 1$. This enables us to reduce our discussion in this case also to that of left Φ-modules.

4. Application of the theory of $\mathfrak{A}$-modules. Let $\mathfrak{R}$ be an arbitrary representation space of the algebra $\mathfrak{A}$. If $\mathfrak{J}$ is a left Φ-ideal, then $\mathfrak{J}x$, the set of vectors bx, x fixed in $\mathfrak{R}$ and b variable in $\mathfrak{J}$, is an invariant subspace of $\mathfrak{R}$. Similarly, the space $\mathfrak{J}\mathfrak{S}$ of vectors $b_1x_1 + \cdots + b_rx_r$, where the x_i range over a set $\mathfrak{S}$ and the b_i range over $\mathfrak{J}$, is an invariant subspace. By the argument of **12** Chapter 4, we may prove that if $\mathfrak{R}$ is irreducible and $\mathfrak{N}$ is the radical, then $\mathfrak{N}\mathfrak{R} = 0$. Hence in this case $\mathfrak{R}$ is actually a representation space of the semi-simple algebra $\bar{\mathfrak{A}} = \mathfrak{A} - \mathfrak{N}$. Moreover, the irreducibility of $\mathfrak{R}$ assures that either $\mathfrak{A}\mathfrak{R} = 0$, or the identity of $\bar{\mathfrak{A}}$ is the identity mapping in $\mathfrak{R}$. In the former case $\mathfrak{R}$ is 1-dimensional and in the latter, by **12** of Chapter 4, $\mathfrak{R}$ is $\mathfrak{A}$-isomorphic to an irreducible left ideal of $\bar{\mathfrak{A}}$. If $\bar{\mathfrak{A}} = \bar{\mathfrak{A}}_1 \oplus \cdots \oplus \bar{\mathfrak{A}}_t$ where the $\bar{\mathfrak{A}}_i$ are irreducible two-sided ideals, then $\mathfrak{R}$ is annihilated by all the $\bar{\mathfrak{A}}_i$ except, say, $\bar{\mathfrak{A}}_1$. Thus $\mathfrak{R}$ is a left $\bar{\mathfrak{A}}_1$-module. If we recall that the number of irreducible left $\mathfrak{A}$-modules $\mathfrak{R}$ such that $\mathfrak{A}\mathfrak{R} \neq 0$ is the number t of components $\bar{\mathfrak{A}}_i$ of $\bar{\mathfrak{A}}$, we may state the following

THEOREM 1. *Let $\mathfrak{N}$ be the radical of $\mathfrak{A}$ and $\bar{\mathfrak{A}} = \mathfrak{A} - \mathfrak{N} = \bar{\mathfrak{A}}_1 \oplus \cdots \oplus \bar{\mathfrak{A}}_t$ where the $\bar{\mathfrak{A}}_i$ are simple. Then any irreducible representation $a \rightarrow A$ is either the 0-representation $(a \rightarrow 0)$ or it is similar to the representation obtained by using one of the irreducible left ideals of $\bar{\mathfrak{A}}$ as a representation space. The number of classes of similar irreducible representations $\neq 0$ is the number of components $\bar{\mathfrak{A}}_i$.*

We recall also that if $\mathfrak{A}$ is semi-simple, any left $\mathfrak{A}$-module in which $1x = x$ for all x, is completely reducible. Now if $\mathfrak{R}$ is an arbitrary representation space of $\mathfrak{A}$, we write $\mathfrak{R} = \mathfrak{S} \oplus \mathfrak{Z}$ where $1y = y$ for all y in $\mathfrak{S}$ and $1z = 0$ for all z in $\mathfrak{Z}$. Since $\mathfrak{S}$ is a left $\mathfrak{A}$-module in which $1y = y$, $\mathfrak{S}$ is completely reducible. Moreover, we may decompose $\mathfrak{Z}$ into 1-dimensional subspaces. This proves

THEOREM 2. *Any representation of a semi-simple algebra is completely reducible.*

As a special case of these theorems, we see that if $\mathfrak{A}$ is a simple algebra which is not a zero algebra, then its representations by matrices are completely reducible. The irreducible representations $\neq 0$ of such an algebra are all similar. If, in particular, $\mathfrak{A} = \Phi_r$, the irreducible representations $\neq 0$ are all similar to the original representation $A \to A$. This can also be seen by noting that $\Phi_r e_{11}$ is an irreducible left ideal, where e_{ij} is a matrix basis. A Φ-basis for this ideal is $x_1 = e_{11}, \cdots, x_r = e_{r1}$ and if $A = \Sigma e_{ij}\alpha_{ij}$, then $Ax_i = \Sigma x_j \alpha_{ji}$. Hence the representation determined by this ideal is the original one, $A \to A$.

5. Representation of an algebra by matrices with elements in a simple algebra. If $\mathfrak{B}$ is an arbitrary algebra, we define a *representation* (*anti-representation*) of $\mathfrak{A}$ *by matrices with elements in* $\mathfrak{B}$ as a homomorphism (anti-homomorphism) between $\mathfrak{A}$ and a sub-algebra of a matrix algebra $\mathfrak{B}_N$. As in the special case where $\mathfrak{B} = \Phi$, we call the representations $a \to A_1$ and $a \to A_2$ in the same $\mathfrak{B}_N$ *similar* if there exists a matrix M independent of a such that $A_2 = M^{-1}A_1M$. The representation $a \to A$ is *reducible* if it is similar to one of the form (1) and *decomposable* if it is similar to one of the form (2). It is *completely reducible* if it is similar to one of the form (3) where the blocks * are 0 and the representations $a \to A_i$ are irreducible. We shall restrict our attention to the study of the representations of an algebra with an identity by matrices with elements in an algebra with an identity. Moreover, we assume that the identity of $\mathfrak{A}$ is mapped into the identity matrix. As we shall see, the theory of anti-representations is somewhat more natural in this case than the theory of ordinary representations. Hence we shall keep the former in the foreground indicating only where necessary the modifications required for the ordinary theory.

We wish to obtain a module formulation of the representation problem. For this purpose it is necessary to recall the theory of free modules discussed in Chapter 3 (**3**). We shall simplify our former terminology by now calling a free $\mathfrak{B}$-module a $\mathfrak{B}$-*space*. Since $\mathfrak{B}$ satisfies the ascending chain condition for ideals, the dimensionality of any $\mathfrak{B}$-space $\mathfrak{R}$ is an invariant. If $x_1, \cdots, x_N$ and $y_1, \cdots, y_N$ are two bases for $\mathfrak{R}$, we write $y_i = \Sigma x_j b_{ji}$ and $x_i = \Sigma y_j c_{ji}$, $B = (b_{ij})$ and $C = (c_{ij})$ in $\mathfrak{B}_N$. Then $BC = CB = 1$ so that $C = B^{-1}$. Conversely if $x_1, \cdots, x_N$ is a basis and B is a unit in $\mathfrak{B}_N$, then the $y_i = \Sigma x_j b_{ji}$ form a second basis.

Now let a be a $\mathfrak{B}$-endomorphism of $\mathfrak{R}$. We set $x_i a = \Sigma x_j a_{ji}$, $A = (a_{ij})$ in $\mathfrak{B}_N$. Then A is uniquely determined by a. Thus we have a single-valued correspondence between the algebra $\mathfrak{L}$ of $\mathfrak{B}$-endomorphisms of $\mathfrak{R}$ and a set of matrices in $\mathfrak{B}_N$. As in the case where $\mathfrak{B}$ is a division ring, we may show that the correspondence is an anti-isomorphism between the algebra of $\mathfrak{B}$-endomorphisms and the algebra $\mathfrak{B}_N$.

We now define an *anti-representation* $\mathfrak{B}$*-space* of $\mathfrak{A}$ as a commutative group $\mathfrak{R}$ that satisfies the following conditions:

1. $\mathfrak{R}$ is a $\mathfrak{B}$-space.
2. $\mathfrak{R}$ is an $\mathfrak{A}$-module such that $x1 = x$ for all x and 1 the identity of $\mathfrak{A}$.
3. $(xa)b = (xb)a$ if $x \epsilon \mathfrak{R}$, $a \epsilon \mathfrak{A}$ and $b \epsilon \mathfrak{B}$.
4. 1α in $\mathfrak{A}$ is mapped into the same endomorphism as 1α in $\mathfrak{B}$.

Now by 3. the endomorphism corresponding to a is a $\mathfrak{B}$-endomorphism. Hence if $x_1, \cdots, x_N$ is a $\mathfrak{B}$-basis for $\mathfrak{R}$ and $ax_i = \Sigma x_j a_{ji}$, then the correspondence between a and the matrix $A = (a_{ij})$ in $\mathfrak{B}_N$ is a ring anti-homomorphism. By 4. to $a\alpha = a(1\alpha)$ there corresponds the matrix $A(1\alpha) = A\alpha$ and so we have an algebra anti-homomorphism between $\mathfrak{A}$ and a subalgebra of $\mathfrak{B}_N$. It follows that each anti-representation $\mathfrak{B}$-space of $\mathfrak{A}$ determines an anti-representation and conversely. Again, as in the case where $\mathfrak{B} = \Phi$, a second basis for $\mathfrak{R}$ defines an anti-representation similar to the anti-representation $a \to A$. The anti-representation spaces $\mathfrak{R}_1$ and $\mathfrak{R}_2$ are $(\mathfrak{A}, \mathfrak{B})$-isomorphic if and only if they determine the same similarity class of anti-representations.

We consider now the algebra $\mathfrak{A}_\mathfrak{B} = \mathfrak{A} \times \mathfrak{B}$. We have seen that if $x_1, \cdots, x_n$ is a basis for $\mathfrak{A}$ over Φ, then each element of $\mathfrak{A}_\mathfrak{B}$ is expressible in one and only one way in the form $x_1b_1 + \cdots + x_nb_n$ where the $b_i \epsilon \mathfrak{B}$. Thus $\mathfrak{A}_\mathfrak{B}$ is a $\mathfrak{B}$-space of rank n relative to the right multiplications $x \to xb$ as module operation. The algebra $\mathfrak{A}_\mathfrak{B}$ is also an $\mathfrak{A}$-module relative to the right multiplications $x \to xa$. Hence $\mathfrak{A}_\mathfrak{B}$ is an anti-representation $\mathfrak{B}$-space of $\mathfrak{A}$. We shall show next that any anti-representation $\mathfrak{B}$-space $\mathfrak{R}$ is an $\mathfrak{A}_\mathfrak{B}$-module. For let $\bar{\mathfrak{A}}$ denote the set of endomorphisms corresponding to the elements of $\mathfrak{A}$ and $\bar{\mathfrak{B}}$ the set corresponding to the elements of $\mathfrak{B}$. Since $a \to a$ of $\bar{\mathfrak{A}}$ and $b \to b$ of $\bar{\mathfrak{B}}$ are homomorphisms, the correspondence between the element $\Sigma a_i b_i$ of $\mathfrak{A}_\mathfrak{B}$ and the endomorphism $\Sigma a_i b_i$ of $\bar{\mathfrak{A}}\bar{\mathfrak{B}} = \bar{\mathfrak{B}}\bar{\mathfrak{A}}$ is a homomorphism. Thus $\mathfrak{R}$ is an $\mathfrak{A}_\mathfrak{B}$-module. On the other hand, any $\mathfrak{A}_\mathfrak{B}$-module which is a $\mathfrak{B}$-space when regarded relative to $\mathfrak{B}$ is an anti-representation $\mathfrak{B}$-space of $\mathfrak{A}$.

In a similar manner, we may show that the theory of ordinary representations is equivalent to a theory of *representation* $\mathfrak{B}$*-spaces* where these are defined by the conditions 1., 4. and

2′. $\mathfrak{R}$ is a left $\mathfrak{A}$-module such that $1x = x$ for all x and 1 the identity of $\mathfrak{A}$.
3′. $(ax)b = a(xb)$ if $x \epsilon \mathfrak{R}$, $a \epsilon \mathfrak{A}$ and $b \epsilon \mathfrak{B}$.

We introduce the algebra $\mathfrak{A}'$ anti-isomorphic to $\mathfrak{A}$. Then we may regard $\mathfrak{R}$ as an $\mathfrak{A}'$-module relative to the product $xa' \equiv ax (a \leftrightarrow a'$ in the anti-isomorphism). Thus $\mathfrak{R}$ is an anti-representation $\mathfrak{B}$-space of $\mathfrak{A}'$ and is therefore an $\mathfrak{A}'_\mathfrak{B}$-module. Conversely, any $\mathfrak{A}'_\mathfrak{B}$-module which is a $\mathfrak{B}$-space is a representation $\mathfrak{B}$-space of $\mathfrak{A}$.

We suppose now that $\mathfrak{B}$ is simple. Then we recall that $\mathfrak{B}$ is a direct sum of, say, m $\mathfrak{B}$-isomorphic irreducible right ideals $\mathfrak{J}$ and that $\mathfrak{B} = \mathfrak{D}_m$ where $\mathfrak{D}$ is a division algebra. $\mathfrak{B}$ itself is a free cyclic module with 1 as a basis. Any $\mathfrak{B}$-module $\mathfrak{R}$ in which $x1 = x$ for all x is a direct sum of irreducible modules $\mathfrak{B}$-isomorphic to the irreducible right ideals $\mathfrak{J}$. Hence $\mathfrak{R}$ is a free cyclic module if and only if it is a direct sum of m irreducible submodules and $\mathfrak{R}$ is a $\mathfrak{B}$-space if and only if it is a direct sum of $h = Nm$ irreducible $\mathfrak{B}$-modules. Then if $\mathfrak{S}$ is any subspace of $\mathfrak{R}$, $\mathfrak{R} = \mathfrak{S} \oplus \mathfrak{S}'$ where $\mathfrak{S}'$ is also a subspace. Thus if $y_1, \cdots, y_u$ is a basis for $\mathfrak{S}$, there exists a basis for $\mathfrak{R}$ that includes the y_i. Using this result we may prove, as in the case where $\mathfrak{B} = \Phi$, that the condition that an anti-representation be reducible is that the anti-representation space $\mathfrak{R}$ contains a proper $\mathfrak{B}$-subspace invariant relative to the endomorphisms a. The condition that the anti-representation be decomposable is that $\mathfrak{R} = \mathfrak{R}_1 \oplus \mathfrak{R}_2$ where the $\mathfrak{R}_i$ are anti-representation subspaces of $\mathfrak{R}$. A sufficient condition for complete reducibility is that $\mathfrak{R}$ be a completely reducible $\mathfrak{A}_{\mathfrak{B}}$-module. As we have seen in Chapter 4, if $\mathfrak{A}_{\mathfrak{B}}$ is semi-simple, then any $\mathfrak{A}_{\mathfrak{B}}$-module such that $x1 = x$, for all x, is completely reducible. Hence if $\mathfrak{A}_{\mathfrak{B}}$ is semi-simple, any anti-representation of $\mathfrak{A}$ by matrices with elements in $\mathfrak{B}$ is completely reducible.

If $\mathfrak{B} = \mathfrak{D}$ is a division algebra, any irreducible $\mathfrak{D}$-module such that $\mathfrak{R}\mathfrak{D} \neq 0$ is a free cyclic module. The $\mathfrak{D}$-spaces defined in this section are simply the vector spaces over $\mathfrak{D}$ that we have considered before. Hence in this case any $\mathfrak{A}_{\mathfrak{D}}$-module $\mathfrak{R}$ in which $x1 = x$ for all x and $(\mathfrak{R}:\mathfrak{D})$ is finite, is an anti-representation $\mathfrak{D}$-space of $\mathfrak{A}$. As above, the condition $(\mathfrak{R}:\mathfrak{D})$ finite is equivalent to the condition that $\mathfrak{R}$ be finitely generated relative to $\mathfrak{A}_{\mathfrak{D}}$. In particular, the irreducible $\mathfrak{A}_{\mathfrak{D}}$-modules are anti-representation $\mathfrak{D}$-spaces. These modules are therefore the irreducible anti-representation $\mathfrak{D}$-spaces of $\mathfrak{A}$. As we have seen, any irreducible $\mathfrak{A}_{\mathfrak{D}}$-module is $\mathfrak{A}_{\mathfrak{D}}$-isomorphic to an irreducible right ideal $\mathfrak{J}$ of $\mathfrak{A}_{\mathfrak{D}} - \mathfrak{N}$, $\mathfrak{N}$ the radical. The size of the matrices determined by $\mathfrak{J}$ is the dimensionality (or rank) of $\mathfrak{J}$ over $\mathfrak{D}$. The number of non-isomorphic irreducible $\mathfrak{A}_{\mathfrak{D}}$-modules in which $x1 = x$ for all x, and hence the number of classes of irreducible anti-representations $\neq 0$, is equal to the number of simple two-sided ideals in $\mathfrak{A}_{\mathfrak{D}} - \mathfrak{N}$.

6. Direct products and composites of fields. As an application of the above theory we shall now obtain the structure of $\mathfrak{A}_{\mathfrak{B}}$ for $\mathfrak{A}$ a separable field over Φ and $\mathfrak{B}$ an arbitrary field over Φ. Suppose first that $\mathfrak{B}$ contains a subfield isomorphic to the least normal field over Φ containing $\mathfrak{A}$. Then if $(\mathfrak{A}:\Phi) = n$, it is well known that there exist precisely n distinct isomorphisms $a \to a^{(i)}$, $i = 1, \cdots, n$, between $\mathfrak{A}$ and subfields of $\mathfrak{B}$.[3] Thus we obtain n anti-homomorphisms between $\mathfrak{A}$ and matrices with elements in $\mathfrak{B}$, and, since these are one dimensional, they are irreducible and dissimilar. It follows from the general theory that $\mathfrak{A}_{\mathfrak{B}} - \mathfrak{N}$, $\mathfrak{N}$ the radical, is a direct sum of at least n ideals. Since the dimensionalities of these ideals over $\mathfrak{B}$ is $\geqq 1$ and $(\mathfrak{A}_{\mathfrak{B}}:\mathfrak{B}) = n$, it follows that $\mathfrak{N} = 0$ and that there are exactly n simple ideals in $\mathfrak{A}_{\mathfrak{B}}$, each one dimensional over $\mathfrak{B}$. Now if $\mathfrak{B}$ is arbitrary, we take a field $\mathfrak{C} \geqq \mathfrak{B}$ and containing a field isomorphic to the least normal field containing $\mathfrak{A}$. Since $(\mathfrak{A}_{\mathfrak{B}})_{\mathfrak{C}} = \mathfrak{A}_{\mathfrak{C}}$, $\mathfrak{A}_{\mathfrak{B}}$ is semi-simple.

THEOREM 3. *If $\mathfrak{A}$ is a separable field over Φ, $(\mathfrak{A}:\Phi) = n$, and $\mathfrak{B}$ is any field over Φ, then $\mathfrak{A}_{\mathfrak{B}}$ is semi-simple. If $\mathfrak{B}$ contains a subfield isomorphic to the least normal field containing $\mathfrak{A}$, the irreducible representations of $\mathfrak{A}$ by matrices in $\mathfrak{B}$ are all one-rowed.*

[3] Cf. van der Waerden's *Moderne Algebra*, vol. 1, p. 115 or 2nd. ed., p. 102.

By the structure theory of semi-simple rings, $\mathfrak{A}_{\mathfrak{B}}$ is a direct sum of fields, say $\mathfrak{F}_1 \oplus \cdots \oplus \mathfrak{F}_t$. If $1 = e_1 + \cdots + e_t$ is the corresponding decomposition of the identity of $\mathfrak{A}_{\mathfrak{B}}$ into the identities of the $\mathfrak{F}_i$, then the set $e_i\mathfrak{A}$ of elements e_ia, a in $\mathfrak{A}$, is a subfield of $\mathfrak{F}_i$ isomorphic to $\mathfrak{A}$. Similarly, $\mathfrak{F}_i$ contains the subfield $e_i\mathfrak{B}$ isomorphic to $\mathfrak{B}$. Since $(e_i\mathfrak{A})(e_i\mathfrak{B}) = e_i\mathfrak{A}\mathfrak{B}e_i = \mathfrak{F}_i$, the field $\mathfrak{F}_i$ is generated by these two fields.

Suppose now that we have any two fields $\mathfrak{A}$ and $\mathfrak{B}$ over Φ and two isomorphisms $a \rightarrow a^S$ and $b \rightarrow b^T$ of $\mathfrak{A}$ and $\mathfrak{B}$, respectively, into subfields $\mathfrak{A}^S$ and $\mathfrak{B}^T$ of a third field $\mathfrak{F}$. Then we call the system $(\mathfrak{F}, S, T)$ a *composite* of $\mathfrak{A}$ and $\mathfrak{B}$ provided that $\mathfrak{F} = [\mathfrak{A}^S, \mathfrak{B}^T]$, the smallest subfield of $\mathfrak{F}$ containing $\mathfrak{A}^S$ and $\mathfrak{B}^T$.[4] We shall regard the two composites $(\mathfrak{F}, S, T)$ and $(\mathfrak{F}', S', T')$ as *equivalent* if the isomorphism $a^S \rightarrow a^{S'}$, $b^T \rightarrow b^{T'}$ may be extended to an isomorphism between $\mathfrak{F}$ and $\mathfrak{F}'$. It is evident that such an extension, if it exists, is uniquely determined.

We have seen that if $\mathfrak{A}$ is separable, then $\mathfrak{A}_{\mathfrak{B}} = \mathfrak{F}_1 \oplus \cdots \oplus \mathfrak{F}_s$. The mappings $a \rightarrow a^{S_i} \equiv ae_i$, for a in $\mathfrak{A}$, is an isomorphism between $\mathfrak{A}$ and the subfield $\mathfrak{A}^{S_i}$ of $\mathfrak{F}_i$. Similarly $b \rightarrow b^{T_i} \equiv be_i$ is an isomorphism between $\mathfrak{B}$ and $\mathfrak{B}^{T_i}$. Moreover, $\mathfrak{F}_i = (\mathfrak{A}^{S_i})(\mathfrak{B}^{T_i}) = [\mathfrak{A}^{S_i}, \mathfrak{B}^{T_i}]$, and therefore $(\mathfrak{F}_i, S_i, T_i)$ is a composite of $\mathfrak{A}$ and of $\mathfrak{B}$.

We wish to prove the following

THEOREM 4. *The composites $(\mathfrak{F}_i, S_i, T_i)$, $i = 1, \cdots, t$, are inequivalent. Any composite of the separable field $\mathfrak{A}$ and the field $\mathfrak{B}$ is equivalent to one of the $(\mathfrak{F}_i, S_i, T_i)$.*

To prove that $(\mathfrak{F}_i, S_i, T_i)$ and $(\mathfrak{F}_j, S_j, T_j)$ are inequivalent if $i \neq j$, we note that e_i has the form $a_1b_1 + \cdots + a_rb_r$, a_k in $\mathfrak{A}$ and b_k in $\mathfrak{B}$, and, since $e_i^2 = e_i$, $e_i = (a_1e_i)(b_1e_i) + \cdots + (a_re_i)(b_re_i) = a_1^{S_i}b_1^{T_i} + \cdots + a_r^{S_i}b_r^{T_i}$. If $(\mathfrak{F}_i, S_i, T_i)$ were equivalent to $(\mathfrak{F}_j, S_j, T_j)$, the required isomorphism would map e_i into $a_1^{S_j}b_1^{T_j} + \cdots + a_r^{S_j}b_r^{T_j} = (a_1b_1 + \cdots + a_rb_r)e_j = e_ie_j = 0$ and this is impossible. Now suppose that $(\mathfrak{F}, S, T)$ is any composite of $\mathfrak{A}$ and $\mathfrak{B}$. Then the mapping $\Sigma ab \rightarrow \Sigma a^Sb^T$ is a homomorphism between $\mathfrak{A}_{\mathfrak{B}}$ and the subalgebra $\mathfrak{A}^S\mathfrak{B}^T$ of $\mathfrak{F}$. Since the only ideals of $\mathfrak{A}_{\mathfrak{B}}$ are the ideals $\mathfrak{F}_{i_1} \oplus \cdots \oplus \mathfrak{F}_{i_r}$ and since $\mathfrak{F}$ has no zero-divisors, the ideal mapped into 0 by the homomorphism is one of the form $\mathfrak{F}_1 \oplus \cdots \oplus \mathfrak{F}_{i-1} \oplus \mathfrak{F}_{i+1} \oplus \cdots \oplus \mathfrak{F}_t = \mathfrak{L}_i$. Hence $\mathfrak{A}^S\mathfrak{B}^T$ is isomorphic to $\mathfrak{F} - \mathfrak{L}_i$ and thus to $\mathfrak{F}_i$. This implies that $\mathfrak{A}^S\mathfrak{B}^T$ is a field and so $\mathfrak{A}^S\mathfrak{B}^T = [\mathfrak{A}^S, \mathfrak{B}^T] = \mathfrak{F}$. Moreover, the isomorphism defined by our homomorphism is the mapping $\Sigma a^Sb^T \rightarrow \Sigma(a^{S_i})(b^{T_i})$. Hence $(\mathfrak{F}, S, T)$ and $(\mathfrak{F}_i, S_i, T_i)$ are equivalent.

We have seen that if $\mathfrak{B}$ contains a field isomorphic to the least normal extension of $\mathfrak{A}$, then $t = n$ and each $\mathfrak{F}_i$ is one dimensional over $\mathfrak{B}$. Hence $\mathfrak{F}_i = \mathfrak{B}^{S_i}$ is isomorphic to $\mathfrak{B}$.

THEOREM 5. *If $\mathfrak{A}$ is separable over Φ, $(\mathfrak{A}:\Phi) = n$ and $\mathfrak{B}$ contains a field isomorphic to the least normal extension of $\mathfrak{A}$ over Φ, then $\mathfrak{A}_{\mathfrak{B}} = \mathfrak{B}_1 \oplus \cdots \oplus \mathfrak{B}_n$ where $\mathfrak{B}_i \cong \mathfrak{B}$.*

[4] If both fields $\mathfrak{A}$ and $\mathfrak{B}$ contain transcendental elements, this definition requires modification. Cf. Chevalley **[9]**.

That these theorems do not hold when both fields are inseparable may be seen from the following example: Let $\mathfrak{A} = \Phi(x)$ where Φ has characteristic p and $x^p = \xi$ is in Φ, but x is not in Φ. Suppose that $\mathfrak{B}$ is the field $\Phi(y)$, $y^p = \xi$. Then $\mathfrak{A}_\mathfrak{B}$ contains the element $z = x - y \neq 0$ which is nilpotent. Since $\mathfrak{A}_\mathfrak{B}$ is commutative, z generates a nilpotent ideal and so $\mathfrak{A}_\mathfrak{B}$ is not semi-simple.

7. Central simple algebras. We take up now the main topic of this chapter, namely, the theory of simple algebras. Throughout our discussion we shall exclude the trivial zero algebras. With this agreement we may state the fundamental structure theorem in the following way.

THEOREM 6 (Wedderburn). *Any simple algebra $\mathfrak{A}$ over Φ is a direct product $\Phi_m \times \mathfrak{D}$ where $\mathfrak{D}$ is a division algebra and conversely, any algebra of this form is simple. If $\mathfrak{A} = \Phi_m \times \mathfrak{D} = \Phi_{m'} \times \mathfrak{F}$ where $\mathfrak{F}$ is a division algebra, then $m = m'$ and $\mathfrak{D}$ and $\mathfrak{F}$ are isomorphic.*

We shall also require

THEOREM 7. $\Phi_{rs} = \Phi_r \times \Phi_s$.

This is an immediate consequence of the computations of **6**, Chapter 2.

A simple algebra $\mathfrak{A}$ is *central* if its center consists of the multiples 1α, α in Φ.[5] For example, Φ_m is central simple. A central algebra is in a sense the opposite of a commutative algebra and we shall see that the theory of direct products for these algebras is considerably simpler than that for commutative algebras indicated in the preceding section.

If $\mathfrak{A} = \Phi_m \times \mathfrak{D} = \mathfrak{D}_m$ where $\mathfrak{D}$ is a division algebra, we have seen that the center $\mathfrak{C}$ of $\mathfrak{A}$ is contained in $\mathfrak{D}$. Hence $\mathfrak{A}$ is central if and only if $\mathfrak{D}$ is central. If $\mathfrak{A}$ is any simple algebra, $\mathfrak{C}$, the center, is a field and $\mathfrak{A}$ may be regarded as an algebra over $\mathfrak{C}$. Obviously $\mathfrak{A}$ is central over $\mathfrak{C}$.

Suppose now that $\mathfrak{A}$ is an arbitrary algebra with an identity and that $\mathfrak{B}$ is a central simple algebra. We wish to show that the two-sided ideals of $\mathfrak{A}_\mathfrak{B}$ may be put into $(1 - 1)$ correspondence with those of $\mathfrak{A}$. First, let $\mathfrak{J}_0$ be a two-sided ideal of $\mathfrak{A}$. Then $\mathfrak{J} = \mathfrak{J}_0\mathfrak{B} = \mathfrak{J}_{0\mathfrak{B}}$ is a two-sided ideal of $\mathfrak{A}_\mathfrak{B}$. Let $x_1, \cdots, x_n$ be a basis for $\mathfrak{A}$ over Φ such that $x_1, \cdots, x_r$ is one for $\mathfrak{J}_0$ over Φ. Then if $\sum_1^n x_ib_i \in \mathfrak{A}$, $b_i = \beta_i$ is in Φ. If $\sum_1^n x_ib_i \in \mathfrak{J}$, $b_{r+1} = \cdots = b_n = 0$. Hence $(\mathfrak{A} \wedge \mathfrak{J})$ consists of the elements $\sum_1^r x_i\beta_i$ and $(\mathfrak{A} \wedge \mathfrak{J}) = \mathfrak{J}_0$. It follows that $\mathfrak{J}_{0\mathfrak{B}} = \bar{\mathfrak{J}}_{0\mathfrak{B}}$ if and only of $\mathfrak{J}_0 = \bar{\mathfrak{J}}_0$.

Now let $\mathfrak{J}$ be an arbitrary two-sided ideal in $\mathfrak{A} \times \mathfrak{B}$, $\mathfrak{J}_0 = (\mathfrak{A} \wedge \mathfrak{J})$, and let $x_1, \cdots, x_n$ be a basis for $\mathfrak{A}$ such that $x_1, \cdots, x_r$ is one for $\mathfrak{J}_0$. Evidently $\mathfrak{J}_0$ is a two-sided ideal in $\mathfrak{A}$ and $\mathfrak{J}_{0\mathfrak{B}} \leqq \mathfrak{J}$. Suppose that $\mathfrak{J}_{0\mathfrak{B}} < \mathfrak{J}$ and let $x_1b_1 + \cdots + x_nb_n$ be an element of $\mathfrak{J}$ not contained in $\mathfrak{J}_{0\mathfrak{B}}$. Then $x_{r+1}b_{r+1} + \cdots + x_nb_n$ has this property also, and so at least one of the b_j, $j = r + 1, \cdots, n$, is

[5] I am indebted to Professor Albert for suggesting this term as a substitute for the overworked term "normal" formerly used in this connection. The term "centralizer" that we shall use later is also due to Albert.

$\neq 0$. Now let $x_{i_1}b_{i_1} + \cdots + x_{i_s}b_{i_s}$, $b_{i_j} \neq 0$, $i_j = r + 1, \cdots, n$, be an element of $\mathfrak{J}$ for which s has the least positive value. The elements

$$b(x_{i_1}b_{i_1} + \cdots + x_{i_s}b_{i_s}), \qquad (x_{i_1}b_{i_1} + \cdots + x_{i_s}b_{i_s})b$$

are in $\mathfrak{J}$ if b is any element of $\mathfrak{B}$. It follows that the first components b_{i_1} of these elements together with 0 form a two sided ideal $\neq 0$ in $\mathfrak{B}$ and hence, since $\mathfrak{B}$ is simple, b_{i_1} is arbitrary. Thus $\mathfrak{J}$ contains $x_{i_1} + x_{i_2}b_2' + \cdots + x_{i_s}b_s'$ and hence it contains

$$b(x_{i_1} + x_{i_2}b_2' + \cdots + x_{i_s}b_s') - (x_{i_1} + x_{i_2}b_2' + \cdots + x_{i_s}b_s')b = \sum_{j=2}^{s} x_{i_j}(bb_j' - b_j'b).$$

Since s is minimal, $bb_j' = b_j'b$ and so, by the centrality, $b_j' = \beta_j \in \Phi$. Thus $\mathfrak{J}$ contains $x_{i_1} + x_{i_2}\beta_2 + \cdots + x_{i_s}\beta_s$ which is evidently in $\mathfrak{A}$. This contradicts the fact that $x_1, \cdots, x_r$ is a basis for $\mathfrak{J}_0 = (\mathfrak{A} \wedge \mathfrak{J})$. We have therefore proved

THEOREM 8. *If $\mathfrak{A}$ is an arbitrary algebra with an identity and $\mathfrak{B}$ is a central simple algebra, then the correspondence $\mathfrak{J}_0 \rightarrow \mathfrak{J}_{0\mathfrak{B}}$ is $(1 - 1)$ between the two-sided ideals of $\mathfrak{A}$ and the two-sided ideals of $\mathfrak{A}_\mathfrak{B}$.*

COROLLARY 1. *If $\mathfrak{A}$ is simple and $\mathfrak{B}$ is central simple, then $\mathfrak{A}_\mathfrak{B}$ is simple.*

If $\mathfrak{N}$ is the radical of $\mathfrak{A}_\mathfrak{B}$, $\mathfrak{N}_0 = \mathfrak{A} \wedge \mathfrak{N}$ is a nilpotent ideal in $\mathfrak{A}$ and is therefore contained in the radical $\mathfrak{N}_0'$ of $\mathfrak{A}$. On the other hand, $\mathfrak{N}_{0\mathfrak{B}}'$ is a nilpotent ideal in $\mathfrak{A}_\mathfrak{B}$ so that $\mathfrak{N}_{0\mathfrak{B}}' \leq \mathfrak{N}$. Hence $\mathfrak{N}_0' = \mathfrak{N}_0$. This implies in particular

COROLLARY 2. *If $\mathfrak{A}$ is semi-simple and $\mathfrak{B}$ is central simple, then $\mathfrak{A}_\mathfrak{B}$ is semi-simple.*

Now let $c = x_1b_1 + \cdots + x_nb_n$ be an element of $\mathfrak{A}_\mathfrak{B} = \mathfrak{A} \times \mathfrak{B}$ commutative with every b in $\mathfrak{B}$. Then $\Sigma x_i(bb_i - b_ib) = 0$ and $bb_i = b_ib$. Hence $b_i \in \Phi$ and $c \in \mathfrak{A}$. It follows that the center of $\mathfrak{A} \times \mathfrak{B}$ coincides with the center of $\mathfrak{A}$. If $\mathfrak{A}$ is central simple, $\mathfrak{A} \times \mathfrak{B}$ is central simple.

THEOREM 9. *If $\mathfrak{A}$ is an algebra with an identity and $\mathfrak{B}$ is central simple, the only elements of $\mathfrak{A} \times \mathfrak{B}$ that commute with all the elements of $\mathfrak{B}$ are the elements of $\mathfrak{A}$. If $\mathfrak{A}$ is central simple, $\mathfrak{A} \times \mathfrak{B}$ is central simple.*

8. Representation of a semi-simple algebra by matrices with elements in a central simple algebra. As before we restrict the discussion to the representations of an algebra $\mathfrak{A}$ in which the identity of $\mathfrak{A}$ is mapped into the identity matrix. If $\mathfrak{A}$ is semi-simple and $\mathfrak{B}$ is central simple, then $\mathfrak{A}_\mathfrak{B}$ is semi-simple. As we have seen in **5** this implies the following

THEOREM 10. *If $\mathfrak{A}$ is a semi-simple algebra, any anti-representation (ordinary representation) of $\mathfrak{A}$ by matrices with elements in a central simple algebra is completely reducible.*

We assume now that $\mathfrak{A}$ is simple. Let $\mathfrak{B}$ be a direct sum of m isomorphic irreducible right ideals. Thus $\mathfrak{B} = \mathfrak{D}_m$ where $\mathfrak{D}$ is a division algebra. We have seen that $\mathfrak{A}_{\mathfrak{B}}$ is simple. Hence $\mathfrak{A}_{\mathfrak{B}}$ is a direct sum of r isomorphic irreducible right ideals and $\mathfrak{A}_{\mathfrak{B}} = \mathfrak{E}_r$ where $\mathfrak{E}$ is a division algebra. Any irreducible right ideal of $\mathfrak{A}_{\mathfrak{B}}$ is a $\mathfrak{B}$-module and is therefore a direct sum of, say, h $\mathfrak{B}$-isomorphic irreducible $\mathfrak{B}$-modules. It follows that $\mathfrak{A}_{\mathfrak{B}}$ is a direct sum of rh irreducible $\mathfrak{B}$-modules. On the other hand, if $(\mathfrak{A}:\Phi) = n$, $\mathfrak{A}_{\mathfrak{B}}$ is a $\mathfrak{B}$-space of rank n. Since any $\mathfrak{B}$-space of rank 1 is a sum of m $\mathfrak{B}$-isomorphic irreducible $\mathfrak{B}$-modules, $\mathfrak{A}_{\mathfrak{B}}$ is a direct sum of mn irreducible $\mathfrak{B}$-modules. Thus $rh = mn$.

Now let $\mathfrak{R}$ be an arbitrary irreducible anti-representation $\mathfrak{B}$-space of $\mathfrak{A}$. Then $\mathfrak{R} = \mathfrak{S}_1 \oplus \cdots \oplus \mathfrak{S}_{\bar{m}}$ where the $\mathfrak{S}_i$ are isomorphic irreducible $\mathfrak{A}_{\mathfrak{B}}$-modules. Each $\mathfrak{S}_i$ is a direct sum of h isomorphic irreducible $\mathfrak{B}$-modules. Hence $\mathfrak{R}$ is a direct sum of $\bar{m}h$ irreducible $\mathfrak{B}$-modules. Since $\mathfrak{R}$ is a $\mathfrak{B}$-space, it follows that $h\bar{m} \equiv 0(m)$. Now if m'' is an integer $\leqq \bar{m}$ such that $hm'' \equiv 0(m)$, then the direct sum of m'' of the $\mathfrak{S}_i$ is a $\mathfrak{B}$-space. It therefore coincides with $\mathfrak{R}$. This implies that $m'' = \bar{m}$ so that $h\bar{m} = m\bar{h}$ is the least common multiple of h and m. The equation $h\bar{m} = m\bar{h}$ shows also that the rank of $\mathfrak{R}$ over $\mathfrak{B}$ is $\bar{h}$. Hence the size of the matrices determined by $\mathfrak{R}$ is $\bar{h}$. If $\mathfrak{R}'$ is a second irreducible anti-representation $\mathfrak{B}$-space of $\mathfrak{A}$, $\mathfrak{R}'$, too, is a direct sum of $\bar{m} = h^{-1}[h, m]$ irreducible $\mathfrak{A}_{\mathfrak{B}}$-modules. It follows that $\mathfrak{R}'$ and $\mathfrak{R}$ are $\mathfrak{A}_{\mathfrak{B}}$-isomorphic. Thus all the irreducible anti-representations of $\mathfrak{A}$ by matrices are similar. Any anti-representation of $\mathfrak{A}$ is completely reducible into irreducible parts all of which are similar to the representation determined by $\mathfrak{R}$. These results may be stated as the following fundamental

THEOREM 11. *Let $\mathfrak{A}$ be a simple algebra and $\mathfrak{B}$ a central simple algebra. Set $(\mathfrak{A}:\Phi) = n$, $\mathfrak{B} = \mathfrak{D}_m$ and $\mathfrak{A}_{\mathfrak{B}} = \mathfrak{E}_r$ where $\mathfrak{D}$ and $\mathfrak{E}$ are division algebras. Then $r \mid mn$ and if $mn = hr$ and $[h, m] = h\bar{m} = \bar{h}m$, $\mathfrak{A}$ has an anti-representation in $\mathfrak{B}_N$ if and only if $\bar{h} \mid N$. Any two anti-representations of $\mathfrak{A}$ in the same $\mathfrak{B}_N$ are similar.*

In a similar manner we may prove

THEOREM 11′. *Let $\mathfrak{A}$ and $\mathfrak{B}$ be as in Theorem 11 and let $\mathfrak{A}'_{\mathfrak{B}} = \mathfrak{E}'_{r'}$ where $\mathfrak{A}'$ is the algebra anti-isomorphic to $\mathfrak{A}$ and $\mathfrak{E}'$ is a division algebra. Then $r' \mid mn$ and if $mn = h'r'$ and $[h', m] = h'\bar{m}' = \bar{h}'m$, $\mathfrak{A}$ has a representation in $\mathfrak{B}_N$ if and only if $\bar{h}' \mid N$. Any two representations of $\mathfrak{A}$ in the same $\mathfrak{B}_N$ are similar.*

We may obtain a somewhat sharper form of Theorem 11 by first specializing this theorem to the case where $\mathfrak{B} = \mathfrak{D}$ is a division algebra and then extending the result thus obtained to the general case where $\mathfrak{B} = \mathfrak{D}_m$. If $\mathfrak{B} = \mathfrak{D}$ then $m = 1$ and $[h, m] = h$. Hence we have the

COROLLARY. *Let $\mathfrak{A}$ be a simple algebra and $\mathfrak{D}$ a central division algebra. Set $(\mathfrak{A}:\Phi) = n$ and $\mathfrak{A}_{\mathfrak{D}} = \mathfrak{E}_s$. Then $n = hs$ and $\mathfrak{A}$ has an anti-representation in $\mathfrak{D}_N$ if and only if $h \mid N$.*

Now if $\mathfrak{A}_{\mathfrak{D}} = \mathfrak{E}_s$, $\mathfrak{A}_{\mathfrak{B}} = \mathfrak{E}_{sm}$ for $\mathfrak{B} = \mathfrak{D}_m$. Hence the integer r of Theorem 11 is equal to sm and $n = hs$. This proves

THEOREM 12. *Let $\mathfrak{A}$ and $\mathfrak{B}$ be as in Theorem 11 and let $\mathfrak{A}_{\mathfrak{D}} = \mathfrak{E}_s$, $\mathfrak{E}$ a division algebra. Then $s \mid n$ and if $n = sh$ and $[h, m] = h\bar{m} = \bar{h}m$, $\mathfrak{A}$ has an anti-representation in $\mathfrak{B}_N$ if and only if $\bar{h} \mid N$.*

THEOREM 12'. *Let $\mathfrak{A}$ and $\mathfrak{B}$ be as in Theorem* 11 *and let* $\mathfrak{A}'_{\mathfrak{D}} = \mathfrak{E}'_{s'}$, $\mathfrak{E}'$ *a division algebra. Then* $s' \mid n$ *and if* $n = s'h'$ *and* $[h', m] = h'\bar{m}' = \bar{h}'m$, $\mathfrak{A}$ *has a representation in* $\mathfrak{B}_N$ *if and only if* $\bar{h}' \mid N$.

We suppose now that $\mathfrak{A}$ is a division algebra and $\mathfrak{B} = \mathfrak{D}$ is a central division algebra. We may regard $\mathfrak{A} \times \mathfrak{D} = \mathfrak{E}_s$ as an $\mathfrak{A}$-space. Then by a repetition of the argument that led to Theorem 11 we may prove that $s \mid d$, $d = (\mathfrak{D}:\Phi)$. The details are left to the reader.

THEOREM 13. *Let $\mathfrak{A}$ be a division algebra and $\mathfrak{D}$ a central division algebra. Then if* $\mathfrak{A} \times \mathfrak{D} = \mathfrak{E}_s$ *where $\mathfrak{E}$ is a division algebra, s is a common factor of* $(\mathfrak{A}:\Phi) = n$ *and of* $(\mathfrak{D}:\Phi) = d$. *If* $(d, n) = 1$, $\mathfrak{A} \times \mathfrak{D}$ *is a division algebra.*

9. Simple subalgebras of a central simple algebra. The theory of representations may be applied to the study of the subalgebras of a central simple algebra $\mathfrak{A}$. For if $\mathfrak{B}$ is a subalgebra, then $b \rightarrow b$ is a representation of $\mathfrak{B}$ by matrices of one row with elements in $\mathfrak{A}$. If $\mathfrak{B}$ is a simple algebra that contains the identity, we may apply Theorem 12'.[6] Let $\mathfrak{A} = \mathfrak{D}_m$ where $\mathfrak{D}$ is a central division algebra, $(\mathfrak{B}:\Phi) = q$ and $\mathfrak{B}' \times \mathfrak{D} = \mathfrak{B}'_{\mathfrak{D}} = \mathfrak{E}'_{s'}$ where $\mathfrak{E}'$ is a division algebra. Then $q = s'h'$ and if $\bar{h}' = m^{-1}[h', m]$, $\mathfrak{B}$ has a representation only in those $\mathfrak{A}_N$ for which $\bar{h}' \mid N$. Since $\mathfrak{B}$ has a one-rowed representation with elements in $\mathfrak{A}$, $\bar{h}' = 1$. Hence $h' \mid m$ and if we write $m = h'l$, we obtain $ms' = ql$. Thus $q \mid ms'$.

THEOREM 14. *If $\mathfrak{B}$ is a simple subalgebra, containing* 1, *of a central simple algebra* $\mathfrak{A} = \mathfrak{D}_m$, $\mathfrak{D}$ *a division algebra, then* $\mathfrak{B}' \times \mathfrak{D} = \mathfrak{E}'_{s'}$ *where $\mathfrak{E}'$ is a division algebra and* $s' \mid q$ *and* $q \mid ms'$.

COROLLARY. *If $\mathfrak{B}$ is a subalgebra, containing* 1, *of a central division algebra* $\mathfrak{D}$, *then* $\mathfrak{B}' \times \mathfrak{D} = \mathfrak{E}'_q$ *where $\mathfrak{E}'$ is a division algebra and* $q = (\mathfrak{B}:\Phi)$.

If $\mathfrak{B}_1$ and $\mathfrak{B}_2$ are isomorphic subalgebras of $\mathfrak{A}$, we may regard these algebras as isomorphic images of the same algebra $\mathfrak{B}$. If $b_1 \rightarrow b_2$ is an isomorphism between $\mathfrak{B}_1$ and $\mathfrak{B}_2$, $b \rightarrow b_1$ and $b \rightarrow b_2$ are representations of $\mathfrak{B}$ by one-rowed matrices with elements in $\mathfrak{A}$. These representations are similar. Hence we have the following

THEOREM 15. *If $\mathfrak{B}_1$ and $\mathfrak{B}_2$ are isomorphic simple subalgebras containing* 1 *of the central simple algebra $\mathfrak{A}$, any isomorphism between $\mathfrak{B}_1$ and $\mathfrak{B}_2$ may be extended to an inner automorphism in $\mathfrak{A}$.*

This, of course, implies

THEOREM 16. *Any automorphism of a central simple algebra is inner.*

10. Derivations. The theorems of **9** have striking analogues in the theory of derivations of an algebra. If $\mathfrak{B}$ is a subalgebra of an algebra $\mathfrak{A}$, a *derivation*

[6] It should be noted that from now on we use a different notation from that of **8**. Here $\mathfrak{A}$ denotes the central simple algebra and $\mathfrak{B}$ the simple algebra that need not be central. This seems desirable since in our applications $\mathfrak{B}$ will usually be a subalgebra of $\mathfrak{A}$.

D of $\mathfrak{B}$ *into* $\mathfrak{A}$ is a mapping of $\mathfrak{B}$ into a part of $\mathfrak{A}$ satisfying the following conditions:

$$(b_1 + b_2)D = b_1D + b_2D, \qquad (b\alpha)D = (bD)\alpha, \qquad (b_1b_2)D = b_1(b_2D) + (b_1D)b_2 .$$

If $\mathfrak{B} = \mathfrak{A}$, we speak simply of a *derivation in* $\mathfrak{A}$. It is readily seen then that if $D_1, D_2 \in \mathfrak{H}$, the set of derivations in $\mathfrak{A}$, then $D\alpha$ and $D_1 \pm D_2 \in \mathfrak{H}$. Since

$$(b_1b_2)D_1D_2 = (b_1D_2)(b_2D_1) + b_1(b_2D_1D_2) + (b_1D_1D_2)b_2 + (b_1D_1)(b_2D_2),$$

D_1D_2 is not in general a derivation. However,

$$(b_1b_2)(D_1D_2 - D_2D_1) = b_1(b_2(D_1D_2 - D_2D_1)) + (b_1(D_1D_2 - D_2D_1))b_2$$

so that $[D_1, D_2] = D_1D_2 - D_2D_1$ is a derivation. For any element d in $\mathfrak{A}$ we may define a derivation by means of the correspondence $x \to [x, d] = xd - dx$. A derivation of this type is called *inner*.

As usual, Leibniz's rule

$$(b_1b_2)D^k = b_1(b_2D^k) + \binom{k}{1}(b_1D)(b_2D^{k-1}) + \cdots + (b_1D^k)b_2$$

is valid. Hence if Φ has characteristic $p \neq 0$,

$$(b_1b_2)D^p = b_1(b_2D^p) + (b_1D^p)b_2$$

so that D^p is a derivation. Similarly we prove by inductior

$$bd^k = d^kb + \binom{k}{1}d^{k-1}b' + \cdots + b^{(k)}$$

where $b' = [b, d]$, $b'' = [[b, d], d]$, etc. Thus for Φ of characteristic $p \neq 0$

$$[b, d^p] = b^{(p)} = [\cdots [[b, \overbrace{d], d], \cdots, d}^{p}].$$

The theory of derivations to a large extent parallels that of isomorphisms. For example, we have the following

THEOREM 17. *If* $\mathfrak{B}$ *is a semi-simple subalgebra, containing* 1, *of a central simple* $\mathfrak{A}$, *then any derivation of* $\mathfrak{B}$ *in* $\mathfrak{A}$ *may be extended to an inner derivation in* $\mathfrak{A}$.

We consider the set of matrices in $\mathfrak{A}_2$ of the form

$$\begin{pmatrix} b & bD \\ 0 & b \end{pmatrix}$$

where b ranges over $\mathfrak{B}$. This set forms an algebra isomorphic to $\mathfrak{B}$ and hence it determines a representation of $\mathfrak{B}$ by matrices with elements in $\mathfrak{A}$. Let $\mathfrak{R}$ be the corresponding representation $\mathfrak{A}$-space. According to the form of the matrices, $\mathfrak{R}$ has a basis x_1, x_2 such that the $\mathfrak{A}$-space $\mathfrak{R}_1 = x_1\mathfrak{A}$ is invariant relative to the endomorphisms b of $\mathfrak{B}$. Since the $\mathfrak{B}'_{\mathfrak{A}}$-module $\mathfrak{R}$ is completely reducible, there exists a second space $\mathfrak{R}_2 = y\mathfrak{A}$ which is also invariant relative to the b and such that $\mathfrak{R} = \mathfrak{R}_1 \oplus \mathfrak{R}_2$. Let $y = x_1a_1 + x_2a_2$ where the $a_i \in \mathfrak{A}$.

Since $x_2 = x_1a_1' + ya_2'$ for suitable a_1' and a_2' in $\mathfrak{A}$, a_2 has the inverse a_2' in $\mathfrak{A}$. We may replace y by ya_2'. Hence we may suppose that $x_2 = x_1d + y$ and $y = x_2 - x_1d$. The matrix relating the $\mathfrak{A}$-basis x_1, y to the $\mathfrak{A}$-basis x_1, x_2 is $\begin{pmatrix} 1 & -d \\ 0 & 1 \end{pmatrix}$. The inverse of this matrix is $\begin{pmatrix} 1 & d \\ 0 & 1 \end{pmatrix}$. Since $\mathfrak{R} = \mathfrak{R}_1 \oplus \mathfrak{R}_2$, the matrix of the endomorphisms b relative to the basis x_1, y has the form $\begin{pmatrix} b_1 & 0 \\ 0 & b_2 \end{pmatrix}$. Hence

$$\begin{pmatrix} 1 & d \\ 0 & 1 \end{pmatrix}\begin{pmatrix} b & bD \\ 0 & b \end{pmatrix}\begin{pmatrix} 1 & -d \\ 0 & 1 \end{pmatrix} = \begin{pmatrix} b_1 & 0 \\ 0 & b_2 \end{pmatrix}$$

A simple computation shows that $b_1 = b_2 = b$ and $bD = [b, d]$ for all b.

As a consequence of Theorem 17 we have

THEOREM 18. *Any derivation of a central simple algebra is inner.*

11. Commuting subalgebras. If $\mathfrak{B}$ is a subalgebra of an algebra $\mathfrak{A}$, we call the subalgebra of $\mathfrak{A}$ of elements commutative with those of $\mathfrak{B}$ the *centralizer* $\mathfrak{A}(\mathfrak{B})$ of $\mathfrak{B}$ in $\mathfrak{A}$. As usual, we denote the algebra of right multiplications in $\mathfrak{A}$ by $\mathfrak{A}_r$ and the algebra of left multiplications in $\mathfrak{A}$ by $\mathfrak{A}_l$. Let $\overline{\mathfrak{B}}_r$ ($\overline{\mathfrak{B}}_l$) be the algebra of right (left) multiplications b_r (b_l) in $\mathfrak{A}$ determined by the elements b of $\mathfrak{B}$. We recall that if $\mathfrak{A}$ has an identity, $\mathfrak{A}_l$ is the algebra of $\mathfrak{A}_r$-endomorphisms and $\mathfrak{A}_r$ is the algebra of $\mathfrak{A}_l$-endomorphisms. Then the algebra of endomorphisms commutative with those of $\mathfrak{A}_l$ and of $\overline{\mathfrak{B}}_r$ is $\overline{\mathfrak{A}(\mathfrak{B})}_r$. For if C is such an endomorphism, $C = c_r$ is a right multiplication. Since $\mathfrak{A}_r$ is isomorphic to $\mathfrak{A}$ under the isomorphism $a \rightarrow a_r$ mapping the elements of b into those of $\overline{\mathfrak{B}}_r$, it follows that $c \in \mathfrak{A}(\mathfrak{B})$. If the subalgebra $\mathfrak{B}$ contains the identity, the algebra of endomorphisms $\mathfrak{A}_l\overline{\mathfrak{B}}_r = \overline{\mathfrak{B}}_r\mathfrak{A}_l$ contains $\mathfrak{A}_l$ and $\overline{\mathfrak{B}}_r$. Hence in this case $\overline{\mathfrak{A}(\mathfrak{B})}_r$ may be characterized as the algebra of $\mathfrak{A}_l\overline{\mathfrak{B}}_r$-endomorphisms acting in $\mathfrak{A}$.

We now suppose that $\mathfrak{A}$ is central simple and that $\mathfrak{B}$ is a simple subalgebra containing the identity of $\mathfrak{A}$. The algebra $\mathfrak{A}_l\overline{\mathfrak{B}}_r$ is a homomorphic image of $\mathfrak{A}' \times \mathfrak{B}$ where $\mathfrak{A}'$ is the algebra anti-isomorphic to $\mathfrak{A}$. We have seen that $\mathfrak{A}' \times \mathfrak{B}$ is simple. Hence this algebra has the form $\mathfrak{E}_r$, $\mathfrak{E}$ a division algebra. It follows that $\mathfrak{A}_l\overline{\mathfrak{B}}_r$ is isomorphic to $\mathfrak{E}_r$ and $\mathfrak{A}_l\overline{\mathfrak{B}}_r = \mathfrak{A}_l \times \overline{\mathfrak{B}}_r = \overline{\mathfrak{E}}_r$ where $\overline{\mathfrak{E}}$ is a division algebra isomorphic to $\mathfrak{E}$. Since $1\mathfrak{A}_l = \mathfrak{A}$, $\mathfrak{A}$ is finitely generated relative to $\overline{\mathfrak{E}}_r$. Hence, by **6** of Chapter 2, the algebra of $\overline{\mathfrak{E}}_r$-endomorphisms has the form $\overline{\mathfrak{E}}_s'$ where $\overline{\mathfrak{E}}'$ is anti-isomorphic to $\mathfrak{E}$ and rs is the dimensionality of $\mathfrak{A}$ over $\overline{\mathfrak{E}}$. Thus $\overline{\mathfrak{A}(\mathfrak{B})}_r = \overline{\mathfrak{E}}_s'$ and $\mathfrak{A}(\mathfrak{B}) = \mathfrak{E}_s'$ where $\mathfrak{E}' \cong \overline{\mathfrak{E}}'$.

We shall now determine $\mathfrak{A}(\mathfrak{A}(\mathfrak{B}))$. Evidently $\mathfrak{A}(\mathfrak{A}(\mathfrak{B}))$ contains $\mathfrak{B}$. On the other hand, if $c \in \mathfrak{A}(\mathfrak{A}(\mathfrak{B}))$, c_r is an $\overline{\mathfrak{E}}_s'$-endomorphism. Since the $\overline{\mathfrak{E}}_s'$-endomorphisms belong to $\overline{\mathfrak{E}}_r = \mathfrak{A}_l\mathfrak{B}_r$, $c_r \in \mathfrak{A}_l \times \overline{\mathfrak{B}}_r$. Since c_r commutes with the elements $\mathfrak{A}_l$, by Theorem 9, $c_r \in \overline{\mathfrak{B}}_r$. Hence $c \in \mathfrak{B}$ so that $\mathfrak{A}(\mathfrak{A}(\mathfrak{B})) = \mathfrak{B}$. This equation implies that $(\mathfrak{B} \wedge \mathfrak{A}(\mathfrak{B}))$ is the center of $\mathfrak{B}$ and the center of $\mathfrak{A}(\mathfrak{B})$.

Let $(\mathfrak{A}:\Phi) = n$, $(\mathfrak{E}:\Phi) = e$. Then $n = (\mathfrak{A}:\overline{\mathfrak{E}})(\mathfrak{E}:\Phi) = rse$. Since $\mathfrak{A}(\mathfrak{B}) = \mathfrak{E}_s'$, $(\mathfrak{A}(\mathfrak{B}):\Phi) = es^2$. Moreover, $\mathfrak{A}' \times \mathfrak{B} = \mathfrak{E}_r$ so that $(\mathfrak{A}' \times \mathfrak{B}:\Phi) = n(\mathfrak{B}:\Phi) = er^2$.

Hence

$$n(\mathfrak{B}:\Phi)(\mathfrak{A}(\mathfrak{B}):\Phi) = e^2r^2s^2 = n^2 = (\mathfrak{A}:\Phi)^2$$

and $(\mathfrak{B}:\Phi)(\mathfrak{A}(\mathfrak{B}):\Phi) = (\mathfrak{A}:\Phi)$.

THEOREM 19. *Let $\mathfrak{A}$ be a central simple algebra and $\mathfrak{B}$ a simple subalgebra containing* 1. *Then if $\mathfrak{A}(\mathfrak{B})$ is the centralizer of $\mathfrak{B}$ and $\mathfrak{A}'$ is the algebra anti-isomorphic to $\mathfrak{A}$, the following statements hold:*

1. *$\mathfrak{A}(\mathfrak{B})$ is simple and contains* 1.
2. *$\mathfrak{A}(\mathfrak{A}(\mathfrak{B})) = \mathfrak{B}$.*
3. *If $\mathfrak{B} \times \mathfrak{A}' = \mathfrak{E}_r$ where $\mathfrak{E}$ is a division algebra, then $\mathfrak{A}(\mathfrak{B}) = \mathfrak{E}'_s$, $\mathfrak{E}'$ anti-isomorphic to $\mathfrak{E}$.*
4. *$(\mathfrak{A}:\Phi) = (\mathfrak{B}:\Phi)(\mathfrak{A}(\mathfrak{B}):\Phi)$.*

12. Subfields and splitting fields. We now let $\mathfrak{B} = \mathfrak{F}$, a field, in the above theorem. Then $\mathfrak{A}(\mathfrak{F}) \geqq \mathfrak{F}$ so that $(\mathfrak{A}:\Phi) = (\mathfrak{F}:\Phi)(\mathfrak{A}(\mathfrak{F}):\Phi) \geqq (\mathfrak{F}:\Phi)^2$. We assume next that $\mathfrak{A} = \mathfrak{D}$ is a central division algebra. Then we may embed $\mathfrak{F}$ in a field $\bar{\mathfrak{F}}$ such that $\mathfrak{D}(\bar{\mathfrak{F}}) = \bar{\mathfrak{F}}$. For if $\mathfrak{D}(\mathfrak{F}) > \mathfrak{F}$, we may choose an element b in $\mathfrak{D}(\mathfrak{F})$ not in $\mathfrak{F}$ and obtain the field $\mathfrak{F}_1 = \mathfrak{F}(b)$ properly containing $\mathfrak{F}$. If $\mathfrak{D}(\mathfrak{F}_1) > \mathfrak{F}_1$, we may repeat this process. Eventually we obtain a field $\bar{\mathfrak{F}}$ with the required property. Our argument shows also that if $\mathfrak{D}(\mathfrak{F}) > \mathfrak{F}$, $\mathfrak{F}$ is not a maximal subfield. Conversely if $\mathfrak{F}$ is not maximal, then $\mathfrak{F} < \mathfrak{F}_1$ a larger field and hence $\mathfrak{D}(\mathfrak{F}) > \mathfrak{F}$. Since $\mathfrak{D}(\bar{\mathfrak{F}}) = \bar{\mathfrak{F}}$, $(\mathfrak{D}:\Phi) = (\bar{\mathfrak{F}}:\Phi)^2$. This proves

THEOREM 20. *The dimensionality of any central division algebra $\mathfrak{D}$ is a square. If $(\mathfrak{D}:\Phi) = \delta^2$, then δ is the dimensionality of any maximal subfield of $\mathfrak{D}$.*

If $(\mathfrak{D}:\Phi) = \delta^2$, δ is the *degree* or the index of $\mathfrak{D}$ and if $\mathfrak{A} = \mathfrak{D}_m$, δ is the *index* of $\mathfrak{A}$. Evidently the dimensionality of $\mathfrak{A}$ is a square $n = (\delta m)^2$. Now if $\mathfrak{F}$ is a subfield of $\mathfrak{A}$ containing 1 such that $(\mathfrak{F}:\Phi) = \delta m$, then $\mathfrak{A}(\mathfrak{F}) = \mathfrak{F}$ and so $\mathfrak{F}$ is a maximal subfield of $\mathfrak{A}$.

We now apply Theorem 14 to $\mathfrak{B} = \mathfrak{F}$. According to this result $\mathfrak{F}' \times \mathfrak{D} = \mathfrak{E}'_{s'}$ where $\mathfrak{E}'$ is a division algebra and $q = \delta m$ is a factor of ms'. Thus $\delta \mid s'$ and $\mathfrak{F}' \times \mathfrak{A} = \mathfrak{E}'_{s'm}$ where $\delta m \mid s'm$. We have seen that the center of $\mathfrak{F}' \times \mathfrak{A}$ is $\mathfrak{F}'$ and that $\mathfrak{F}' \leqq \mathfrak{E}'$. Hence

$$((\mathfrak{F}' \times \mathfrak{A}):\Phi) = (\mathfrak{F}:\Phi)(\delta m)^2 = (\mathfrak{E}':\Phi)(s'm)^2.$$

Since $s' \geqq \delta$ and $(\mathfrak{E}':\Phi) \geqq (\mathfrak{F}:\Phi)$, it follows from the above equation that $s' = \delta$ and $(\mathfrak{E}':\Phi) = (\mathfrak{F}:\Phi)$. Hence $\mathfrak{E}' = \mathfrak{F}'$ and $\mathfrak{F}' \times \mathfrak{A} = \mathfrak{F}'_n$. Since $\mathfrak{F}$ is commutative, we may also write $\mathfrak{F} \times \mathfrak{A} = \mathfrak{A}_{\mathfrak{F}} = \mathfrak{F}_n$.

Now we shall call a field $\mathfrak{F}$ over Φ a *splitting field* for a central simple algebra $\mathfrak{A} = \mathfrak{D}_m$ if $\mathfrak{A}_{\mathfrak{F}} = \mathfrak{F}_n$. Since $(\mathfrak{F}_n)_{\mathfrak{K}} = \mathfrak{K}_n$ if $\mathfrak{K} \geqq \mathfrak{F}$, any extension field of a splitting field is a splitting field. If $\mathfrak{D}_{\mathfrak{F}} = \mathfrak{E}_s$, $\mathfrak{A}_{\mathfrak{F}} = (\mathfrak{D}_m)_{\mathfrak{F}} = \mathfrak{E}_{sm}$. Hence by the uniqueness part of Wedderburn's theorem, if $\mathfrak{F}$ is a splitting field for $\mathfrak{A}$, it is one for $\mathfrak{D}$. The converse, that $\mathfrak{F}$ splits $\mathfrak{A}$ if it splits $\mathfrak{D}$, is clear. For if $\mathfrak{D}_{\mathfrak{F}} = \mathfrak{F}_\delta$, then $\mathfrak{A}_{\mathfrak{F}} = \mathfrak{F}_{\delta m}$. The result that we obtained in the last paragraph is the sufficiency part of the following

THEOREM 21. *Necessary and sufficient conditions that a field* $\mathfrak{F}$ *be a splitting field of a central division algebra* $\mathfrak{D}$ *of degree* δ *are that* $f = (\mathfrak{F}:\Phi)$ *be a multiple* $m\delta$ *of* δ *and that* $\mathfrak{F}$ *be isomorphic to a subalgebra containing* 1 *of* $\mathfrak{D}_m$.

To prove the necessity of the conditions we use the Corollary to Theorem 11. Since $\mathfrak{D}_{\mathfrak{F}} = \mathfrak{F} \times \mathfrak{D} = \mathfrak{F}_{\mathfrak{D}} = \mathfrak{F}_\delta$, $f = m\delta$ and $\mathfrak{F}$ has an anti-representation in $\mathfrak{D}_m$. Since $\mathfrak{F}$ is a field, $\mathfrak{F}$ is isomorphic to a subfield containing 1 of $\mathfrak{D}_m$.

The existence of a splitting field of a central simple algebra implies

THEOREM 22. *If* $\mathfrak{A}$ *is central simple and* Γ *is any field over* Φ (*not necessarily of finite dimensionality*), *then* $\mathfrak{A}_\Gamma$ *is central simple.*

For let $\mathfrak{F}$ be a finite dimensional splitting field. There exists a field Σ containing $\mathfrak{F}$ and Γ.[7] Then $\mathfrak{A}_\Sigma = (\mathfrak{A}_\mathfrak{F})_\Sigma = \Sigma_n$. Hence $(\mathfrak{A}_\Gamma)_\Sigma = \Sigma_n$. Since the extension of any ideal in $\mathfrak{A}_\Gamma$ is an ideal in $(\mathfrak{A}_\Gamma)_\Sigma$, $\mathfrak{A}_\Gamma$ is simple. Similarly $\mathfrak{A}_\Gamma$ is central.

13. The Brauer group. We have seen that the direct product of any two central simple algebras is a central simple algebra. We shall consider now the structure of the direct product $\mathfrak{A}' \times \mathfrak{A}$ where $\mathfrak{A}$ is central simple and $\mathfrak{A}'$ is anti-isomorphic to $\mathfrak{A}$. For this purpose we apply Theorem 14 to the case where $\mathfrak{B} = \mathfrak{A}$. We then obtain that $\mathfrak{A}' \times \mathfrak{A} = \mathfrak{E}'_{s'm}$ where $n = (\mathfrak{A}:\Phi)$ is a factor of $s'm$. By comparing the dimensionalities over Φ, we see that $s'm = n$ and $\mathfrak{E}' = \Phi$. Hence we have proved

THEOREM 23. *If* $\mathfrak{A}$ *is a central simple algebra and* $\mathfrak{A}'$ *is the algebra anti-isomorphic to* $\mathfrak{A}$, *then* $\mathfrak{A}' \times \mathfrak{A} = \Phi_n$.

A second proof of this theorem that is more direct is the following: Let $\mathfrak{A}_r$ and $\mathfrak{A}_l$, respectively, denote the algebras of right and of left multiplications in $\mathfrak{A}$. Consider $\mathfrak{A}_r\mathfrak{A}_l = \mathfrak{A}_l\mathfrak{A}_r$. The elements of this algebra are linear transformations in $\mathfrak{A}$ regarded as an n-dimensional space over Φ. We note also that the algebra $\mathfrak{A}' \times \mathfrak{A}$ is homomorphic to $\mathfrak{A}_l\mathfrak{A}_r$ and since $\mathfrak{A}' \times \mathfrak{A}$ is simple, these algebras are isomorphic. It follows that $\mathfrak{A}_l\mathfrak{A}_r$ contains n^2 linearly independent elements. Hence $\mathfrak{A}_l\mathfrak{A}_r$ is isomorphic to Φ_n and this holds for $\mathfrak{A}' \times \mathfrak{A}$.

This result enables us to define a remarkable group first discovered by R. Brauer. We consider the set $\mathfrak{S}$ of central simple algebras over a fixed field Φ. Two elements $\mathfrak{A}$ and $\mathfrak{B}$ of $\mathfrak{S}$ are said to be *similar* ($\mathfrak{A} \sim \mathfrak{B}$) if their division algebras $\mathfrak{D}$, $\bar{\mathfrak{D}}$ in the representation $\mathfrak{A} = \mathfrak{D}_m$, $\mathfrak{B} = \bar{\mathfrak{D}}_{\bar{m}}$ are isomorphic. Since $\mathfrak{D}$ is determined in the sense of isomorphism by $\mathfrak{A}$, the relation of similarity is well-defined. Evidently this relation has the properties of an equivalence and hence it determines a decomposition of $\mathfrak{S}$ into non-overlapping sets $\{\mathfrak{A}\}$, $\{\mathfrak{B}\}$, $\cdots$. ($\{\mathfrak{A}\}$ denotes the set of algebras similar to a fixed $\mathfrak{A}$.) The elements of the Brauer group $\mathfrak{G}(\Phi)$ are the sets $\{\mathfrak{A}\}$. Multiplication is defined by $\{\mathfrak{A}\}\{\mathfrak{B}\} = \{\mathfrak{A} \times \mathfrak{B}\}$. This is single-valued. For if $\mathfrak{A} = \mathfrak{D}^{(1)}_{n_1} \sim \bar{\mathfrak{A}} = \bar{\mathfrak{D}}^{(1)}_{m_1}$ and $\mathfrak{B} = \mathfrak{D}^{(2)}_{n_2} \sim \bar{\mathfrak{B}} = \bar{\mathfrak{D}}^{(2)}_{m_2}$, then $\mathfrak{A} \times \mathfrak{B} = (\mathfrak{D}^{(1)} \times \mathfrak{D}^{(2)})_{n_1 n_2}$ and $\bar{\mathfrak{A}} \times \bar{\mathfrak{B}} = (\bar{\mathfrak{D}}^{(1)} \times \bar{\mathfrak{D}}^{(2)})_{m_1 m_2}$. The central simple algebras $\mathfrak{D}^{(1)} \times \mathfrak{D}^{(2)}$ and $\bar{\mathfrak{D}}^{(1)} \times \bar{\mathfrak{D}}^{(2)}$ are isomorphic and hence their division algebras are isomorphic. The class of

[7] More precisely, containing subfields isomorphic to $\mathfrak{F}$ and Γ.

matrix algebras ($\mathfrak{A} \sim 1$) acts as an identity in $\mathfrak{G}$. By the above theorem $\{\mathfrak{A}\}\{\mathfrak{A}'\} = \{1\}$ so that $\{\mathfrak{A}'\} = \{\mathfrak{A}\}^{-1}$. Since direct multiplication is commutative and associative, $\mathfrak{G}(\Phi)$ is a commutative group.

If P is a field over Φ the mapping $\{\mathfrak{A}\} \rightarrow \{\mathfrak{A}_P\}$ is a homomorphism between $\mathfrak{G}(\Phi)$ and a subgroup of $\mathfrak{G}(P)$. For $(\mathfrak{A} \times \mathfrak{B})_P = \mathfrak{A}_P \times \mathfrak{B}_P$.

Our principal objective in **14–16** is the theorem that every element of $\mathfrak{G}(\Phi)$ has finite order. Thus far we have had to refer only to results obtained elsewhere in this book. However, we must now call on a part of the theory of commutative fields. In particular we shall use the results of **6** which have until now served only as illustrations of the theory of direct products.

14. Separable subfields. We suppose that Φ has characteristic $p \neq 0$ and that $\mathfrak{D}$ is a central division algebra of degree p. If a is any element of $\mathfrak{D}$ not in Φ, $\Phi(a)$ is a subfield of dimensionality p over Φ, and if b is an element of $\mathfrak{D}$ not in $\Phi(a)$, then the algebra generated by a and b is $\mathfrak{D}$. Now suppose that a is not separable over Φ so that $a^p = \alpha \in \Phi$. Consider the derivation $x \rightarrow x' = [x, a]$. If $x \notin \Phi(a)$, $x' \neq 0$. However, $x^{(p)} = [x, a^p] = 0$. Hence there is a $k \geqq 1$ such that $x^{(k)} \neq 0$ but $x^{(k+1)} = 0$. If we set $b = x^{(k-1)}(x^{(k)})^{-1}$, we obtain $b' = 1$ and if we set $c = ab$, then $c' = a$. Thus $[c^p, a] = \overbrace{[c[\cdots[c, a]\cdots]]}^{p} = a$. Hence $[c^p - c, a] = 0$ and $c^p = c + g(a)$ where $g(a) \in \Phi(a)$. Evidently $g(a)$ commutes with c and with a and hence $g(a) = \gamma \in \Phi$, and c is a separable element.

LEMMA. *If $\mathfrak{D}$ is a central division algebra of degree p and characteristic p and if $\mathfrak{D}$ contains an inseparable element over Φ, then $\mathfrak{D}$ also contains a separable element not in Φ.*

We note that the element b satisfies an equation of the form $b^p = \beta$ since $[b^p, a] = 0$. The elements $b^i a^j$, $i, j = 0, \cdots, p - 1$, form a basis for $\mathfrak{D}$ and the following relations determine the multiplication:

$$a^p = \alpha, \qquad b^p = \beta, \qquad ba - ab = 1.$$

Similarly, we may use a and c as generators with the following relations

$$a^p = \alpha, \qquad c^p = c + \gamma, \qquad a^{-1}ca = c + 1.$$

Since $c + 1, c + 2, \cdots, c + (p - 1)$ satisfy the equation $t^p = t + \gamma$, $\Phi(c)$ is a cyclic field over Φ with the generating automorphism $c \rightarrow c + 1$.

The above lemma may be used to prove the following

THEOREM 24. *Any central division algebra $\mathfrak{D}$ over Φ contains a maximal separable subfield.*

If Φ has characteristic 0, there is nothing to prove. Hence we suppose that Φ has characteristic $p \neq 0$. Let a_1 in $\mathfrak{D}$ be separable and $(\Phi(a_1):\Phi) = r_1 > 0$. If $\mathfrak{B}$ is the algebra of elements commutative with those of $\Phi(a_1)$, then $(\mathfrak{D}:\Phi) = \delta^2 = (\Phi(a_1):\Phi)(\mathfrak{B}:\Phi) = r_1 b$ for $b = (\mathfrak{B}:\Phi)$. The field $\Phi(a_1)$ is the center of $\mathfrak{B}$. If $\mathfrak{B}$ contains an element a_2 separable over $\Phi(a_1)$ and $(\Phi(a_1, a_2):\Phi(a_1)) = r_2 > 0$, then $\Phi(a_1, a_2)$ is separable of dimensionality $r_1 r_2$ over Φ. This process leads to a maximal subfield separable over Φ or to a central division algebra whose subfields, properly containing the center, are all purely inseparable. Let $\mathfrak{D}$ be such an algebra and Φ its center. If $\mathfrak{F} = \Phi(a_1, a_2, \cdots, a_k)$ is a maximal subfield of $\mathfrak{D}$, each a_i satisfies an equation of the form $a_i^{p^{m_i}} = \alpha_i$. Hence $\mathfrak{F}$ contains a subfield $\mathfrak{F}_0$ such that $(\mathfrak{F}:\mathfrak{F}_0) = p$ and $\mathfrak{F} = \mathfrak{F}_0(a)$, $a^p \in \mathfrak{F}_0$. The elements commutative with those of $\mathfrak{F}_0$ form a central division algebra $\mathfrak{B}$ of degree p over $\mathfrak{F}_0$. Since $\mathfrak{B}$ contains the element a such that $a \notin \mathfrak{F}_0$ but $a^p \in \mathfrak{F}_0$, it follows that $\mathfrak{B}$ contains an element c such that $c^p - c = g(a) \in \mathfrak{F}_0$ but $c \notin \mathfrak{F}_0$. Then

$$(c^p - c)^{p^m} = c^{p^{m+1}} - c^{p^m} = (c^{p^m})^p - (c^{p^m}) = g(a)^{p^m}$$

is in Φ if m is sufficiently large. Hence c^{p^m} is separable over Φ and $c^{p^m} \notin \Phi$ since $c^{p^m} = c^{p^{m-1}} + g(a)^{p^{m-1}}$, $c^{p^{m-1}} = c^{p^{m-2}} + g(a)^{p^{m-2}}, \cdots$ implies that $\Phi(c^{p^m}, \mathfrak{F}_0) = \Phi(c, \mathfrak{F}_0) > \mathfrak{F}_0$. Thus the assumption that a $\mathfrak{D}$ possessing only purely inseparable subfields exists leads to a contradiction, and hence the theorem is proved.

15. Crossed products. If $\mathfrak{D}$ is a central division algebra of degree δ, let $\mathfrak{F}$ be a maximal separable subfield of $\mathfrak{D}$. Then $\mathfrak{F}$ may be extended to a field $\mathfrak{K}$ normal, separable and of dimensionality $\nu = \delta m$ over Φ. We have seen that $\mathfrak{K}$ is a splitting field and hence is contained in $\mathfrak{D}_m$, a central simple algebra similar to $\mathfrak{D}$. Furthermore, $\mathfrak{K}$ is a maximal subfield of $\mathfrak{D}_m$. Let $1, S, \cdots, V$ be the elements of the Galois group $\mathfrak{G}$ of $\mathfrak{K}$ over Φ. Since the automorphism $k \to k^S$ in $\mathfrak{K}$ may be extended to an inner automorphism in $\mathfrak{D}_m$, there is a non-singular element u_S in $\mathfrak{D}_m$ such that $u_S^{-1}ku_S = k^S$, or $ku_S = u_Sk^S$ for all k in $\mathfrak{K}$. The element $u_{ST}^{-1}u_Su_T$ commutes with all k and hence $u_Su_T = u_{ST}\rho_{S,T}$, ρ in $\mathfrak{K}$. By the associative law we obtain

$$\rho_{S,TU}\rho_{T,U} = \rho_{ST,U}\rho_{S,T}^{U}$$

so that $\rho = \{\rho_{S,T}\}$ is a factor set. Consider now the crossed product $\mathfrak{K}(\mathfrak{G}, \rho)$ of $\mathfrak{K}$ with its Galois group $\mathfrak{G}$, and having the factor set ρ.[8] Evidently it is homomorphic to the subalgebra $\mathfrak{B}$ of $\mathfrak{D}_m$ consisting of the elements Σu_Sk_S. Since $\mathfrak{K}(\mathfrak{G}, \rho)$ is simple, $\mathfrak{K}(\mathfrak{G}, \rho)$ is isomorphic to $\mathfrak{B}$ and since $(\mathfrak{K}(\mathfrak{G}, \rho):\Phi) = \nu^2$, $\mathfrak{B} = \mathfrak{D}_m$.

THEOREM 25. *Any central simple algebra is similar to a crossed product* $\mathfrak{K}(\mathfrak{G}, \rho)$.

This theorem enables us to apply the theory of crossed products to that of central simple algebras. We recall the definition that the factor sets ρ and σ are associates if there exist elements μ_S in $\mathfrak{K}$ such that $\rho_{S,T} = \sigma_{S,T}\dfrac{\mu_{ST}}{\mu_S^T\,\mu_T}$. Thus if v_S is a second set of elements in $\mathfrak{D}_m = \mathfrak{K}(\mathfrak{G}, \rho)$ such that $v_S^{-1}kv_S = k^S$, then $u_S^{-1}v_S \in \mathfrak{K}$ and $v_S = u_S\mu_S$. Then if $v_Sv_T = v_{ST}\sigma_{S,T}$, ρ and σ are associates $(\rho \sim \sigma)$. With these definitions we have the following

THEOREM 26. $\mathfrak{K}(\mathfrak{G}, \rho) \sim 1$ *if and only if* $\rho \sim 1$.

[8] Since the correspondence is the identity, we may use this simplification of the notation of Chapter 4.

If the conditions are satisfied, we replace u_S by elements $v_S = u_S\mu_S$ to obtain $v_S v_T = v_{ST}$. It follows from **18**, Chapter 4, that $\mathfrak{R}(\mathfrak{G}, \rho)$ is isomorphic to Φ_ν. Conversely, suppose that $\mathfrak{R}(\mathfrak{G}, \rho) \sim 1$. Then $\mathfrak{R}(\mathfrak{G}, \rho)$ is isomorphic to $\bar{\mathfrak{R}}(\bar{\mathfrak{G}}, 1)$ and hence $\mathfrak{R}(\mathfrak{G}, \rho)$ contains a field $\mathfrak{R}_1$ isomorphic to $\mathfrak{R}$ and elements v_{S_1} such that every element has the form $\Sigma v_{S_1} k_{S_1}^{(1)}$, $k^{(1)}$ in $\mathfrak{R}_1$ and

$$k^{(1)} v_{S_1} = v_{S_1} k^{(1)S_1}, \qquad v_{S_1} v_{T_1} = v_{S_1 T_1}, \tag{4}$$

S_1 in the Galois group of $\mathfrak{R}_1$. We may suppose that S_1 is the automorphism $k_1 \to (k^S)_1$ where $k \to k_1$ is a particular isomorphism between $\mathfrak{R}$ and $\mathfrak{R}_1$. This isomorphism may be extended to an automorphism $a \to a_1$ in $\mathfrak{R}(\mathfrak{G}, \rho)$. If $u_{S_1} = (u_S)_1$, we have

$$k^{(1)} u_{S_1} = u_{S_1} k^{(1)S_1}, \qquad u_{S_1} u_{T_1} = u_{S_1 T_1} \rho_{S,T}^{(1)}.$$

If we compare with (4), we obtain $\rho^{(1)} \sim 1$ and hence $\rho \sim 1$.

We consider now two crossed products $\mathfrak{R}_1(\mathfrak{G}_1, \rho_1)$ and $\mathfrak{R}_2(\mathfrak{G}_2, \sigma_2)$ where $\mathfrak{R}_1 \cong \mathfrak{R}_2$ and, say, $k_1 \to k_2$ is an isomorphism between $\mathfrak{R}_1$ and $\mathfrak{R}_2$. Let S_1 and S_2 be the corresponding automorphisms of the Galois groups $\mathfrak{G}_1$ and $\mathfrak{G}_2$ in the sense that $k_1^{S_1} \to k_2^{S_2}$. We wish to obtain the form of $\mathfrak{R}_1(\mathfrak{G}_1, \rho_1) \times \mathfrak{R}_2(\mathfrak{G}_2, \sigma_2)$.

Evidently $\mathfrak{R}_1(\mathfrak{G}_1, \rho_1) \times \mathfrak{R}_2(\mathfrak{G}_2, \sigma_2)$ contains $\mathfrak{R}_1 \times \mathfrak{R}_2$. The latter is a direct sum of ν fields isomorphic to $\mathfrak{R}_1$. If e_i is the identity of one of the components, the elements of the component have the form $e_i k_2$, k_2 in $\mathfrak{R}_2$, i.e. $\mathfrak{R}_1 \times \mathfrak{R}_2 = e_1\mathfrak{R}_2 \oplus \cdots \oplus e_\nu\mathfrak{R}_2$. Similarly $e_i\mathfrak{R}_2 = e_i\mathfrak{R}_1$. The correspondence $k_1 \to k_2^{(i)}$ obtained by writing $e_i k_1 = e_i k_2^{(i)}$ is an (anti-) representation of $\mathfrak{R}_1$ in $\mathfrak{R}_2$ and we have seen that we obtain in this way all of the irreducible representations (in the sense of similarity, which in this case means identity) of $\mathfrak{R}_1$ in $\mathfrak{R}_2$. On the other hand, we have the ν distinct representations $k_1 \to k_2^{S_2}$ where S_2 ranges over $\mathfrak{G}_2$. Hence $k_2^{(i)} = k_2^{S_2}$ and the idempotent elements are in $(1 - 1)$ correspondence with the elements of $\mathfrak{G}_2$. We may therefore denote e_i as e_{S_2} and note that

$$\Sigma e_{S_2} = 1, \qquad e_{S_2} e_{T_2} = 0 \text{ if } S_2 \neq T_2, \qquad e_{S_2}^2 = e_{S_2},$$
$$e_{S_2}(k_1 - k_2^{S_2}) = 0.$$

The mappings $x \to v_{T_2}^{-1} x v_{T_2}$ and $x \to u_{T_1}^{-1} x u_{T_1}$ are automorphisms of $\mathfrak{R}_1 \times \mathfrak{R}_2$. Since the simple components of a semi-simple algebra are uniquely determined, the elements $v_{T_2}^{-1} e_{S_2} v_{T_2}$ and $u_{T_1}^{-1} e_{S_2} u_{T_1}$ are again e's. Since v_{T_2} commutes with k_1 and $v_{T_2}^{-1} k_2 v_{T_2} = k_2^{T_2}$,

$$v_{T_2}^{-1} e_{S_2} v_{T_2}(k_1 - k_2^{S_2 T_2}) = 0.$$

Hence $v_{T_2}^{-1} e_{S_2} v_{T_2} = e_{S_2 T_2}$ and similarly, $u_{T_1}^{-1} e_{S_2} u_{T_1} = e_{T_2^{-1} S_2}$.

We now define $e_{S,T} = v_{S_2}^{-1} v_{T_2} e_{T_2}$ and verify that

$$e_{S,T} e_{U,V} = \delta_{T,U} e_{S,V}, \qquad \Sigma e_{S,S} = 1.$$

These matrix units may be used to write $\mathfrak{R}_1(\mathfrak{G}_1, \rho_1) \times \mathfrak{R}_2(\mathfrak{G}_2, \sigma_2) = \mathfrak{B}_\nu$ where $\mathfrak{B}$ is the set of elements commutative with the $e_{S,T}$ and is isomorphic to $e_{1,1}(\mathfrak{K}_1(\mathfrak{G}_1, \rho_1) \times \mathfrak{K}_2(\mathfrak{G}_2, \sigma_2))e_{1,1} = \bar{\mathfrak{B}}$. By the relations noted above $e_{1,1}u_{S_1}v_{S_2} = u_{S_1}v_{S_2}e_{1,1} = \bar{w}_S \in \bar{\mathfrak{B}}$ and $\bar{k}_1 \equiv k_1e_{1,1} = k_2e_{1,1} \equiv \bar{k}_2 \in \bar{\mathfrak{B}}$. Then

$$\bar{w}_S\bar{w}_T = \bar{w}_{ST}\bar{\rho}_{1S_1,T_1}\bar{\sigma}_{2S_2,T_2}, \qquad \bar{k}\bar{w}_S = \bar{w}_S\bar{k}^S$$

where $\bar{k} \equiv \bar{k}_1$ and $\bar{k}^S = \overline{k_1^{S_1}}$. We have shown that $\bar{\mathfrak{B}}$ contains the crossed product $\bar{\mathfrak{K}}(\bar{\mathfrak{G}}, \bar{\rho}_1\bar{\sigma}_2)$ and since its dimensionality over Φ is ν^2, $\bar{\mathfrak{B}}$ coincides with the crossed product. Hence we have proved the following

THEOREM 27. *Let $\mathfrak{K}_1$ and $\mathfrak{K}_2$ be isomorphic separable normal fields over Φ and $k_1 \to k_2$ an isomorphism between them. Then $\mathfrak{K}_1(\mathfrak{G}_1, \rho_1) \times \mathfrak{K}_2(\mathfrak{G}_2, \sigma_2) \sim \bar{\mathfrak{K}}(\bar{\mathfrak{G}}, \bar{\rho}_1\bar{\sigma}_2)$ where $\bar{\mathfrak{K}}$ is isomorphic to the $\mathfrak{K}_i$ and $k_i \to \bar{k}_i$ is an isomorphism between $\mathfrak{K}_i$ and $\bar{\mathfrak{K}}$ such that $\bar{k}_1 = \bar{k}_2$.*

This theorem has the following significance: Let $\rho_{S,T}$ and $\sigma_{S,T}$ be factor sets and define $\tau_{S,T} = \rho_{S,T}\sigma_{S,T}$ as the product factor set. The set of factor sets forms a commutative group relative to this multiplication. The factor sets of the form $\rho_{S,T} = \dfrac{\mu_{ST}}{\mu_S^T \mu_T}$ form a subgroup and two factor sets that belong to the same coset relative to this subgroup are associates. If $\mathfrak{H}_{\mathfrak{K}}$ is the factor group whose elements are the classes of associate factor sets and $\mathfrak{G}_{\mathfrak{K}}(\Phi)$ the subgroup of the Brauer group over Φ consisting of those classes of central simple algebras that have $\mathfrak{K}$ as a splitting field, then, by our theorem, $\mathfrak{H}_{\mathfrak{K}}$ and $\mathfrak{G}_{\mathfrak{K}}(\Phi)$ are isomorphic. We prove next

THEOREM 28. *If $\mathfrak{K}(\mathfrak{G}, \rho)$ has index δ, then $\rho^\delta \sim 1$.*

Let $\mathfrak{K}(\mathfrak{G}, \rho) = \mathfrak{D}_m$ where $\mathfrak{D}$ is a central division algebra. Then $\mathfrak{D}_m = \mathfrak{J}_1 \oplus \cdots \oplus \mathfrak{J}_m$ where the $\mathfrak{J}$'s are isomorphic irreducible right ideals and hence have the same dimensionality when regarded as vector spaces over $\mathfrak{K}$. Since $(\mathfrak{D}_m:\mathfrak{K}) = \nu$ and $(\mathfrak{D}_m:\Phi) = \nu^2 = \delta^2m^2$, $(\mathfrak{J}:\mathfrak{K})m = \nu$ and $(\mathfrak{J}:\mathfrak{K}) = \delta$. The elements u_S define semi-linear transformations in $\mathfrak{J}$ over $\mathfrak{K}$. Hence if $x_1, \cdots, x_\delta$ is a basis for $\mathfrak{J}$ over $\mathfrak{K}$ and $x_iu_S = \Sigma x_j\mu_{jiS}$, μ in $\mathfrak{K}$, then the matrices $M_S = (\mu_{jiS})$ satisfy the equations $M_TM_S^T = M_{ST}\rho_{S,T}$. If we set $\det M_S = \mu_S$, we obtain $\rho_{S,T}^\delta\mu_{S,T} = \mu_S^T\mu_T$. Hence $\rho^\delta \sim 1$.

16. The exponent of a central simple algebra. The results of the last section imply

THEOREM 29. *If $\mathfrak{A}$ is a central simple algebra of index δ, then $\{\mathfrak{A}\}^\delta = 1$, i.e. the direct product of δ algebras isomorphic to $\mathfrak{A}$ is of the form $\Phi_{m\delta}$.*

For we have seen that $\mathfrak{A} \sim \mathfrak{K}(\mathfrak{G}, \rho)$ and $\mathfrak{A}_1 \times \cdots \times \mathfrak{A}_\delta \sim \mathfrak{K}(\mathfrak{G}, \rho^\delta)$ if $\mathfrak{A}_i \cong \mathfrak{A}$. Since $\rho^\delta \sim 1$, $\mathfrak{A}_1 \times \cdots \times \mathfrak{A}_\delta \sim 1$.

Thus the order of each element of the Brauer group is finite. If e is the order of $\{\mathfrak{A}\}$, we call e the *exponent* of the central simple algebra $\mathfrak{A}$. The above theorem implies that the exponent is a divisor of the index of $\mathfrak{A}$.

THEOREM 30. *If p is a prime factor of the index of $\mathfrak{A}$, then p is a divisor of the exponent of $\mathfrak{A}$.*

If $\mathfrak{F}$ is a field containing Φ, we have seen that the correspondence between $\{\mathfrak{A}\}$ and $\{\mathfrak{A}_\mathfrak{F}\}$ is a homomorphism between the Brauer group over Φ and a subgroup of the Brauer group over $\mathfrak{F}$. Hence the exponent $e_\mathfrak{F}$ of $\mathfrak{A}_\mathfrak{F}$ is a divisor of that of $\mathfrak{A}$. Now let $\mathfrak{A} \sim \mathfrak{K}(\mathfrak{G}, \rho) = \mathfrak{D}_m$ where $(\mathfrak{K}:\Phi) = \nu = \delta_m$. Let p^s be the highest power of p dividing ν and $\mathfrak{G}_p$ a Sylow subgroup of order p^s of $\mathfrak{G}$. Corresponding to $\mathfrak{G}_p$ there is a subfield $\mathfrak{F}$ of $\mathfrak{K}$ such that $(\mathfrak{K}:\mathfrak{F}) = p^s$. Consider $\mathfrak{K}(\mathfrak{G}, \rho)_{\mathfrak{F}_1}$ where $\mathfrak{F}_1 \cong \mathfrak{F}$. Since $\mathfrak{K}(\mathfrak{G}, \rho)_{\mathfrak{F}_1}$ has the splitting field $\mathfrak{K}_1 \cong \mathfrak{K}$ and $(\mathfrak{K}_1:\mathfrak{F}_1) = p^s$, the degree of $\mathfrak{K}(\mathfrak{G}, \rho)_{\mathfrak{F}_1}$ is p^t, $t \leq s$, and hence $e_{\mathfrak{F}_1} = p^u$ where $u \leq t$. Thus $e \equiv 0(e_{\mathfrak{F}_1})$, $\equiv 0(p)$, unless $u = 0$. Now, if $u = 0$, $\mathfrak{K}(\mathfrak{G}, \rho)_{\mathfrak{F}_1} \sim 1$ and hence $(\mathfrak{F}_1:\Phi)$ is divisible by δ. Since $(\mathfrak{F}_1:\Phi) = \dfrac{\nu}{p^s}$ and $\left(\dfrac{\nu}{p^s}, p\right) = 1$, this is impossible.

We prove finally the following theorem which in most considerations of central division algebras yields a reduction to the case of prime power degree.

Theorem 31. *If $\mathfrak{D}$ is a central division algebra of degree $\delta = p_1^{s_1} \cdots p_l^{s_l}$ where the p_i are distinct primes, then $\mathfrak{D} = \mathfrak{D}_1 \times \cdots \times \mathfrak{D}_l$ where $\mathfrak{D}_i$ has degree $p_i^{s_i}$ and is uniquely determined in the sense of isomorphism by $\mathfrak{D}$.*

Let $e = p_1^{t_1} \cdots p_l^{t_l}$, $0 < t_i \leq s_i$, be the exponent of $\mathfrak{D}$. By the usual group theoretic argument, $\{\mathfrak{D}\} = \{\mathfrak{D}_1\} \cdots \{\mathfrak{D}_l\}$ where $\{\mathfrak{D}_i\}^{p^{t_i}} = 1$. We may suppose that $\mathfrak{D}_i$ is a division algebra. Then its degree is $p_i^{s_i'}$, $s_i' \geq t_i$. Since the degrees of the $\mathfrak{D}_i$ are relatively prime, the direct product $\mathfrak{D}_1 \times \cdots \times \mathfrak{D}_l$ is a division algebra and since it is similar to the division algebra $\mathfrak{D}$, $\mathfrak{D} \cong \mathfrak{D}_1 \times \cdots \times \mathfrak{D}_l$ and $s_i' = s_i$. Now if $\mathfrak{D} = \mathfrak{D}_1 \times \cdots \times \mathfrak{D}_l = \mathfrak{E}_1 \times \cdots \times \mathfrak{E}_l$ where $\mathfrak{E}_i$ has degree $p_i^{s_i}$, then $\mathfrak{D}_i^{q_i} \sim \mathfrak{E}_i^{q_i}$ if $q_i = e p_i^{-t_i}$. Since $(q_i, p_i^{t_i}) = 1$, there exist integers a_i, b_i such that $q_i a_i + p_i^{t_i} b_i = 1$. Then $(\mathfrak{D}_i^{q_i})^{a_i} \sim (\mathfrak{E}_i^{q_i})^{a_i}$, $\mathfrak{D}_i \sim \mathfrak{E}_i$ and since both are division algebras, $\mathfrak{D}_i \cong \mathfrak{E}_i$.

17. Central division algebras over special fields. If a is any element of an algebra $\mathfrak{A}$, a satisfies an equation $\varphi(a) = 0$ where $\varphi(t)$ is a polynomial $\neq 0$ in $\Phi[t]$. For let $a^0 = 1$ if $\mathfrak{A}$ has an identity and $a^0 = 0$ otherwise, and consider the sequence $a^0, a^1, a^2, \cdots$. There are only a finite number of linearly independent elements in this sequence. Hence there exists an m, $0 < m \leq n$ the dimensionality of $\mathfrak{A}$, such that $a^m = a^{m-1}\alpha_1 + \cdots + a^0\alpha_m$. Thus $\varphi(a) = 0$ for $\varphi(t) = t^m - t^{m-1}\alpha_1 - \cdots - t^0\alpha_m$ where $t^0 = 1$ or 0 according as $a^0 = 1$ or 0. Now let m be minimal. Then the a^i with $i < m$ are linearly independent and so the α_i used in expressing a^m in terms of the a^i, $i < m$, are uniquely determined. It follows that the corresponding polynomial $\varphi(t) \equiv \mu_a(t)$ is the only polynomial of degree m with leading coefficient 1 having a as a root. Moreover, it is clear that m is the least degree for the polynomials $\varphi(t) \neq 0$ such that $\varphi(a) = 0$. By the division process we may show also that $\mu_a(t)$ is a factor of any $\varphi(t)$ such that $\varphi(a) = 0$. We shall call $\mu_a(t)$ the *minimum polynomial of a*.

If $\mathfrak{A}$ is a division algebra, $\mu_a(t)$ is irreducible. For if $\mu_a(t) = \mu^{(1)}(t)\mu^{(2)}(t)$, then $\mu^{(1)}(a)\mu^{(2)}(a) = 0$ and so either $\mu^{(1)}(a) = 0$ or $\mu^{(2)}(a) = 0$. Because of the minimality of the degree of $\mu_a(t)$ either $\mu^{(1)}(t)$ or $\mu^{(2)}(t)$ is of degree 0. Now if

Φ is algebraically closed, the irreducible polynomials are all linear and so for any a, we have $a - 1\alpha = 0$ or $a = 1\alpha$. This proves

THEOREM 32. *The only division algebra over an algebraically closed field Φ is Φ itself.*

Suppose that Φ is a real closed field. It is well known that the only algebraic extensions of Φ are Φ itself and $\Phi(i)$, $i^2 = -1$.[9] Let $\mathfrak{A} \neq \Phi$ be a central division algebra over Φ. If $(\mathfrak{A}:\Phi) = m^2$, $m > 1$, there is a maximal subfield Σ of $\mathfrak{A}$ such that $(\Sigma:\Phi) = m$. Hence $\Sigma = \Phi(i)$ and $m = 2$. Since $\Phi(i)$ is normal, $\mathfrak{A}$ is a crossed product and so there is a second element j in $\mathfrak{A}$ such that $j^{-1}ij = -i$, $j^2 = 1\beta$. The element β is negative and j may be normalized so that $j^2 = -1$. Hence $\mathfrak{A}$ has the basis $1, i, j, k = ij$ with

$$i^2 = -1, \qquad j^2 = -1, \qquad ij = -ji,$$

and $\mathfrak{A}$ is Hamilton's quaternion algebra. As is well known, an $\mathfrak{A}$ of this form is a division algebra. If $\mathfrak{A}$ is a division algebra over Φ that is not central, the center of $\mathfrak{A}$ is the algebraically closed field $\Phi(i)$. Hence by Theorem 32, $\mathfrak{A} = \Phi(i)$.

THEOREM 33 (Frobenius). *The only division algebras over a real closed field Φ are Φ, $\Phi(i)$ and the quaternion algebra $\Phi(i, j)$.*

Now let Φ be a finite field and $\mathfrak{A}$ a central division algebra over Φ. We denote the multiplicative group of elements $\neq 0$ of $\mathfrak{A}$ by $\mathfrak{A}'$. If Σ is a maximal subfield, Σ', the set of elements $\neq 0$ of Σ, is a subgroup of $\mathfrak{A}'$. Any element $b \neq 0$ may be embedded in a maximal subfield and since any maximal subfield has the form $u^{-1}\Sigma u$, $b \in u^{-1}\Sigma' u$ for a suitable u. Thus $\mathfrak{A}'$ is a sum of subgroups conjugate to Σ'. Now the conjugates of a subgroup of a finite group include all the elements of the group only if the subgroup is the entire group. Hence $\Sigma' = \mathfrak{A}'$ and $\mathfrak{A}$ is commutative. Thus $\mathfrak{A} = \Phi$.

THEOREM 34 (Wedderburn). *The only central division algebra over a finite field Φ is Φ itself.*

This, of course, means that every division algebra over a finite field is commutative. Moreover, since any division ring may be regarded as an algebra over its center, this theorem holds also for arbitrary finite division rings.

18. Minimum polynomial of an algebra. In the remainder of this chapter we consider algebras that are not necessarily simple. We shall define a special class of semi-simple algebras called separable, and shall give a constructive criterion for an algebra to belong to this class. If Φ has characteristic 0, every semi-simple algebra is separable so that in this case our criterion will be one for semi-simplicity. We shall also obtain a structure theorem, due to Wedderburn, which to a certain extent reduces the study of arbitrary algebras to that of semi-simple and of nilpotent algebras.

We consider first the theory of the minimum polynomial of an element a of

[9] See van der Waerden's *Moderne Algebra*, vol. 1 p. 228 or 2nd. ed. p. 237.

an arbitrary algebra $\mathfrak{A}$. Suppose that we have a $(1-1)$ representation $x \to X$ of $\mathfrak{A}$ by matrices in the matrix algebra Φ_N, such that if $\mathfrak{A}$ has an identity, then $1 \to 1$ the identity of Φ_N.[10] Let $\bar{\mathfrak{A}}$ denote the set of matrices representing $\mathfrak{A}$ and let A be the matrix corresponding to a. Then we assert that the minimum polynomial $\mu_a(t)$ of a is the minimum polynomial of the matrix A regarded as an element of the algebra Φ_N. For it is clear that $\mu_a(t)$ is the minimum polynomial of A regarded as an element of $\bar{\mathfrak{A}}$. Our assumptions imply that if $\bar{\mathfrak{A}}$ has an identity, then this identity is the identity of Φ_N. Hence, in this case, the minimum polynomial in $\bar{\mathfrak{A}}$ of A is the same as its minimum polynomial in Φ_N. Now suppose that $\bar{\mathfrak{A}}$ does not have an identity and let $\mu_A(t) = t^m - t^{m-1}\alpha_1 - \cdots - 1\alpha_{m-1}$ be the minimum polynomial of A in Φ_N. Then $1\alpha_m = A^m - A^{m-1}\alpha_1 - \cdots - A\alpha_{m-1}$ belongs to $\bar{\mathfrak{A}}$ and must therefore be 0. Thus the constant term of $\mu_A(t)$ is 0 and hence $\mu_A(t)$ is the minimum polynomial in $\bar{\mathfrak{A}}$ of A. Now we recall that the minimum polynomial of a matrix A is the last invariant factor of the matrix $(1t - A)$ belonging to $\Phi[t]_N$. Hence $\mu_a(t)$ is the last invariant factor of $(1t - A)$, and if $f(t)$ is the characteristic polynomial det $(1t - A)$, then $\mu_a(t)$ is a factor of $f(t)$. Since $f(t)$ is the product of all of the invariant factors, and each invariant factor is a factor of the last one, any irreducible factor of $f(t)$ is a factor of $\mu_a(t)$.

Let $x_1, \cdots, x_n$ be a basis for $\mathfrak{A}$ over Φ, and set $\mathrm{P} = \Phi(\xi_1, \cdots, \xi_n)$ the field obtained from Φ by adjoining the indeterminates ξ_i. We form the algebra $\mathfrak{A}_\mathrm{P}$ and shall call the element $x_1\xi_1 + \cdots + x_n\xi_n$ of $\mathfrak{A}_\mathrm{P}$ a *general element* of $\mathfrak{A}$. Now suppose that $x_i \to X_i$ in the $(1-1)$ representation of $\mathfrak{A}$ in Φ_N. Then $\Sigma x_i\gamma_i \to \Sigma X_i\gamma_i$, γ in P, is a $(1-1)$ representation of $\mathfrak{A}_\mathrm{P}$ in P_N satisfying the condition that $1 \to 1$ if $\mathfrak{A}_\mathrm{P}$ has an identity.[11] We may therefore apply the above considerations to $\Sigma x_i\xi_i$. We see then that $m(t, \xi)$, the last invariant factor of $(1t - \Sigma X_i\xi_i)$, is the minimum polynomial of $\Sigma x_i\xi_i$ and is a divisor of the characteristic polynomial

$$f(t, \xi) = \det (t - \Sigma X_i\xi) = t^N - t^{N-1}\varphi_1(\xi) + \cdots + (-1)^N\varphi_N(\xi).$$

Since the coefficients of $f(t, \xi)$ are polynomials in the ξ_i, it follows from Gauss' lemma that the coefficients of $m(t, \xi)$ are polynomials in the ξ's.[12] We have shown also that $m(t, \xi)$ and $f(t, \xi)$ have the same irreducible factors in $\mathrm{P}[t]$, differing at most in the multiplicities of these factors. From the definition of $m(t, \xi)$ as the minimum polynomial of $\Sigma x_i\xi_i$ in $\mathfrak{A}_\mathrm{P}$, it is clear that $m(t, \xi)$ depends only on $\Sigma x_i\xi_i$ and not on the particular representation used. We shall call this polynomial a *minimum polynomial of the algebra* $\mathfrak{A}$.

[10] Throughout this discussion we may use anti-representations in place of ordinary representations of the algebra $\mathfrak{A}$.

[11] If an extension $\mathfrak{A}_\mathrm{P}$ of an algebra $\mathfrak{A}$ has an identity, then $\mathfrak{A}$ has an identity. This follows from the well-known theorem that a set of linear equations with coefficients in Φ have a solution in an extension field P if and only if they have a solution in Φ. The details of the proof are left to the reader.

[12] See Albert, *Modern Higher Algebra*, p. 37.

Let $y_1, \cdots, y_n$, $y_i = \Sigma x_j\mu_{ji}$, be a second basis for $\mathfrak{A}$ and let $m'(t, \eta)$ be the minimum polynomial determined by the general element $\Sigma y_i\eta_i$. If we use the field $\Sigma = \Phi(\xi, \eta)$, we may compare $m(t, \xi)$ and $m'(t, \eta)$. Now we recall that if $a_1, \cdots, a_r$ are linearly independent elements in an algebra $\mathfrak{A}$, then they remain linearly independent in any extension $\mathfrak{A}_\Sigma$ of $\mathfrak{A}$. It follows from this that the minimum polynomial of an element of an algebra is unchanged when the field over which the algebra is defined is extended. Hence $m(t, \xi)$ and $m'(t, \eta)$ are the minimum polynomials of $\Sigma x_i\xi_i$ and $\Sigma y_i\eta_i$, respectively, in the algebra $\mathfrak{A}_\Sigma$. Since $y_i = \Sigma x_j\mu_{ji}$, $\Sigma y_i\eta_i = \Sigma x_j\xi_j'$ where $\xi_j' = \Sigma \mu_{ji}\eta_i$. Hence $m(t, \xi') = m'(t, \eta)$. In this sense $m(t, \xi)$ is an invariant of $\mathfrak{A}$.

We write

$$m(t, \xi) = t^r - t^{r-1}T(\xi) + \cdots + (-1)^r N(\xi).$$

If the ξ's are specialized in Φ, say $\xi_i = \alpha_i$, we obtain a polynomial $m_a(t) = m(t, \alpha)$ called *the principal polynomial* associated with the element $a = \Sigma x_i\alpha_i$ of $\mathfrak{A}$. Using the relation $m(t, \xi') = m'(t, \eta)$ we see that $m_a(t)$ does not depend on the choice of the basis. Hence this is true also for the functions $T(a) \equiv T(\alpha)$ and $N(a) \equiv N(\alpha)$ which we call respectively the *principal trace* and the *principal norm* of a. The equation $m(x(\xi), \xi) = 0$ is equivalent to n polynomial identities $\rho_i(\xi) = 0$ obtained by expressing $m(x(\xi), \xi)$ as $\Sigma x_i\rho_i(\xi)$. Hence we have $m(a, \alpha) = 0$. It follows that $m_a(t)$ is divisible by $\mu_a(t)$. Since $f(t, \alpha)$ and $\mu_a(t)$ have the same irreducible factors, $m_a(t)$ and $\mu_a(t)$ have the same irreducible factors.

The matrix $(T(x_ix_j))$ is called a *discriminant matrix* of $\mathfrak{A}$. A change of basis replaces this matrix by a cogredient one ($M'TM$, M non-singular). The det $(T(x_ix_j))$ is a *discriminant* of $\mathfrak{A}$. The discriminants differ by square factors $\neq 0$ in Φ.

We consider now the problem of computing the minimum polynomial $m(t, \xi)$. First let $\mathfrak{A} = \Phi_r$. Here we use the representation of $\mathfrak{A}$ by itself. If the ξ_{ij} are indeterminates, $f(t, \xi) = \det (1t - (\xi_{ij}))$ is irreducible in $P[t]$, $P = \Phi(\xi_{ij})$.[13] Hence $m(t, \xi) = f(t, \xi)$. By a similar argument we treat $\mathfrak{A} = \Phi_{r_1}^{(1)} \oplus \cdots \oplus \Phi_{r_s}^{(s)}$, $\Phi^{(i)} \cong \Phi$. A general element of $\mathfrak{A}$ is

$$\begin{pmatrix} \xi_{ij}^{(1)} & & \\ & \ddots & \\ & & \xi_{ij}^{(s)} \end{pmatrix}.$$

Using this representation we obtain $m(t, \xi) = \Pi f_h(t, \xi^{(h)})$ where f_j is the characteristic polynomial of $(1t - (\xi^{(h)}))$.

Now let $\mathfrak{A}$ be arbitrary. We note that $m(t, \xi)$ is unchanged if $\mathfrak{A}$ is replaced by $\mathfrak{A}_\Gamma$, Γ an extension of Φ; for $\Sigma x_i\xi_i$ is also a general element for $\mathfrak{A}_\Gamma$. Hence it suffices to determine $m(t, \xi)$ for $\mathfrak{A}_\Gamma$ where Γ is the algebraic closure of Φ. We

[13] See L. E. Dickson, *Algebras and Their Arithmetics*, p. 115.

suppose that $x \to X$ is a $(1-1)$ representation of $\mathfrak{A}_\Gamma$ such that $1 \to 1$ if $\mathfrak{A}_\Gamma$ has an identity. We may take this representation to have the form

$$(5) \qquad \begin{pmatrix} X^{(1)} & & * \\ & \ddots & \\ 0 & & X^{(s)} \end{pmatrix}$$

where $x \to X^{(h)}$ are irreducible representations some of which may be the 0 representation. Now we recall that if $\mathfrak{N}$ is the radical of $\mathfrak{A}_\Gamma$, then $\mathfrak{A}_\Gamma - \mathfrak{N} = \Gamma_{r_1}^{(1)} \oplus \cdots \oplus \Gamma_{r_s}^{(s)}$ where the $\Gamma_{r_i}^{(i)}$ are division algebras. Since Γ is algebraically closed, it follows from Theorem 32 that $\Gamma^{(i)} \cong \Gamma$. The representation $x \to X^{(h)}$ is a representation of one of the Γ_r's. Hence if this representation is $\neq 0$, the set of matrices $X^{(h)}$ is the complete set of matrices having the same number of rows and columns as $X^{(h)}$. Hence it is possible to express the matrix units $e_{ij}^{(h)}$ of the h-th block as linear combinations of the matrices $X^{(h)}$. It follows that the characteristic polynomial $f_h(t, \xi)$ of $(1t - \Sigma X_i^{(h)}\xi_i)$ is irreducible, and consequently, $m(t, \xi) = \Pi f_{h_j}(t, \xi)$, a product of certain of the f_h's. Since $m(t, \xi)$ is the last invariant factor of $(1t - X)$, $m(t, \xi)$ is divisible by each of the distinct f_h's. We wish to show now that the representations $x \to X^{(h)}$ include all of the irreducible representations $\neq 0$ of $\mathfrak{A}_\Gamma$. For this purpose we recall that if $\mathfrak{J}^{(k)}$ is a particular irreducible left ideal in $\Gamma_{r_k}^{(k)}$, then the representations of $\mathfrak{A}_\Gamma$ determined by the s ideals $\mathfrak{J}^{(k)}$, $k = 1, \cdots, s$, constitute a complete set of inequivalent irreducible representations $\neq 0$ of $\mathfrak{A}_\Gamma$. If $\mathfrak{A}^{(k)}$ denotes the two-sided ideal in $\mathfrak{A}_\Gamma$ mapped into $\Gamma_{r_k}^{(k)}$, the representation of $\mathfrak{A}^{(k)}$ determined by $\mathfrak{J}^{(l)}$, $l \neq k$, is the 0 representation. Hence if the representation determined by $\mathfrak{J}^{(k)}$ does not occur among the constituents $x \to X^{(h)}$, then the elements of $\mathfrak{A}^{(k)}$ are represented in (5) by matrices whose diagonal blocks are all 0. Evidently such matrices form a nilpotent algebra and since the representation of $\mathfrak{A}^{(k)}$ is $(1-1)$, $\mathfrak{A}^{(k)}$ is nilpotent contrary to the relation $\mathfrak{A}^{(k)} - (\mathfrak{N} \wedge \mathfrak{A}^{(k)}) \cong \Gamma_{r_k}^{(k)}$. We remark finally that if $\mathfrak{A}$ has an identity, then none of the representations $x \to X^{(h)}$ are 0. Hence $m(t, \xi)$ is divisible by t if and only if $\mathfrak{A}$ does not have an identity.

If $f_h(t, \xi) = t^{n_h} - t^{n_h - 1}T^{(h)}(\xi) + \cdots + (-1)^{n_h}N^{(h)}(\xi)$, $T^{(h)}(\xi)$ is the trace of $\Sigma X_i^{(h)}\xi_i$ and $N^{(h)}(\xi)$ is the determinant of this matrix. Evidently, $T(\xi) = \sum_j T^{(h_j)}(\xi)$ and $N(\xi) = \Pi N^{(h_j)}(\xi)$. Using the properties of $T^{(h)}$ and $N^{(h)}$ and the fact that $\Sigma x_i\alpha_i \to \Sigma X_i^{(h)}\alpha_i$ is a homomorphism, we obtain the following important relations for the principal trace and the principal norm:

$$T(a + b) = T(a) + T(b), \qquad T(a\alpha) = T(a)\alpha, \qquad T(ab) = T(ba)$$

$$N(ab) = N(a)N(b), \qquad N(a\alpha) = N(a)\alpha^r.$$

Of course, $N(a) = 0$ if $\mathfrak{A}$ does not have an identity.

Examples. 1) If $\mathfrak{A}$ is a separable field, let P be the minimum normal extension of $\mathfrak{A}$. Then $\mathfrak{A}_P = P^{(1)} \oplus \cdots \oplus P^{(n)}$ where $n = (\mathfrak{A}:\Phi)$. Hence $m(t, \xi)$ has degree n and so it coincides with the characteristic polynomial of the matrix of the general element in the regular representation.

2) Let $\mathfrak{A}$ be the purely inseparable field of characteristic $p \neq 0$ of the form $\Phi(x_1, \cdots, x_m)$ where $x_i^p = \gamma_i$ in Φ and $(\mathfrak{A}:\Phi) = p^m$. Then the elements $x_1^{k_1} \cdots x_m^{k_m}$, $0 \leqq k_i < p$, form a basis for $\mathfrak{A}$ and if $x = \Sigma x_1^{k_1} \cdots x_m^{k_m} \xi_{k_1 \cdots k_m}$ then $m(t, \xi) = t^p - \Sigma \gamma_1^{k_1} \cdots \gamma_m^{k_m} \xi_{k_1 \cdots k_m}^p$.

19. Separable algebras. If $\mathfrak{A}$ is a separable field over Φ, $(\mathfrak{A}:\Phi)$ finite, we have seen that $\mathfrak{A}_\Gamma$ is semi-simple for any extension field Γ of Φ. Now let $\mathfrak{A}$ be inseparable of characteristic p, a an inseparable element and $\varphi(t) = (t^p)^r + (t^p)^{r-1}\beta_1 + \cdots + \beta_r$ its minimum polynomial. Since $1, a, \cdots, a^{p^r-1}$ are linearly independent in $\mathfrak{A}$, they are linearly independent in $\mathfrak{A}_\Gamma$. Hence $b = a^r + a^{r-1}\gamma_1 + \cdots + \gamma_r \neq 0$ for any γ_i in Γ. We suppose that Γ is algebraically closed and choose $\gamma_i = \beta_i^{\frac{1}{p}}$. Then $b^p = 0$ and so $\mathfrak{A}_\Gamma$ is not semi-simple. These facts lead us to define a *separable algebra over* Φ as an algebra $\mathfrak{A}$ over Φ such that $\mathfrak{A}_\Gamma$ is semi-simple for arbitrary extension fields Γ of Φ. As an extension of our result on fields, we have

THEOREM 35. *A necessary and sufficient condition that $\mathfrak{A}$ be separable over Φ is that $\mathfrak{A} = \mathfrak{A}_1 \oplus \cdots \oplus \mathfrak{A}_t$ where $\mathfrak{A}_i$ is simple and has a separable center $\mathfrak{C}_i$ over Φ.*

Necessity. By definition, if $\mathfrak{A}$ is separable, $\mathfrak{A}$ is semi-simple and hence $\mathfrak{A} = \mathfrak{A}_1 \oplus \cdots \oplus \mathfrak{A}_t$ where each $\mathfrak{A}_i$ is simple. The center $\mathfrak{C}_i$ of $\mathfrak{A}_i$ is separable. For, otherwise, one of the $\mathfrak{C}_i$, say $\mathfrak{C}_1$, contains an inseparable element and hence if Γ is the algebraic closure of Φ, then $\mathfrak{C}_{1\Gamma}$ contains a nilpotent element $b \neq 0$. Since b is in the center of $\mathfrak{A}_\Gamma$, the principal ideal $b\mathfrak{A}_\Gamma$ is nilpotent, contrary to hypothesis.

Sufficiency. Since $\mathfrak{A}_\Gamma = \mathfrak{A}_{1\Gamma} \oplus \cdots \oplus \mathfrak{A}_{t\Gamma}$, it suffices to consider the case where $\mathfrak{A} = \mathfrak{A}_1$ is simple and its center $\mathfrak{C}$ is separable. Let P be a field isomorphic to the least normal extension of $\mathfrak{C}$. We have seen that $\mathfrak{C} \times \mathrm{P} = \mathrm{P}^{(1)} \oplus \cdots \oplus \mathrm{P}^{(r)}$ where $\mathrm{P}^{(j)}$ is isomorphic to P and $r = (\mathfrak{C}:\Phi)$. Let $1 = e_1 + \cdots + e_r$ where $e_j \in \mathrm{P}^{(j)}$. Then $\mathrm{P}^{(j)} = e_j\mathrm{P}$ and so for any c in $\mathfrak{C}$, $e_i c = e_i \rho^{(i)}$ where $\rho^{(i)} \in \mathrm{P}$ and where the correspondences $c \rightarrow \rho^{(i)}$ are the distinct, irreducible (anti-) representations of $\mathfrak{C}$ in P. Now let $x_1, \cdots, x_n$ $(n = \nu^2)$ be a basis of $\mathfrak{A}$ over $\mathfrak{C}$ and

$$x_i x_{i'} = \Sigma x_k c_{kii'}, \tag{6}$$

$c_{kii'}$ in $\mathfrak{C}$. Then if $c_1, \cdots, c_r$ is a basis for $\mathfrak{C}$ over Φ, the elements $x_i c_j$ form a basis for $\mathfrak{A}$ over Φ. Thus every element of $\mathfrak{A} \times \Gamma$ has the form $\Sigma x_i c_j \gamma_{ij}$, γ_{ij} in P. If we express the elements c_j in terms of the e_j, we obtain a unique expression $\Sigma x_i e_j \gamma_{ij}$ for each element of $\mathfrak{A} \times \mathrm{P}$. Since $e_j e_k = 0$ for $j \neq k$, $\mathfrak{A} \times \mathrm{P} = \mathfrak{A}_1 \oplus \cdots \oplus \mathfrak{A}_r$ where $\mathfrak{A}_j$ is a two-sided ideal with the basis $x_1^{(j)} = x_1 e_j; \cdots, x_n^{(j)} = x_n e_j$ over P. By (6), $x_i^{(j)} x_{i'}^{(j)} = \Sigma x_k^{(j)} \gamma_{kii'}^{(j)}$. Thus $\mathfrak{A}_j$ is a central simple algebra over P, isomorphic as an algebra over Φ to $(\mathfrak{A} \text{ over } \mathfrak{C})_\mathrm{P}$. Now if Γ is any extension of Φ, we form an extension Σ containing P and Γ. Then $\mathfrak{A}_\Sigma = (\mathfrak{A}_\mathrm{P})_\Sigma = \mathfrak{A}_{1\Sigma} \oplus \cdots \oplus \mathfrak{A}_{r\Sigma}$ and the $\mathfrak{A}_{j\Sigma}$ are central simple. Since $\mathfrak{A}_\Sigma = (\mathfrak{A}_\Gamma)_\Sigma$, $\mathfrak{A}_\Gamma$ is semi-simple.

A second criterion for separability of an algebra is given by

THEOREM 36. *A necessary and sufficient condition that $\mathfrak{A}$ be separable is that its discriminants Δ be $\neq 0$.*

We note first that $\Delta = \det (T(x_i x_j)) = 0$ if and only if there exists an element $z \neq 0$ in $\mathfrak{A}$ such that $T(za) = 0$ for all a. For if $\Delta = 0$, the equations $\Sigma T(x_i x_j)\zeta_j = 0$ have a solution $(\xi_1, \cdots, \xi_n) \neq (0, \cdots, 0)$, and hence the element $z = \Sigma x_i \xi_i \neq 0$ satisfies $T(x_i z) = T(z x_i) = 0$ for $i = 1, \cdots, n$. It follows that $T(za) = 0$ for all a. The converse is evident: If $T(za) = 0$ for all a then $T(x_i z) = 0$ and so the equations $\Sigma T(x_i x_j)\zeta_j = 0$ have a non-trivial solution. Then $\Delta = 0$. Now suppose that $\mathfrak{A}$ is not separable. Then there exists a field Γ such that $\mathfrak{A}_\Gamma$ has a radical $\mathfrak{N}$. If $z \in \mathfrak{N}$, $za \in \mathfrak{N}$ for all a in $\mathfrak{A}_\Gamma$. Now let $x \rightarrow X$ be a $(1-1)$ representation of $\mathfrak{A}_\Gamma$ by matrices. Then if $z \rightarrow Z$, $a \rightarrow A$, the matrix $ZA = X$ is nilpotent. If we use the form (5), we see that each X_i is nilpotent and hence each $T^{(i)}(X) = 0$. Then $T(X) = \Sigma T^{h_i}(X) = 0$ and so $T(za) = 0$ for all a and $\Delta = 0$. Now let $\mathfrak{A}$ be separable and let Γ be the algebraic closure of Φ. Then $\mathfrak{A}_\Gamma = \Gamma_{n_1}^{(1)} \oplus \cdots \oplus \Gamma_{n_t}^{(t)}$ where $\Gamma^{(i)} \cong \Gamma$. Then $\mathfrak{A}_\Gamma$ is the algebra of matrices

$$x = \begin{pmatrix} (\xi_{ij}^{(1)}) & & \\ & \ddots & \\ & & (\xi_{kl}^{(t)}) \end{pmatrix}$$

where the ξ's are arbitrary and, as we have seen, $T(X)$ is the ordinary trace of x. If we use the basis $e_{i_k j_k}^{(k)}$, $i_k, j_k = 1, \cdots, n_k$; $k = 1, \cdots, t$, for $\mathfrak{A}_\Gamma$ where $e_{i_k j_k}^{(k)}$ is a matrix basis for $\Gamma_{n_k}^{(k)}$, we obtain by a simple computation $\det (T(e_{i_k j_k}^{(k)} e_{i_l j_l}^{(l)})) = \pm 1$.

20. A theorem of Wedderburn.

THEOREM 37. *Let $\mathfrak{A}$ be an algebra with the radical $\mathfrak{N}$. Then if $\bar{\mathfrak{A}} = \mathfrak{A} - \mathfrak{N}$ is separable, there exists a sub-algebra $\mathfrak{S}$ of $\mathfrak{A}$ such that $\bar{\mathfrak{A}} = \mathfrak{N} + \mathfrak{S}$, $\mathfrak{N} \wedge \mathfrak{S} = 0$.*[14]

Suppose first that $\mathfrak{N}^2 \neq 0$. Then $(\mathfrak{A} - \mathfrak{N}^2 : \Phi) < (\mathfrak{A} : \Phi)$. Since $(\mathfrak{A} - \mathfrak{N}^2) - (\mathfrak{N} - \mathfrak{N}^2) \cong \mathfrak{A}$, $(\mathfrak{N} - \mathfrak{N}^2)$ is the radical of $(\mathfrak{A} - \mathfrak{N}^2)$ and the latter algebra satisfies the hypothesis of the theorem. We may assume that the theorem has already been established for algebras of dimensionality $< (\mathfrak{A} : \Phi)$. Hence there exists a sub-algebra $\mathfrak{S}_1$ of $\mathfrak{A}$ that contains $\mathfrak{N}^2$ and such that

$$(\mathfrak{A} - \mathfrak{N}^2) = (\mathfrak{S}_1 - \mathfrak{N}^2) + (\mathfrak{N} - \mathfrak{N}^2), \qquad (\mathfrak{S}_1 - \mathfrak{N}^2) \wedge (\mathfrak{N} - \mathfrak{N}^2) = 0.$$

These equations are equivalent to

$$\mathfrak{A} = \mathfrak{S}_1 + \mathfrak{N}, \qquad \mathfrak{S}_1 \wedge \mathfrak{N} = \mathfrak{N}^2. \tag{7}$$

[14] This theorem is due to Wedderburn for fields Φ of characteristic 0. Moreover, the proof in the general case is a rather trivial modification of Wedderburn's argument ([8] or Dickson [2]).

Since $\mathfrak{A} - \mathfrak{N} = (\mathfrak{S}_1 + \mathfrak{N}) - \mathfrak{N} \cong \mathfrak{S}_1 - (\mathfrak{S}_1 \wedge \mathfrak{N}) = \mathfrak{S}_1 - \mathfrak{N}^2$, the radical of $\mathfrak{S}_1$ is $\mathfrak{N}^2$ and $\mathfrak{S}_1$ satisfies the condition of the theorem. Since $\mathfrak{S}_1 \wedge \mathfrak{N} = \mathfrak{N}^2$, $(\mathfrak{S}_1:\Phi) < (\mathfrak{A}:\Phi)$ and so there exists a subalgebra $\mathfrak{S}'$ of $\mathfrak{S}_1$ such that $\mathfrak{S}_1 = \mathfrak{S} + \mathfrak{N}^2$, $\mathfrak{S} \wedge \mathfrak{N}^2 = 0$. Then by (7), $\mathfrak{A} = \mathfrak{S} + \mathfrak{N}$ and $\mathfrak{S} \wedge \mathfrak{N} = 0$.

Suppose next that $\bar{\mathfrak{A}} = \bar{\Phi}_{n_1}^{(1)} \oplus \cdots \oplus \bar{\Phi}_{n_t}^{(t)}$ where $\bar{\Phi}^{(i)} \cong \Phi$ and let $\bar{u}_k$ be the identity of $\bar{\Phi}_{n_k}^{(k)}$. We may choose an idempotent element u_k in $\bar{u}_k$ so that $u_k u_l = 0$ if $k \neq l$. Then $u_k\mathfrak{A}u_k - (u_k\mathfrak{A}u_k \wedge \mathfrak{N}) \cong (u_k\mathfrak{A}u_k + \mathfrak{N}) - \mathfrak{N} = \Phi_{n_k}^{(k)}$. Hence $u_k\mathfrak{A}u_k$ is primary and its radical is $u_k\mathfrak{A}u_k \wedge \mathfrak{N} = u_k\mathfrak{N}u_k$. It follows that $u_k\mathfrak{A}u_k$ contains a subalgebra $\mathfrak{S}_k \cong \bar{\Phi}_{n_k}^{(k)}$. Since $u_k u_l = 0$ if $k \neq l$, $\mathfrak{S}_k\mathfrak{S}_l = 0$ and so $\mathfrak{S} \equiv \mathfrak{S}_1 + \cdots + \mathfrak{S}_t = \mathfrak{S}_1 \oplus \cdots \oplus \mathfrak{S}_t$ is semi-simple and its dimensionality is Σn_k^2. Hence $\mathfrak{S} \wedge \mathfrak{N} = 0$ and by comparing dimensionalities we see that $\mathfrak{S} + \mathfrak{N} = \mathfrak{A}$.

Finally, let $\mathfrak{A}$ be any algebra for which the assumption of the theorem holds and for which $\mathfrak{N}^2 = 0$. If Γ is any extension of Φ, $\mathfrak{A}_\Gamma - \mathfrak{N}_\Gamma \cong (\bar{\mathfrak{A}})_\Gamma$ and the latter algebra is semi-simple. Hence $\mathfrak{N}_\Gamma$ is the radical of $\mathfrak{A}_\Gamma$. If Γ is the algebraic closure of Φ, $\mathfrak{A}_\Gamma - \mathfrak{N}_\Gamma = \bar{\Gamma}_{n_1}^{(1)} \oplus \cdots \oplus \bar{\Gamma}_{n_r}^{(r)}$ where $\bar{\Gamma}^{(k)} \cong \Gamma$. The matrix units of the simple components of $\bar{\mathfrak{A}}_\Gamma$ are expressible in terms of a basis $\bar{y}_1, \cdots, \bar{y}_r$ of $\bar{\mathfrak{A}}$ as $\Sigma \bar{y}_i w_i$, w_i in Γ. Since there are only a finite number of w's involved in these expressions and each w is algebraic over Φ, they generate a finite extension P of Φ. Evidently $\bar{\mathfrak{A}}_P$ is a direct sum of matrix algebras over P and $\mathfrak{A}_P - \mathfrak{N}_P = \bar{P}_{n_1}^{(1)} \oplus \cdots \oplus \bar{P}_{n_t}^{(t)}$, and by what has already been proved, $\mathfrak{A}_P = \mathfrak{N}_P + \tilde{\mathfrak{S}}$, $\tilde{\mathfrak{S}} \wedge \mathfrak{N}_P = 0$ where $\tilde{\mathfrak{S}}$ is a subalgebra of $\mathfrak{A}_P$. Let $\rho_0 = 1, \rho_1, \cdots, \rho_s$ be a basis for P over Φ and $x_1, \cdots, x_n$ a basis for $\mathfrak{A}$ over Φ (P) such that $x_{r+1}, \cdots, x_n$ is a basis for $\mathfrak{N}$ over Φ (P). Then $x_i = y_i - z_i$ where $y_i \in \tilde{\mathfrak{S}}$ and $z_i \in \mathfrak{N}_P$. The elements $y_1, \cdots, y_r$ form a basis for $\tilde{\mathfrak{S}}$ and $y_i = x_i + \Sigma z_{ij}\rho_j$ where $z_{ij} \in \mathfrak{N}$. We set $x_i' = x_i + z_{i0}$. Then $x_1', \cdots, x_r', x_{r+1}, \cdots, x_n$ is a basis for $\mathfrak{A}$ over Φ and $y_i = x_i' + z_i'$ where $z_i' = \sum_{k=1}^{s} z_{ik}\rho_k$. Hence $x_i'x_j' = \Sigma x_k'\gamma_{kij} + v_{ij}$, γ_{ijk} in Φ and v_{ij} in $\mathfrak{N}$. It follows that $y_i y_j = \Sigma y_k \gamma_{kij} + u_{ij}$, u_{ij} in $\mathfrak{N}_P$, and since the $y_i \in \tilde{\mathfrak{S}}$, $u_{ij} = 0$. If we substitute the expressions $x_i' + z_i'$ for y_i, this relation becomes

$$x_i'x_j' + x_i'z_j' + z_i'x_j' = \Sigma(x_k' + z_k')\gamma_{kij}.$$

If we express each term as $\Sigma a_i\rho_i$ and compare the coefficients of $\rho_0 = 1$, we obtain $x_i'x_j' = \Sigma x_k'\gamma_{kij}$. Thus the totality of elements $\Sigma x_i'\alpha_i$, α in Φ, is the required algebra $\mathfrak{S}$.

CHAPTER 6

MULTIPLICATIVE IDEAL THEORY

1. Quotient rings. It is a well-known discovery of Emmy Noether's that the fundamental factorization theorem for the ideals of a maximal domain of algebraic numbers may be deduced from some very simple properties of these domains. These properties are embodied in the theorem: If $\mathfrak{o}$ is a commutative domain of integrity, then the ideals ($\neq 0$, $\mathfrak{o}$) are factorable in one and only one way as products of prime ideals if and only if $\mathfrak{o}$ satisfies the following conditions:

N1. $\mathfrak{o}$ is integrally closed (in its quotient field).

N2. The descending chain condition holds for the ideals containing any fixed ideal $\neq 0$.

N3. The ascending chain condition holds for all ideals.

In this chapter we shall consider the extension of this result to non-commutative rings. The contents of this theory are due mainly to Speiser, Brandt, Artin, Hasse and Deuring; the axiomatic foundations to Asano. Many of the results of this chapter have been anticipated in our discussion of principal ideal domains. We shall also need to refer to the theory of principal ideal rings that we have developed in **15–16** of Chapter 4.

We begin with a ring $\mathfrak{o}$ having an identity. An element a of $\mathfrak{o}$ will be called *regular* if it is neither a left nor a right zero-divisor. The first restriction that we shall impose on $\mathfrak{o}$ is that it have a (right) *quotient ring*, i.e. a ring $\mathfrak{A}$ containing $\mathfrak{o}$ such that 1) every regular element of $\mathfrak{o}$ has an inverse in $\mathfrak{A}$ and 2) every element of $\mathfrak{A}$ has the form ab^{-1} where a and b are in $\mathfrak{o}$. It is a simple matter to obtain a condition on $\mathfrak{o}$ that $\mathfrak{A}$ exist. For this purpose we consider any pair of elements a and b in $\mathfrak{o}$ with b regular. Then b^{-1} is in $\mathfrak{A}$ and hence $b^{-1}a$ has the form $a_1b_1^{-1}$, a_1 and b_1 in $\mathfrak{o}$. Then $ba_1 = ab_1$. A necessary condition for the existence of $\mathfrak{A}$ is therefore that for any pair of elements a, b in $\mathfrak{o}$, b regular, there exists a common right multiple $m = ab_1 = ba_1$ such that b_1 is regular.

Conversely, suppose that this condition holds. As in Chapter 3, we consider the pairs (a, b) of elements a, b in $\mathfrak{o}$ such that b is regular. If (c, d) is a second pair of this type and m is any multiple of the form $db_1 = bd_1$ such that b_1 (and hence d_1) is regular, then we regard (a, b) as equivalent to (c, d) $((a, b) \sim (c, d))$ if $ad_1 = cb_1$. We note that if this condition holds for a particular m, it holds for any $n = db_2 = bd_2$ such that b_2 and d_2 are regular. For we may determine regular elements e_2 and e_1 such that $b_1e_2 = b_2e_1$. Then $d_1e_2 = d_2e_1$ and $ad_2 = cb_2$. It follows directly from this that the relation $\sim$ is symmetric, reflexive and transitive. As usual, we denote the set of pairs equivalent to (a, b) as a/b.

If $m = db_1 = bd_1$ and b_1 and d_1 are regular, we define $a/b + c/d$ as $(ad_1 + cb_1)/m$, and if $n = bc_1 = cb_1$, b_1 regular, we define $(a/b)(c/d) = ac_1/db_1$. The functions thus defined are single-valued and they turn the set $\mathfrak{A}$ of "fractions"

a/b into a ring. We leave the verification of these facts to the reader. The ring $\mathfrak{A}$ has an identity $1/1$ and contains a subring, consisting of the elements $a/1$, that is isomorphic to $\mathfrak{o}$. We shall identify this subring with $\mathfrak{o}$ and write a for $a/1$. Then if a is not a zero-divisor in $\mathfrak{o}$, it has the inverse $a^{-1} = 1/a$ in $\mathfrak{A}$. Since any element of $\mathfrak{A}$ has the form $(a/1)(1/b) = ab^{-1}$, $\mathfrak{A}$ is a right quotient ring of $\mathfrak{o}$.

We note next that the quotient ring is uniquely determined in the sense of isomorphism. For it is readily verified that if $\mathfrak{o}$ and $\mathfrak{o}'$ are isomorphic under the correspondence $a \rightarrow a'$, then their quotient rings $\mathfrak{A}$ and $\mathfrak{A}'$ are isomorphic under the correspondence $ab^{-1} \rightarrow a'(b')^{-1}$. Finally, we remark that if $\mathfrak{A}$ is a quotient ring of a ring $\mathfrak{o}$, any regular element of $\mathfrak{A}$ has an inverse in $\mathfrak{A}$.

2. Orders and ideals. Once the condition that $\mathfrak{o}$ have a quotient ring has been determined, it is more convenient to shift our attention from $\mathfrak{o}$ to $\mathfrak{A}$. Thus we suppose that $\mathfrak{A}$ is given as any ring with an identity in which every regular element has an inverse. For example, $\mathfrak{A}$ may be any ring satisfying the descending chain condition for left and for right ideals.[1] We consider the subrings $\mathfrak{o}$ of $\mathfrak{A}$ defined by the following

Definition 1. An *order* $\mathfrak{o}$ in $\mathfrak{A}$ is a subring containing 1 and having the property that every element of $\mathfrak{A}$ has the form ab^{-1} for suitable elements a and b in $\mathfrak{o}$.

It may be remarked that this definition is one-sided since we do not require that the elements of $\mathfrak{A}$ be representable in the form $b^{-1}a$, b, a in $\mathfrak{o}$. The latter condition will be satisfied, however, for the orders with which we shall be primarily concerned in the sequel.

The orders $\mathfrak{o}_1$ and $\mathfrak{o}_2$ are *equivalent* if there exist regular elements a_1, b_1, a_2, b_2 in $\mathfrak{A}$ such that $a_1\mathfrak{o}_1b_1 \leqq \mathfrak{o}_2$ and $a_2\mathfrak{o}_2b_2 \leqq \mathfrak{o}_1$. Evidently this relation is symmetric, reflexive and transitive. *We shall restrict our attention to the orders that are equivalent to a fixed order* $\mathfrak{o}_0$, and for simplicity, we use the term "order" in place of "order equivalent to $\mathfrak{o}_0$." We remark that in order to prove that a subring $\mathfrak{o}'$ containing the identity is an order it suffices to show that there exists an order $\mathfrak{o}$ and regular elements a, b, a' and b' such that $a\mathfrak{o}b \leqq \mathfrak{o}'$ and $a'\mathfrak{o}b' \leqq \mathfrak{o}$. For if z is any element of $\mathfrak{A}$, there exist elements p, q in $\mathfrak{o}$ such that $a^{-1}za = pq^{-1}$ and so $z = (apb)(aqb)^{-1}$.

Definition 2. A subset $\mathfrak{a}$ of $\mathfrak{A}$ is a (fractional) *right* $\mathfrak{o}$-*ideal* if 1) $\mathfrak{a}\mathfrak{o} \leqq \mathfrak{a}$, 2) $\mathfrak{a}$ contains a regular element and 3) there exists a regular element a in $\mathfrak{A}$ such that $a\mathfrak{a} \leqq \mathfrak{o}$.

Left $\mathfrak{o}$-*ideals* are defined in a similar fashion. If $\mathfrak{a}$ is both a right $\mathfrak{o}$-ideal and a left $\mathfrak{o}$-ideal, it is a *two-sided* $\mathfrak{o}$-*ideal*. If a is any regular element, the set $\mathfrak{a} = a\mathfrak{o}$ is a right $\mathfrak{o}$-ideal. For 1) and 2) are clear and 3) holds since $a^{-1}\mathfrak{a} = \mathfrak{o}$. An ideal of this type is called *principal*. In terms of this definition the conditions 2) and 3) may be replaced respectively by 2′) $\mathfrak{a}$ contains a principal right $\mathfrak{o}$-ideal and 3′) $\mathfrak{a}$ is contained in a principal right $\mathfrak{o}$-ideal.

[1] For in this case, if a is not a left zero-divisor then the mapping $x \rightarrow ax$ is an $\mathfrak{A}_r$-isomorphism between $\mathfrak{A}$ and the right ideal $a\mathfrak{A}$. By the descending chain condition for right ideals, $a\mathfrak{A} = \mathfrak{A}$. Hence there is an element b such that $ab = 1$. Similarly, an element b' exists such that $b'a = 1$. It follows that $b = b'$.

Now let $\mathfrak{o}$ be an order and let $\mathfrak{a}$ be a right $\mathfrak{o}$-ideal. Consider the set $\mathfrak{o}_l$ of elements x such that $x\mathfrak{a} \leqq \mathfrak{a}$.[2] Evidently, $\mathfrak{o}_l$ is a subring of $\mathfrak{A}$ containing the identity. Since there exist regular elements a and b such that $b\mathfrak{o} \leqq \mathfrak{a} \leqq a\mathfrak{o}$, $b\mathfrak{o}a^{-1} \leqq \mathfrak{o}_l$ and since $\mathfrak{o}_l(b\mathfrak{o}) \leqq a\mathfrak{o}$, $\mathfrak{o}_l b \leqq a\mathfrak{o}$ and $\mathfrak{o}_l \leqq a\mathfrak{o}b^{-1}$. Thus $\mathfrak{o}_l$ is an order. Since $a^{-1}\mathfrak{a} \leqq \mathfrak{o}$, $\mathfrak{a}a^{-1}\mathfrak{a} \leqq \mathfrak{a}$ and so $\mathfrak{a}a^{-1} \leqq \mathfrak{o}_l$, $\mathfrak{a} \leqq \mathfrak{o}_l a$. Finally, b is in $\mathfrak{a}$ so that $\mathfrak{o}_l b \leqq \mathfrak{a}$. This shows that $\mathfrak{a}$ is a left $\mathfrak{o}_l$-ideal.

In a similar manner if $\mathfrak{a}$ is a left $\mathfrak{o}$-ideal, the set $\mathfrak{o}_r$ of elements y such that $\mathfrak{a}y \leqq \mathfrak{a}$ is an order and $\mathfrak{a}$ is a right $\mathfrak{o}_r$-ideal. Hence if we begin with a right $\mathfrak{o}$-ideal, we may determine first $\mathfrak{o}_l$ and then use the fact that $\mathfrak{a}$ is a left $\mathfrak{o}_l$-ideal to show that the set $\mathfrak{o}_r$ of y's such that $\mathfrak{a}y \leqq \mathfrak{a}$ is an order and that $\mathfrak{a}$ is a right $\mathfrak{o}_r$-ideal. Evidently $\mathfrak{o} \leqq \mathfrak{o}_r$.

Theorem 1. *If $\mathfrak{a}$ is a right (left) $\mathfrak{o}$-ideal, the set of elements x such that $x\mathfrak{a} \leqq \mathfrak{a}$ is an order $\mathfrak{o}_l$ and the set of elements y such that $\mathfrak{a}y \leqq \mathfrak{a}$ is an order $\mathfrak{o}_r$. The set $\mathfrak{a}$ is a left $\mathfrak{o}_l$-ideal and a right $\mathfrak{o}_r$-ideal.*

The orders $\mathfrak{o}_l$ and $\mathfrak{o}_r$ are respectively the *left order* and the *right order* of $\mathfrak{a}$. If $\mathfrak{a} \leqq \mathfrak{o}_r$, $\mathfrak{a} \leqq \mathfrak{o}_l$ and conversely. In this case $\mathfrak{a}$ is called *integral*. If $\mathfrak{a}$ is a principal $\mathfrak{o}$-ideal $a\mathfrak{o}$, $a\mathfrak{o}_r \leqq a\mathfrak{o}$ and so $\mathfrak{o}_r = \mathfrak{o}$. Similarly $\mathfrak{o}_l = a\mathfrak{o}a^{-1}$.

If $\mathfrak{a}$ is a right ideal with right order $\mathfrak{o}_r$ and left order $\mathfrak{o}_l$, let $\mathfrak{a}^{-1}$ denote the set of elements z in $\mathfrak{A}$ such that $\mathfrak{a}z\mathfrak{a} \leqq \mathfrak{a}$. Evidently the elements z may also be characterized by either of the equations $\mathfrak{a}z \leqq \mathfrak{o}_l$ or $z\mathfrak{a} \leqq \mathfrak{o}_r$. If c and d are regular elements such that $d\mathfrak{o}_r \leqq \mathfrak{a} \leqq c\mathfrak{o}_r$, then $\mathfrak{a}(\mathfrak{o}_r c^{-1})\mathfrak{a} = (\mathfrak{a}\mathfrak{o}_r)(c^{-1}\mathfrak{a}) \leqq (\mathfrak{a}\mathfrak{o}_r)\mathfrak{o}_r = \mathfrak{a}$ and $\mathfrak{a}^{-1}d \leqq \mathfrak{o}_r$. Then $\mathfrak{o}_r c^{-1} \leqq \mathfrak{a}^{-1} \leqq \mathfrak{o}_r d^{-1}$ and since $\mathfrak{a}^{-1}$ is a left $\mathfrak{o}$-module, $\mathfrak{a}^{-1}$ is a left $\mathfrak{o}_r$-ideal. Similarly $\mathfrak{a}^{-1}$ is a right $\mathfrak{o}_l$-ideal.

Theorem 2. *If $\mathfrak{a}$ is an ideal with right order $\mathfrak{o}_r$ and left order $\mathfrak{o}_l$, then the set of elements z in $\mathfrak{A}$ such that $\mathfrak{a}z\mathfrak{a} \leqq \mathfrak{a}$ is a left $\mathfrak{o}_r$-ideal and a right $\mathfrak{o}_l$-ideal.*

The ideal $\mathfrak{a}^{-1}$ is called the *inverse* of $\mathfrak{a}$. Since $\mathfrak{a}^{-1}$ is the set of elements z such that $z\mathfrak{a} \leqq \mathfrak{o}_r$, it follows that if $\mathfrak{b}$ is a second ideal with right order $\mathfrak{o}_r$ (left order $\mathfrak{o}_l$) and $\mathfrak{a} \geqq \mathfrak{b}$, then $\mathfrak{a}^{-1} \leqq \mathfrak{b}^{-1}$. If $\mathfrak{a} = a\mathfrak{o}$, $\mathfrak{o}_r = \mathfrak{o}$ and if z is in $\mathfrak{a}^{-1}$, $za = b$ is in $\mathfrak{o}$. Hence $z = ba^{-1}$ and $\mathfrak{a}^{-1} = \mathfrak{o}a^{-1}$.

3. Bounded orders. We shall see that a fundamental concept of the present theory is one that we have already encountered in our study of principal ideal domains, namely, that of a bounded ideal. If $\mathfrak{a}$ is a right (left) $\mathfrak{o}$-ideal, $\mathfrak{a}$ is *bounded* if it contains a two-sided $\mathfrak{o}$-ideal. It should be remarked that the present definition applies to any ideal, integral or not, and that it is unnecessary to state explicitly that the two-sided ideal is $\neq 0$ since this requirement is contained in our new definition of an ideal. If all ideals in $\mathfrak{o}$ are bounded, we say that $\mathfrak{o}$ itself is *bounded*. In this section we shall investigate some of the properties of bounded orders.

Since any ideal contains a principal ideal, in order that $\mathfrak{o}$ be bounded it clearly suffices that every principal $\mathfrak{o}$-ideal contains a two-sided $\mathfrak{o}$-ideal. Suppose that

[2] Unfortunately, our notation here is the same as that for the ring of left multiplications. For this reason we shall not use our old notation for the latter system in this chapter.

$\mathfrak{o}$ is *maximal* in the sense that there exists no order (equivalent to $\mathfrak{o}$) properly containing $\mathfrak{o}$.[3] Then we have the following theorem.

THEOREM 3. *If $\mathfrak{o}$ is a maximal order, the following conditions on $\mathfrak{o}$ are equivalent*: 1) *$\mathfrak{o}$ is bounded,* 2) *every integral right (or left) $\mathfrak{o}$-ideal contains a two-sided $\mathfrak{o}$-ideal,* 3) *for every regular c, $\mathfrak{o}c\mathfrak{o}$ is a two-sided $\mathfrak{o}$-ideal and* 4) *for every regular element a there exists a regular element b such that $\mathfrak{o}b \leqq a\mathfrak{o}$ (or $b\mathfrak{o} \leqq a\mathfrak{o}$).*

Suppose that for any regular element a in $\mathfrak{o}$, $a\mathfrak{o}$ contains a two-sided $\mathfrak{o}$-ideal $\mathfrak{a}$. Since $\mathfrak{o}$ is maximal, both orders of $\mathfrak{a}$ coincide with $\mathfrak{o}$. Hence $(a\mathfrak{o})^{-1} = \mathfrak{o}a^{-1} \leqq \mathfrak{a}^{-1}$ and $\mathfrak{o}a^{-1}\mathfrak{o} \leqq \mathfrak{a}^{-1}$. This implies that $\mathfrak{o}a^{-1}\mathfrak{o}$ is contained in a right (left) principal $\mathfrak{o}$-ideal and so $\mathfrak{o}a^{-1}\mathfrak{o}$ is a two-sided $\mathfrak{o}$-ideal. If c is an arbitrary regular element, c has the form ba^{-1} with b and a in $\mathfrak{o}$. Then $\mathfrak{o}c\mathfrak{o} = \mathfrak{o}ba^{-1}\mathfrak{o} \leqq \mathfrak{o}a^{-1}\mathfrak{o}$ and $\mathfrak{o}c\mathfrak{o}$ is a two-sided $\mathfrak{o}$-ideal. We have therefore proved that 2) implies 3). Now suppose that 3) holds. If a is an arbitrary regular element, the two-sided $\mathfrak{o}$-ideal $\mathfrak{a} = \mathfrak{o}a^{-1}\mathfrak{o}$ contains $\mathfrak{o}a^{-1}$ and $a^{-1}\mathfrak{o}$. Hence $\mathfrak{a}^{-1} \leqq (\mathfrak{o}a^{-1})^{-1} = a\mathfrak{o}$. Since $\mathfrak{a}^{-1}$ is a two-sided $\mathfrak{o}$-ideal, it contains a principal left $\mathfrak{o}$-ideal $\mathfrak{o}b$. Similarly $\mathfrak{o}a$ contains $\mathfrak{a}^{-1}$ and a suitable $b'\mathfrak{o}$, b' regular. Thus 3) implies 1) and 4). Since 2) is an obvious consequence of 1), the conditions 1), 2) and 3) are equivalent. Finally we prove that 4) implies 1). Let $a\mathfrak{o}$ be an arbitrary principal ideal and b a regular element such that $a\mathfrak{o} \geqq \mathfrak{o}b$. Then $a\mathfrak{o} \geqq \mathfrak{o}b\mathfrak{o}$ and the latter is a right $\mathfrak{o}$-ideal. Since $\mathfrak{o}$ is maximal, the left order of $\mathfrak{o}b\mathfrak{o}$ is $\mathfrak{o}$ and so $\mathfrak{o}b\mathfrak{o}$ is also a left $\mathfrak{o}$-ideal. This shows that 1) holds and the theorem is completely proved.

The following theorem shows that if $\mathfrak{o}$ is a bounded maximal order, then the condition that every x of $\mathfrak{A}$ has the form ab^{-1} may be replaced by the dual requirement that every x has the form $c^{-1}d$, c and d in $\mathfrak{o}$.

THEOREM 4. *If $\mathfrak{o}$ is a bounded maximal order, every x in $\mathfrak{A}$ has the form $c^{-1}d$ with c and d in $\mathfrak{o}$.*

We know that any x of $\mathfrak{A}$ may be expressed as ab^{-1}, a and b in $\mathfrak{o}$. Since b is regular, $b\mathfrak{o} \geqq \mathfrak{a} \geqq \mathfrak{o}c$ where $\mathfrak{a}$ is a two-sided $\mathfrak{o}$-ideal and c is regular. Since $\mathfrak{o}$ is maximal, the left and the right orders of $\mathfrak{a}$ are $\mathfrak{o}$ and hence $\mathfrak{o}b^{-1} \leqq \mathfrak{a}^{-1} \leqq c^{-1}\mathfrak{o}$. Thus $ab^{-1} = c^{-1}d$ for a suitable d in $\mathfrak{o}$.

THEOREM 5. *Suppose that $\mathfrak{o}$ is a bounded maximal order. Then if $\mathfrak{o}'$ is any order and $\mathfrak{M}$ is a set such that $a'\mathfrak{M}b' \leqq \mathfrak{o}'$ for suitable regular elements a', b', there exist regular elements c and d such that $c\mathfrak{M} \leqq \mathfrak{o}$ and $\mathfrak{M}d \leqq \mathfrak{o}$.*

Since $\mathfrak{o}'$ is equivalent to $\mathfrak{o}$, there exist regular elements a and b such that $a\mathfrak{M}b \leqq \mathfrak{o}$. Hence $a\mathfrak{M} \leqq \mathfrak{o}b^{-1} \leqq \mathfrak{o}b^{-1}\mathfrak{o} \leqq c'\mathfrak{o}$. Thus $\mathfrak{M} \leqq c\mathfrak{o}$ for $c = a^{-1}c'$. Similarly $\mathfrak{M} \leqq \mathfrak{o}d$ for a suitable regular d.

We use this result to prove

THEOREM 6. *If $\mathfrak{o}$ is a bounded maximal order, every maximal order $\mathfrak{o}'$ (equivalent to $\mathfrak{o}$) is bounded.*

Let a be any regular element in $\mathfrak{o}'$ and suppose that b and c are regular elements such that $b\mathfrak{o}c \leqq \mathfrak{o}'$. Then $a\mathfrak{o}' \geqq (ab)\mathfrak{o}c$ and $(ab)\mathfrak{o} \geqq \mathfrak{o}b'$ for a certain regular b'.

[3] Examples of bounded maximal orders will be given later.

Hence $a\mathfrak{o}' \geqq \mathfrak{o}b'c$. If we apply the preceding theorem to $\mathfrak{M} = \mathfrak{o}'$, we obtain the existence of a regular element c' such that $\mathfrak{o}'c' \leqq \mathfrak{o}$. Then $a\mathfrak{o}' \geqq \mathfrak{o}'c'b'c$ and $\mathfrak{o}'$ is bounded as a consequence of Theorem 3.

4. The axioms. We now impose the following conditions on our order $\mathfrak{o}$:

I. $\mathfrak{o}$ is maximal.

II. The descending chain conditions hold for the integral one-sided $\mathfrak{o}$-ideals that contain any fixed integral two-sided $\mathfrak{o}$-ideal.

III. The ascending chain condition holds for integral two-sided $\mathfrak{o}$-ideals.

IV. $\mathfrak{o}$ is bounded.

If $\mathfrak{A}$ is a field, any ideal in the old sense, which is not zero, is an ideal in the sense of this chapter. Hence the conditions II and III are equivalent respectively to Noether's conditions N2 and N3. We recall now the meaning of N1. Suppose that $\mathfrak{A}$ is an algebra with an identity, of finite dimensionality over the underlying field Φ, and let $\mathfrak{g}$ be an order in Φ. Then $\mathfrak{g}$ is any subring of Φ containing 1 and having Φ as its minimal containing field. An element a of $\mathfrak{A}$ is called $\mathfrak{g}$-*integral* if it is a root of a polynomial in $\mathfrak{g}[t]$ having leading coefficient 1. If $\mathfrak{A} = \Phi$, $\mathfrak{o} = \mathfrak{g}$ is *integrally closed* if every element of $\mathfrak{A}$ which is $\mathfrak{o}$-integral belongs to $\mathfrak{o}$. In order to discuss the relation between this property and I we require the following general condition.

THEOREM 7. *If $\mathfrak{g}$ satisfies the ascending chain condition for ideals, then a necessary and sufficient condition that an element a in $\mathfrak{A}$ be $\mathfrak{g}$-integral is that all of the powers a^k be contained in the same finitely generated $\mathfrak{g}$-module (in $\mathfrak{A}$).*

If a is integral, $a^m = a^{m-1}\gamma_1 + \cdots + 1\gamma_m$ where the γ_i are in $\mathfrak{g}$. It follows that all a^k belong to the $\mathfrak{g}$-module $(1, a, \cdots, a^{m-1})$ generated by a^i, $0 \leqq i \leqq m - 1$. Conversely let $\mathfrak{M}$ be a finitely generated $\mathfrak{g}$-module containing all a^k. Since the ascending chain condition holds for the ideals of $\mathfrak{g}$, $\mathfrak{M}$ satisfies the ascending chain condition for $\mathfrak{g}$-modules. Hence for the chain $(1) \leqq (1, a) \leqq (1, a, a^2) \leqq \cdots$ there exists an integer m such that $(1, a, \cdots, a^{m-1}) = (1, a, \cdots, a^m)$. It follows that $a^m = a^{m-1}\gamma_1 + \cdots + 1\gamma_m$ for suitable γ_i in $\mathfrak{g}$.

We may now show that if III holds in an order $\mathfrak{o}$ of a field $\mathfrak{A} = \Phi$, then the conditions I and N1 are equivalent. For let a be integral. Then all the powers of a and hence $\mathfrak{o}[a]$, the set of polynomials in a with coefficients in $\mathfrak{o}$, belong to an $\mathfrak{o}$-module $(a_1, \cdots, a_r)$ generated by elements a_i of $\mathfrak{A}$. We may write $a_i = b_i d^{-1}$ where b_i, d are in $\mathfrak{o}$. Hence $\mathfrak{o}[a] \leqq \mathfrak{o}d^{-1}$ and so $\mathfrak{o}[a]$ is an order (equivalent to $\mathfrak{o}$). Since $\mathfrak{o}[a] \geqq \mathfrak{o}$ our condition I implies that $\mathfrak{o}[a] = \mathfrak{o}$, i.e. $a \in \mathfrak{o}$. Thus $\mathfrak{o}$ is integrally closed. Suppose, conversely, that $\mathfrak{o}$ is integrally closed. Then if $\mathfrak{o}'$ is any order, $\mathfrak{o}' \leqq \mathfrak{o}b$ for a suitable b in $\mathfrak{A}$. Hence the elements of $\mathfrak{o}'$ are $\mathfrak{o}$-integral and so $\mathfrak{o}' \leqq \mathfrak{o}$. We have therefore proved that $\mathfrak{o}$ is maximal. Hence if $\mathfrak{A}$ is a field, the assumptions I, II and III are equivalent to N1, N2 and N3.

We return to the general case of an arbitrary $\mathfrak{A}$ and consider some consequences of our axioms. First, we remark that any ideal of $\mathfrak{o}$ in the old sense, which contains a two-sided $\mathfrak{o}$-ideal $\mathfrak{a}$, is an integral $\mathfrak{o}$-ideal. Since these ideals are in $(1 - 1)$ correspondence with the ideals of the difference ring $\mathfrak{o} - \mathfrak{a}$, condition

II is equivalent to II′: If $\mathfrak{a}$ is an integral two-sided $\mathfrak{o}$-ideal, the descending chain conditions hold for the (ordinary) one-sided ideals of $\mathfrak{o} - \mathfrak{a}$.

We recall that for any ring with an identity the ascending chain conditions for one-sided ideals are consequences of the descending chain conditions. Hence the former hold for $\mathfrak{o}$ and consequently we have

V. The ascending chain conditions hold for the integral one-sided $\mathfrak{o}$-ideals that contain any fixed integral two-sided $\mathfrak{o}$-ideal.

Evidently this implies III. However, we have preferred to state III as a separate assumption since in many important instances, it holds while II fails.

If $\mathfrak{a}$ and $\mathfrak{b}$ are integral right (left, two-sided) $\mathfrak{o}$-ideals, it is clear that $\mathfrak{a} + \mathfrak{b}$ is also of this type. If $\mathfrak{a}$ and $\mathfrak{b}$ are integral two-sided $\mathfrak{o}$-ideals, $\mathfrak{ab}$ contains a regular element and so $\mathfrak{ab}$ is an integral two-sided $\mathfrak{o}$-ideal. Since the intersection $\mathfrak{a} \wedge \mathfrak{b} \geqq \mathfrak{ab}$, it follows that $\mathfrak{a} \wedge \mathfrak{b}$ is an integral two-sided $\mathfrak{o}$-ideal.

Definition 3. An integral two-sided $\mathfrak{o}$-ideal $\mathfrak{p} \neq \mathfrak{o}$ is a *prime ideal* if for any pair of integral two-sided $\mathfrak{o}$-ideals $\mathfrak{a}$, $\mathfrak{b}$ such that $\mathfrak{ab} \equiv 0(\mathfrak{p})$, we have either $\mathfrak{a} \equiv 0(\mathfrak{p})$ or $\mathfrak{b} \equiv 0(\mathfrak{p})$.

If $\mathfrak{ab} \equiv 0(\mathfrak{p})$, then $\mathfrak{a}'\mathfrak{b}' \equiv 0(\mathfrak{p})$ for $\mathfrak{a}' = \mathfrak{a} + \mathfrak{p}$ and $\mathfrak{b}' = \mathfrak{b} + \mathfrak{p}$. Since $\mathfrak{a}' \equiv 0(\mathfrak{p})$ implies $\mathfrak{a} \equiv 0(\mathfrak{p})$, in order to ascertain whether or not $\mathfrak{p}$ is prime it is sufficient to test the integral ideals $\mathfrak{a}'$, $\mathfrak{b}'$ containing $\mathfrak{p}$. Thus if $\mathfrak{p}$ is *maximal* in the sense that no two-sided $\mathfrak{o}$-ideal $\neq \mathfrak{o}$, $\mathfrak{p}$ exists between $\mathfrak{o}$ and $\mathfrak{p}$, then $\mathfrak{p}$ is prime. This remark is the trivial part of the following important theorem:

VI. An integral two-sided $\mathfrak{o}$-ideal $\mathfrak{p} \neq \mathfrak{o}$ is prime if and only if it is maximal. When the condition holds, $\mathfrak{o} - \mathfrak{p}$ is a matrix ring over a division ring.

The property of maximality is equivalent to the property of simplicity of the difference ring $\bar{\mathfrak{o}} = \mathfrak{o} - \mathfrak{p}$. Now suppose that $\mathfrak{p}$ is prime. Since $\bar{\mathfrak{o}}$ satisfies the descending chain condition for left ideals, it has a radical $\bar{\mathfrak{r}}$ and there exists a two-sided $\mathfrak{o}$-ideal $\mathfrak{r}$ in $\mathfrak{o}$ such that $\bar{\mathfrak{r}} = \mathfrak{r} - \mathfrak{p}$. The ring $\mathfrak{o} - \mathfrak{r}$ is isomorphic to $\bar{\mathfrak{o}} - \bar{\mathfrak{r}}$. Hence $\mathfrak{o} - \mathfrak{r}$ is semi-simple. Now $\bar{\mathfrak{r}}^s = 0$ for a suitable s. Hence $\mathfrak{r}^s \equiv 0(\mathfrak{p})$ and since $\mathfrak{p}$ is prime, $\mathfrak{r} \equiv 0(\mathfrak{p})$. This shows that $\bar{\mathfrak{o}}$ is semi-simple. Either $\bar{\mathfrak{o}}$ is simple or there exist two two-sided ideals $\bar{\mathfrak{a}}$, $\bar{\mathfrak{b}} \neq 0$ in $\bar{\mathfrak{o}}$ such that $\bar{\mathfrak{a}}\bar{\mathfrak{b}} = 0$. This implies that there exist two-sided $\mathfrak{o}$-ideals $\mathfrak{a}$, $\mathfrak{b}$ contained in $\mathfrak{o}$ such that $\mathfrak{ab} \equiv 0(\mathfrak{p})$ but neither $\mathfrak{a}$ nor $\mathfrak{b} \equiv 0(\mathfrak{p})$. Hence $\bar{\mathfrak{o}}$ is simple and $\mathfrak{p}$ is maximal. The second part of the theorem is, of course, a consequence of the fundamental structure theorem for simple rings.

If $\mathfrak{o} - \mathfrak{p} = \mathfrak{d}_k$, $\mathfrak{d}$ a division ring, then k is the *capacity* of the prime ideal $\mathfrak{p}$.

5. Orders in an algebra. Let $\mathfrak{A}$ be a separable algebra over Φ, $(\mathfrak{A}:\Phi) = n$, and suppose that $\mathfrak{g}$ is an order in Φ. We consider the problem of embedding $\mathfrak{g}$ in an order $\mathfrak{o}$ of $\mathfrak{A}$.[4] If $\mathfrak{o}$ is such an order, let $\mathfrak{B} = \mathfrak{o}\Phi$ be the smallest sub-algebra of $\mathfrak{A}$ containing $\mathfrak{o}$. Then if b is an element of $\mathfrak{o}$ which is regular in $\mathfrak{A}$, b is regular in $\mathfrak{B}$ and hence its inverse is in $\mathfrak{B}$. It is therefore clear from the definition of an order that $\mathfrak{B} = \mathfrak{A}$, and so $\mathfrak{o}$ contains a basis $u_1, \cdots, u_n$ of $\mathfrak{A}$ over Φ. On the other hand, if $\mathfrak{o}$ is any subring of $\mathfrak{A}$ that contains $\mathfrak{g}$ and contains a basis

[4] For the present the word "order" is used in the original sense. The choice of the equivalence class of orders will be made later.

$u_1, \cdots, u_n$ of $\mathfrak{A}$, then $\mathfrak{o}$ is an order. For, any element of $\mathfrak{A}$ has the form $(\Sigma u_i\gamma_i)\gamma^{-1}$ where the γ_i, γ are in $\mathfrak{g}$ and since $\Sigma u_i\gamma_i$ is in $\mathfrak{o}$, this element has the form $b\gamma^{-1}$, b, γ in $\mathfrak{o}$.

THEOREM 8. *A necessary and sufficient condition that a subring of $\mathfrak{A}$ containing $\mathfrak{g}$ be an order is that it contain a basis of $\mathfrak{A}$ over Φ.*

We may, of course, take $u_1 = 1$ to be one of the elements of the basis. We now assume that $\mathfrak{g}$ satisfies the conditions I and III (or N1, N3) and we suppose that $\mathfrak{o}$ is an order having only $\mathfrak{g}$-integral elements. In order to investigate orders of this type we require the following theorems.

THEOREM 9. *If a and b are $\mathfrak{g}$-integral and $ab = ba$, then $a \pm b$ and ab are $\mathfrak{g}$-integral.*

For if $a^m = \sum_0^{m-1} a^i\gamma_i$ and $b^{m'} = \sum_0^{m'-1} b^j\eta_j$ with the γ's and the η's in $\mathfrak{g}$, then all of the powers $(a + b)^k$ are contained in the $\mathfrak{g}$-module generated by a^ib^j, $0 \leqq i \leqq m - 1, 0 \leqq j \leqq m' - 1$. Hence by Theorem 7, $(a + b)$ is integral. A similar argument applies to $a - b$ and ab.

As an immediate corollary we have the result that if $\mathfrak{A}$ is commutative, the set of $\mathfrak{g}$-integral elements is a subring of $\mathfrak{A}$. This remark may be used in the general case to prove

THEOREM 10. *If a is a $\mathfrak{g}$-integral element, $\mathfrak{g}$ satisfying* I *and* III, *then the minimum polynomial $\mu_a(t)$, the principal polynomial $m_a(t)$, and the characteristic polynomial $f_a(t)$ in any* $(1 - 1)$ *representation of $\mathfrak{A}$ all belong to $\mathfrak{g}[t]$.*

Let $\varphi(t)$ in $\mathfrak{g}[t]$ be a polynomial with leading coefficient 1 having a for a root, and suppose that $\mathfrak{B}$ is a root field over Φ of $\varphi(t)$.[5] Then $\varphi(t) = \Pi(t - a_i)$ in $\mathfrak{B}[t]$ and since $\mu_a(t)$ is a factor of $\varphi(t)$, $\mu_a(t) = \Pi'(t - a_j)$ a product of certain of the factors $(t - a_j)$. The elements a_j are $\mathfrak{g}$-integral in $\mathfrak{B}$ and hence the coefficients of $\mu_a(t)$ are $\mathfrak{g}$-integral. Since $\mathfrak{g}$ is integrally closed, $\mu_a(t)$ is in $\mathfrak{g}[t]$. The results for $m_a(t)$ and $f_a(t)$ follow from this since the roots of these polynomials are the same except for multiplicities as those of $\mu_a(t)$.

Suppose, as before, that $\mathfrak{o}$ is an order containing $\mathfrak{g}$ and consisting of $\mathfrak{g}$-integral elements. Let $u_1 = 1, u_2, \cdots, u_n$ be a basis of $\mathfrak{A}$ contained in $\mathfrak{o}$. Then by replacing the u_i, $i > 1$, by certain multiples $u_i\gamma_i$, γ_i in $\mathfrak{g}$ and returning to the original notation, we may suppose that the constants of multiplication γ_{kij} in $u_iu_j = \Sigma u_k\gamma_{kij}$ are in $\mathfrak{g}$. It follows that the totality of elements $\Sigma u_i\mu_i$, μ_i in $\mathfrak{g}$, is an order $\mathfrak{o}_0 \leqq \mathfrak{o}$.[6] Suppose that $d = \Sigma u_i\delta_i$ is in $\mathfrak{o}$. Then by the preceding theorem the principal traces $T(u_id)$ and $T(u_iu_j)$ belong to $\mathfrak{g}$. We have the equations

$$T(u_id) = \Sigma T(u_iu_j)\delta_j$$

[5] $\mathfrak{B} = \Phi(a_1, \cdots, a_m)$ where $\varphi(t) = \Pi(t - a_i)$ in $\mathfrak{B}[t]$. Cf. Albert's *Modern Higher Algebra*, p. 156.

[6] Incidentally, this argument shows that orders of the type considered here do exist. For, we may start with an arbitrary basis u_i, $u_1 = 1$, for which the multiplication constants are in $\mathfrak{g}$ and take $\mathfrak{o}_0$ to be the set of elements $\Sigma u_i\mu_i$, μ_i in $\mathfrak{g}$. Then $\mathfrak{o}_0$ is an order.

which, since $\Delta = \det(T(u_i u_j)) \neq 0$, give δ_j as an element of $\mathfrak{g}\Delta^{-1}$. Thus $\mathfrak{o}$ is a submodule of the finitely generated $\mathfrak{g}$-module $\mathfrak{o}_0\Delta^{-1}$. Since the ascending chain condition holds for the ideals of $\mathfrak{g}$, any submodule of a finitely generated $\mathfrak{g}$-module is finitely generated.[7] In particular $\mathfrak{o}$ has this property. Conversely, if $\mathfrak{o}$ is any order containing $\mathfrak{g}$ and $\mathfrak{o}$ is finitely generated over $\mathfrak{g}$, then by Theorem 7, all the elements of $\mathfrak{o}$ are $\mathfrak{g}$-integral.

THEOREM 11. *Suppose that $\mathfrak{o}$ is an order containing $\mathfrak{g}$. Then a necessary and sufficient condition that $\mathfrak{o}$ contains only $\mathfrak{g}$-integral elements is that it be a finitely generated $\mathfrak{g}$-module.*

Now if $\mathfrak{M}$ is any $\mathfrak{g}$-module generated by a finite number of elements $v_1, \cdots, v_r$, we may write $v_j = (\Sigma u_i \nu_{ij})\nu^{-1}$, ν_{ij}, and ν in $\mathfrak{g}$. Then $\mathfrak{M} \leqq \mathfrak{o}_0\nu^{-1} \leqq \mathfrak{o}\nu^{-1}$. In particular, if $\mathfrak{o}'$ is any order of $\mathfrak{g}$-integral elements and $\mathfrak{o}'$ contains $\mathfrak{g}$, then $\mathfrak{o}' \leqq \mathfrak{o}\nu^{-1}$ for a suitable ν. By symmetry there exists an element μ in $\mathfrak{g}$ such that $\mathfrak{o}'\mu^{-1} \geqq \mathfrak{o}$ and so we have proved the following

THEOREM 12. *Any two orders $\mathfrak{o}$ and $\mathfrak{o}'$ of $\mathfrak{g}$-integral elements containing $\mathfrak{g}$ are equivalent.*

If we refer back to the proof of Theorem 11, we see that the element Δ does not depend on $\mathfrak{o}$ but rather on the basis $u_1, \cdots, u_n$. Hence our argument shows that if $\mathfrak{o}'$ is any order containing $\mathfrak{o}$ and containing only $\mathfrak{g}$-integral elements, then $\mathfrak{o}' \leqq \mathfrak{o}_0\Delta^{-1}$. Moreover if $\mathfrak{o}'$ is any order equivalent to $\mathfrak{o}$, $\mathfrak{o}' \leqq a\mathfrak{o}b$, a finitely generated $\mathfrak{g}$-module, and so all the elements of $\mathfrak{o}'$ are $\mathfrak{g}$-integral. Thus any $\mathfrak{o}'$ containing $\mathfrak{o}$ and equivalent to $\mathfrak{o}$ is contained in $\mathfrak{o}_0\Delta^{-1}$; it follows that there exists a maximal order $\mathfrak{o}'$ equivalent to $\mathfrak{o}$ and containing $\mathfrak{o}$.

THEOREM 13. *If $\mathfrak{o}$ is as in the preceding theorem, then $\mathfrak{o}$ may be embedded in a maximal order $\mathfrak{o}'$ equivalent to $\mathfrak{o}$.*

Let $\mathfrak{G}$ denote the ring of $\mathfrak{g}$-integral elements of the center $\mathfrak{C}$ of $\mathfrak{A}$. Then if $\mathfrak{o}$ is an order containing only $\mathfrak{g}$-integral elements, by Theorem 9, $\mathfrak{o}\mathfrak{G}$ is an order containing $\mathfrak{G}$ (and hence $\mathfrak{g}$) and containing only $\mathfrak{g}$-integral elements. If $\mathfrak{o}$ itself contains $\mathfrak{g}$, we have seen that $\mathfrak{o}$ and $\mathfrak{o}\mathfrak{G}$ are equivalent. Hence if, in addition, $\mathfrak{o}$ is maximal, then $\mathfrak{o} = \mathfrak{o}\mathfrak{G}$ and $\mathfrak{o}$ contains $\mathfrak{G}$.

THEOREM 14. *Any maximal order $\mathfrak{o}$ that contains $\mathfrak{g}$ and contains only $\mathfrak{g}$-integral elements contains all the $\mathfrak{g}$-integral elements of the center of $\mathfrak{A}$.*

Let $\mathfrak{o}'$ be an order equivalent to $\mathfrak{o}$, an order that contains $\mathfrak{g}$ and contains only $\mathfrak{g}$-integral elements. Then we have seen that $\mathfrak{o}' \leqq a\mathfrak{o}b$, a finitely generated $\mathfrak{g}$-module, and consequently that all the elements of $\mathfrak{o}'$ are $\mathfrak{g}$-integral. Now let $\mathfrak{A}$ be commutative and let $\mathfrak{o}$ be the totality of $\mathfrak{g}$-integral elements. Then $\mathfrak{o}$ is an order, and if $\mathfrak{o}'$ is any order equivalent to $\mathfrak{o}$, its elements are $\mathfrak{g}$-integral and so $\mathfrak{o}' \leqq \mathfrak{o}$.

THEOREM 15. *If $\mathfrak{A}$ is commutative, the totality $\mathfrak{o}$ of $\mathfrak{g}$-integral elements of $\mathfrak{A}$ is a maximal order. Any order equivalent to $\mathfrak{o}$ is contained in $\mathfrak{o}$.*

[7] Theorem 3, Chapter 3.

If $\mathfrak{o}' \leqq a\mathfrak{o}b$ and $\mathfrak{o}$ is an order containing $\mathfrak{g}$, then $\mathfrak{o}'\mathfrak{g} \leqq a\mathfrak{o}b$ also. Hence if $\mathfrak{o}'$ is equivalent to $\mathfrak{o}$, $\mathfrak{o}'\mathfrak{g}$ is equivalent to $\mathfrak{o}$ and if $\mathfrak{o}'$ is maximal, then $\mathfrak{o}' = \mathfrak{o}'\mathfrak{g}$. To sum up; if $\mathfrak{g}$ satisfies I and III, the orders $\mathfrak{o}$ that have the properties: 1) $\mathfrak{o}$ contains $\mathfrak{g}$ and 2) $\mathfrak{o}$ contains only $\mathfrak{g}$-integral elements, belong to a single equivalence class. All orders of this class have property 2) and all maximal orders in the class have both properties. For the remainder of this section we assume that $\mathfrak{g}$ satisfies I and III and that $\mathfrak{o}$ satisfies 1) and 2).

If $\mathfrak{a}$ is a right ideal, $\mathfrak{a} \leqq a\mathfrak{o}$ a principal $\mathfrak{o}$-ideal and so $\mathfrak{a}$ is a finitely generated $\mathfrak{g}$-module. Since $\mathfrak{a}$ contains a regular element b, it contains a basis $v_1 = bu_1, \cdots, v_n = bu_n$ of $\mathfrak{A}$. The identity $1 = \Sigma v_i\rho_i$, ρ_i in Φ, and hence there is a relation of the form $\eta = \Sigma v_i\eta_i$ with η, η_i in $\mathfrak{g}$. Thus η is in the intersection $\mathfrak{a} \wedge \mathfrak{g}$ and $\mathfrak{a} \wedge \mathfrak{g} \neq 0$. Evidently $\mathfrak{a} \wedge \mathfrak{g}$ is a $\mathfrak{g}$-ideal. We note also that the ideal $\eta\mathfrak{o} = \mathfrak{o}\eta$ is a two-sided $\mathfrak{o}$-ideal contained in $\mathfrak{a}$ and so $\mathfrak{a}$ is bounded.

THEOREM 16. *Any order $\mathfrak{o}$ in $\mathfrak{A}$ is bounded.*

Suppose next that $\mathfrak{a}$ is an integral two-sided $\mathfrak{o}$-ideal. We consider the difference $\mathfrak{g}$-module $\mathfrak{o} - \mathfrak{a}$ and note that it is finitely generated. Since it is annihilated by $\mathfrak{a}_0 = \mathfrak{a} \wedge \mathfrak{g}$, it may be regarded as a $(\mathfrak{g} - \mathfrak{a}_0)$-module. We have seen that $\mathfrak{a}_0 \neq 0$. Hence if $\mathfrak{g}$ satisfies condition II, $\mathfrak{g} - \mathfrak{a}_0$ satisfies the descending chain condition for ideals and therefore $\mathfrak{o} - \mathfrak{a}$ satisfies the descending chain condition for $(\mathfrak{g} - \mathfrak{a}_0)$-sub-modules. It follows that the descending chain condition holds for the integral $\mathfrak{o}$-ideals containing $\mathfrak{a}$.

THEOREM 17. *If $\mathfrak{g}$ satisfies condition* II, *any order $\mathfrak{o}$ of $\mathfrak{A}$ satisfies this condition also.*

A special type of domain $\mathfrak{g}$ that satisfies our conditions is a principal ideal domain. For we have seen in Chapter 3 that N2 and N3 hold for $\mathfrak{g}$ and N1 may be proved as in the case of the ring of integers by using the unique factorization theorem. Let $\mathfrak{o}$ be an order of $\mathfrak{A}$ containing $\mathfrak{g}$ and containing only $\mathfrak{g}$-integral elements. Then if $u_1, \cdots, u_n$ is a basis for $\mathfrak{A}$ contained in $\mathfrak{o}$ and having an integral multiplication table, we have seen that $\mathfrak{o}$ contains the free $\mathfrak{g}$-module with the basis u_i and $\mathfrak{o}$ is contained in the free $\mathfrak{g}$-module having the basis $u_i\Delta^{-1}$ where $\Delta = \det (T(u_iu_j))$. It follows that $\mathfrak{o}$ is a free $\mathfrak{g}$-module with a basis of n elements.

THEOREM 18. *If $\mathfrak{g}$ is a principal ideal domain, any order $\mathfrak{o}$ has a free basis of $n = (\mathfrak{A}:\Phi)$ elements.*

6. Factorization of two-sided ideals. We return now to the general case of an arbitrary ring $\mathfrak{A}$ and an order $\mathfrak{o}$ in $\mathfrak{A}$ that satisfies conditions I–IV.[8] Our first aim is to prove the existence and the uniqueness of factorization of any integral two-sided $\mathfrak{o}$-ideal as a product of prime ideals. If we examine the argument of **5**, Chapter 3, we see that the decisive step is the theorem that if $\mathfrak{a}$ and $\mathfrak{b}$ are two-sided $\mathfrak{o}$-ideals, then $\mathfrak{a} \leqq \mathfrak{b}$ if and only if there exists an integral

[8] As a matter of fact we shall require in this section only conditions I, III, IV and VII: any prime $\mathfrak{o}$-ideal is maximal.

two-sided $\mathfrak{o}$-ideal $\mathfrak{c}$ such that $\mathfrak{a} = \mathfrak{c}\mathfrak{b}$. This fact will be established here as a consequence of the following

THEOREM 19. *If $\mathfrak{a}$ is a two-sided $\mathfrak{o}$-ideal properly contained in $\mathfrak{o}$, then $\mathfrak{a}^{-1} > \mathfrak{o}$.*

We require a pair of lemmas.

LEMMA 1. *Any integral two-sided $\mathfrak{o}$-ideal $\mathfrak{a}$ is contained in a prime ideal.*

This is clear because of the ascending chain condition and the criterion VI.

LEMMA 2. *Any integral two-sided $\mathfrak{o}$-ideal $\mathfrak{a}$ contains a product of prime ideals.*

If $\mathfrak{a}$ is prime, the lemma holds. Otherwise there exist two integral two-sided $\mathfrak{o}$-ideals $\mathfrak{a}'$ and $\mathfrak{a}''$ containing $\mathfrak{a}$ such that $\mathfrak{a}'\mathfrak{a}'' \equiv 0\ (\mathfrak{a})$ but $\mathfrak{a}' \not\equiv 0$, $\mathfrak{a}'' \not\equiv 0\ (\mathfrak{a})$. Then $\mathfrak{a}' > \mathfrak{a}$, $\mathfrak{a}'' > \mathfrak{a}$. If we repeat this process with $\mathfrak{a}'$ and $\mathfrak{a}''$ and the ideals arising from them, we obtain the lemma as a consequence of the ascending chain condition.

Proof of theorem. Let $\mathfrak{p}$ be a prime ideal containing $\mathfrak{a}$. Then if $\mathfrak{a}^{-1} = \mathfrak{o}$, $\mathfrak{p}^{-1} = \mathfrak{o}$. Let a be a regular element in $\mathfrak{p}$ and consider the right $\mathfrak{o}$-ideal $a\mathfrak{o}$ contained in $\mathfrak{p}$. By the boundedness condition, $a\mathfrak{o}$ contains a two-sided $\mathfrak{o}$-ideal, and so by Lemma 2 $a\mathfrak{o}$ contains a product $\mathfrak{p}_1 \cdots \mathfrak{p}_r$ of prime ideals $\mathfrak{p}_i$. We suppose that the $\mathfrak{p}_i$ have been selected so that r is minimal, i.e. $a\mathfrak{o}$ contains no product of $r - 1$ prime ideals. Since $\mathfrak{p} \geqq a\mathfrak{o} \geqq \mathfrak{p}_1 \cdots \mathfrak{p}_r$, one of the $\mathfrak{p}_i = \mathfrak{p}$ and $\mathfrak{p}_1 \cdots \mathfrak{p}_r = \mathfrak{b}\mathfrak{p}\mathfrak{c}$. Then $a^{-1}\mathfrak{b}\mathfrak{p}\mathfrak{c} \leqq \mathfrak{o}$ and $a^{-1}\mathfrak{b} \leqq (\mathfrak{p}\mathfrak{c})^{-1}$. Since $(\mathfrak{p}\mathfrak{c})(\mathfrak{p}\mathfrak{c})^{-1} \leqq \mathfrak{o}$, $\mathfrak{p}(\mathfrak{c}(\mathfrak{p}\mathfrak{c})^{-1}) \leqq \mathfrak{o}$ and $\mathfrak{c}(\mathfrak{p}\mathfrak{c})^{-1} \leqq \mathfrak{p}^{-1} = \mathfrak{o}$. Thus $(\mathfrak{p}\mathfrak{c})^{-1} \leqq \mathfrak{c}^{-1}$ and since $(\mathfrak{p}\mathfrak{c})^{-1} \geqq \mathfrak{c}^{-1}$, we have $(\mathfrak{p}\mathfrak{c})^{-1} = \mathfrak{c}^{-1}$. This implies that $a^{-1}\mathfrak{b} \leqq \mathfrak{c}^{-1}$, $a^{-1}\mathfrak{b}\mathfrak{c} \leqq \mathfrak{o}$ and $\mathfrak{b}\mathfrak{c} \leqq a\mathfrak{o}$. Since $\mathfrak{b}\mathfrak{c}$ is a product of $r - 1$ prime ideals, we have a contradiction to the minimality of r. An important consequence of this theorem is

THEOREM 20. *If $\mathfrak{a}$ is a right (left) $\mathfrak{o}$-ideal, $\mathfrak{a}^{-1}\mathfrak{a} = \mathfrak{o}$ ($\mathfrak{a}\mathfrak{a}^{-1} = \mathfrak{o}$).*

The set $\mathfrak{a}^{-1}\mathfrak{a}$ is contained in $\mathfrak{o}$ and is therefore an integral two-sided $\mathfrak{o}$-ideal. Since $\mathfrak{o}$ is maximal, the orders of $\mathfrak{a}^{-1}\mathfrak{a}$ are $\mathfrak{o}$. Now $(\mathfrak{a}^{-1}\mathfrak{a})^{-1}(\mathfrak{a}^{-1}\mathfrak{a}) \leqq \mathfrak{o}$ and so $(\mathfrak{a}^{-1}\mathfrak{a})^{-1}\mathfrak{a}^{-1} \leqq \mathfrak{a}^{-1}$. Hence $(\mathfrak{a}^{-1}\mathfrak{a})^{-1} \leqq \mathfrak{o}$ and so by the preceding theorem $\mathfrak{a}^{-1}\mathfrak{a} = \mathfrak{o}$.

We may now prove the important

THEOREM 21. *If $\mathfrak{a}$ is a right $\mathfrak{o}$-ideal contained in a two-sided $\mathfrak{o}$-ideal $\mathfrak{b}$, there exists an integral right $\mathfrak{o}$-ideal $\mathfrak{c}$ such that $\mathfrak{a} = \mathfrak{c}\mathfrak{b}$.*

Since $\mathfrak{a} \leqq \mathfrak{b}$, $\mathfrak{c} \equiv \mathfrak{a}\mathfrak{b}^{-1} \leqq \mathfrak{b}\mathfrak{b}^{-1} = \mathfrak{o}$ and $\mathfrak{c}$ is an integral right $\mathfrak{o}$-ideal. Since $\mathfrak{b}^{-1}\mathfrak{b} = \mathfrak{o}$, $\mathfrak{c}\mathfrak{b} = \mathfrak{a}\mathfrak{b}^{-1}\mathfrak{b} = \mathfrak{a}$.

We may now carry over the discussion of **5**, Chapter 3. We obtain then 1) the commutativity of multiplication of integral two-sided $\mathfrak{o}$-ideals and 2) the unique factorization of any integral two-sided ideal as a product of prime ideals. By Theorem 20 the two-sided $\mathfrak{o}$-ideals form a group $G(\mathfrak{o})$ under multiplication with $\mathfrak{o}$ as the identity and $\mathfrak{a}^{-1}$ as the inverse of $\mathfrak{a}$. Now if $\mathfrak{a}$ is any two-sided $\mathfrak{o}$-ideal, there exists a regular element a in $\mathfrak{o}$ such that $a\mathfrak{a} \leqq \mathfrak{o}$, or $(a\mathfrak{o})\mathfrak{a} \leqq \mathfrak{o}$.[9] The ideal $a\mathfrak{o}$ contains an integral two-sided $\mathfrak{o}$-ideal $\mathfrak{b}$ and so $\mathfrak{b}\mathfrak{a} = \mathfrak{c} \leqq \mathfrak{o}$. Thus

[9] For there exists a regular element $b = a^{-1}c$, a and c in $\mathfrak{o}$, such that $\mathfrak{a} \leqq b\mathfrak{o}$. Hence $\mathfrak{a} \leqq a^{-1}c\mathfrak{o} \leqq a^{-1}\mathfrak{o}$ and $a\mathfrak{a} \leqq \mathfrak{o}$.

$\mathfrak{a} = \mathfrak{b}^{-1}\mathfrak{c}$ where $\mathfrak{b}$ and $\mathfrak{c}$ are integral and $G(\mathfrak{o})$ is generated by the integral ideals contained in it. It follows, of course, that $G(\mathfrak{o})$ is commutative. Every $\mathfrak{a}$ in $G(\mathfrak{o})$ has a unique representation in the form $\mathfrak{p}_1^{g_1} \cdots \mathfrak{p}_s^{g_s}$ where the $g_i \gtrless 0$ and the $\mathfrak{p}_i$ are distinct prime ideals. Hence we have proved the fundamental

THEOREM 22. *The two-sided* $\mathfrak{o}$*-ideals form a commutative group* $G(\mathfrak{o})$ *under multiplication.* $G(\mathfrak{o})$ *is a direct product of the infinite cyclic groups generated by the prime ideals of* $\mathfrak{o}$.

7. The structure of $\mathfrak{o} - \mathfrak{a}$. Let $\mathfrak{a} = \mathfrak{p}_1^{e_1} \cdots \mathfrak{p}_s^{e_s}$ where the $\mathfrak{p}_i$ are distinct primes and the e_i are > 0. Then if we set $\mathfrak{a}_i = \mathfrak{a}\mathfrak{p}_i^{-e_i}$, we obtain $\mathfrak{o} = \mathfrak{a}_1 + \cdots + \mathfrak{a}_s$ and $\mathfrak{a}_i \wedge (\mathfrak{a}_1 + \cdots + \mathfrak{a}_{i-1} + \mathfrak{a}_{i+1} + \cdots + \mathfrak{a}_s) = \mathfrak{a}$. This follows directly from Theorem 21 and the unique factorization theorem. Hence $\mathfrak{o} - \mathfrak{a} = \bar{\mathfrak{a}}_1 \oplus \cdots \oplus \bar{\mathfrak{a}}_s$ where $\bar{\mathfrak{a}}_i = \mathfrak{a}_i - \mathfrak{a}$. We have the relation $\mathfrak{o} - \mathfrak{p}_i^{e_i} = (\mathfrak{a}_i + \mathfrak{p}_i^{e_i}) - \mathfrak{p}_i^{e_i} \cong \mathfrak{a}_i - (\mathfrak{a}_i \wedge \mathfrak{p}_i^{e_i}) = \bar{\mathfrak{a}}_i$. We note also that if $\mathfrak{a} = \mathfrak{p}^e$, $\mathfrak{p}$ a prime, then $\bar{\mathfrak{o}} = \mathfrak{o} - \mathfrak{a}$ contains the nilpotent ideal $\bar{\mathfrak{p}} = \mathfrak{p} - \mathfrak{p}^e$ and $\bar{\mathfrak{o}} - \bar{\mathfrak{p}}$ is isomorphic to $\mathfrak{o} - \mathfrak{p}$ a simple ring. It follows that $\bar{\mathfrak{p}}$ is the radical of $\bar{\mathfrak{o}}$ and $\bar{\mathfrak{o}}$ is a primary ring.

We wish to prove that for arbitrary $\mathfrak{a}$, $\mathfrak{o} - \mathfrak{a}$ is a principal ideal ring. Because of the direct sum decomposition it suffices to take $\mathfrak{a} = \mathfrak{p}^e$, and by Theorem 41 of Chapter 4, our theorem will be proved if we show that the radical $\bar{\mathfrak{p}} = \mathfrak{p} - \mathfrak{p}^e$ of $\mathfrak{o} - \mathfrak{p}^e$ is a principal right ideal and a principal left ideal. Now the ideals of $\bar{\mathfrak{o}}$ have the form $\bar{\mathfrak{b}} = \mathfrak{b} - \mathfrak{p}^e$ when $\mathfrak{b}$ is an integral $\mathfrak{o}$-ideal containing $\mathfrak{p}^e$. Hence by Theorem 21 any right (left) ideal $\bar{\mathfrak{b}}$ of $\bar{\mathfrak{o}}$ that is contained in $\bar{\mathfrak{p}}$ has the form $\bar{\mathfrak{c}}\bar{\mathfrak{p}}$ ($\bar{\mathfrak{p}}\bar{\mathfrak{c}}$) where $\bar{\mathfrak{c}}$ is a right (left) ideal. The theorem will therefore follow from

THEOREM 23. *Let* $\mathfrak{O}$ *be a primary ring and let* $\mathfrak{P}$ *be its radical. Then if every right (left) ideal of* $\mathfrak{O}$ *contained in* $\mathfrak{P}$ *may be written in the form* $\mathfrak{C}\mathfrak{P}$ ($\mathfrak{P}\mathfrak{C}$) *where* $\mathfrak{C}$ *is a right (left) ideal,* $\mathfrak{P}$ *is a principal right (left) ideal.*

We recall that $\mathfrak{O}$ is a matrix ring $\mathfrak{O}_{0k}$ where $\mathfrak{O}_0$ is completely primary. The radical of $\mathfrak{O}_0$ is $\mathfrak{O}_0 \wedge \mathfrak{P} = \mathfrak{P}_0$ and $\mathfrak{P} = \Sigma e_{ij}\mathfrak{P}_0$ if e_{ij} form a matrix basis for $\mathfrak{O} = \Sigma e_{ij}\mathfrak{O}_0$. We suppose first that $\mathfrak{P}^2 = 0$. Let w be an element $\neq 0$ in $\mathfrak{P}_0$ and consider the right ideal $w\mathfrak{O}$. Evidently $w\mathfrak{O} \leqq \mathfrak{P}$ and so $w\mathfrak{O} = \mathfrak{C}\mathfrak{P}$, $\mathfrak{C}$ a right ideal. Since $(\mathfrak{C} + \mathfrak{P})\mathfrak{P} = \mathfrak{C}\mathfrak{P}$, we may suppose that $\mathfrak{C} > \mathfrak{P}$. Consider the simple ring $\bar{\mathfrak{O}} = \mathfrak{O} - \mathfrak{P}$ and the right ideal $\bar{\mathfrak{C}} = \mathfrak{C} - \mathfrak{P}$ in it. We know that $\bar{\mathfrak{C}}$ has the form $\bar{u}\bar{\mathfrak{O}}$ where $\bar{u}$ is an idempotent element $\neq 0$. Now the cosets $\bar{e}_{ij} = e_{ij} + \mathfrak{P}$ form a matrix basis for $\bar{\mathfrak{O}}$ and $\bar{\mathfrak{O}} = \Sigma\bar{e}_{ij}\bar{\mathfrak{O}}_0$ where $\bar{\mathfrak{O}}_0 = (\mathfrak{O}_0 + \mathfrak{P}) - \mathfrak{P}$ is a division ring isomorphic to $\mathfrak{O}_0 - \mathfrak{P}_0$. It follows that there exists a regular element $\bar{q}$ in $\bar{\mathfrak{O}}$ such that $\bar{u} = \bar{q}^{-1}\sum_1^t \bar{e}_{ii}\bar{q}$. Any element q in the coset $\bar{q}$ is regular in $\mathfrak{O}$ and because of the form of $\bar{\mathfrak{C}}$, $\mathfrak{C}$ consists of the elements of the form $(q^{-1}e_tq)x + z$ where x is in $\mathfrak{O}$, z is in $\mathfrak{P}$ and $e_t = \sum_1^t e_{ii}$. Hence $w = \sum_j q^{-1}e_ty_j$ with y_j in $\mathfrak{P}$ and $qw = \sum e_ty_j$. If we write $q = \Sigma e_{ij}q_{ij}$, q_{ij} in $\mathfrak{O}_0$, this equation implies that $q_{ij}w = 0$ for $i = t + 1, \cdots, n$. Since every element of $\mathfrak{O}_0$ that is not a unit is contained in $\mathfrak{P}_0$, these q_{ij} are in $\mathfrak{P}_0$. This contradicts the fact that q is regular, unless $t = k$, i.e. $\bar{u} = \bar{1}$. Then

$\bar{\mathfrak{C}} = \bar{\mathfrak{O}}$, $\mathfrak{C} = \mathfrak{O}$ and $\mathfrak{P} = w\mathfrak{O}$ is a principal right ideal. If $\mathfrak{P}^2 \neq 0$, we consider $\mathfrak{O} - \mathfrak{P}^2$ and note that it satisfies the hypothesis of the theorem. Moreover its radical $\mathfrak{P}^* = \mathfrak{P} - \mathfrak{P}^2$ satisfies $\mathfrak{P}^{*2} = 0$. Hence $\mathfrak{P} - \mathfrak{P}^2$ is principal. This implies that $\mathfrak{P} = w\mathfrak{O} + \mathfrak{P}^2$. Then $\mathfrak{P}^2 = w\mathfrak{P} + \mathfrak{P}^3$, $\mathfrak{P}^3 = w\mathfrak{P}^2 + \mathfrak{P}^4, \cdots$ and so $\mathfrak{P} = w\mathfrak{O}$. In a like manner we prove that $\mathfrak{P}$ is a principal left ideal.

As we have seen, this implies

THEOREM 24. *If $\mathfrak{a}$ is an integral two-sided $\mathfrak{o}$-ideal, $\mathfrak{o} - \mathfrak{a}$ is a principal ideal ring.*

8. Bounded $\mathfrak{o}$-modules. The preceding theorem enables us to obtain the structure of any finitely generated $\mathfrak{o}$-module $\mathfrak{M}$ that is *bounded* in the sense that its annihilating ideal contains a regular element. Then the annihilating ideal $\mathfrak{a}$ is an integral two-sided $\mathfrak{o}$-ideal and $\mathfrak{M}$ may be regarded as an $\bar{\mathfrak{o}}$-module, $\bar{\mathfrak{o}} = \mathfrak{o} - \mathfrak{a}$. Since $\bar{\mathfrak{o}}$ is a principal ideal ring satisfying the descending chain conditions for one-sided ideals, the results of **15–16**, Chapter 4 are directly applicable. We obtain in this way that $\mathfrak{M} = \mathfrak{M}_1 \oplus \cdots \oplus \mathfrak{M}_u$ where $\mathfrak{M}_i$ is an indecomposable cyclic $\mathfrak{o}$-module (or $\bar{\mathfrak{o}}$-module).

If we call the annihilating ideal $\mathfrak{a}$ the *bound* of $\mathfrak{M}$, then in order that $\mathfrak{M}$ be indecomposable it is necessary that its bound be a prime power. This follows directly from the decomposition of $\bar{\mathfrak{o}} = \mathfrak{o} - \mathfrak{a}$ as $\bar{\mathfrak{a}}_1 \oplus \cdots \oplus \bar{\mathfrak{a}}_s$ where $\bar{\mathfrak{a}}_i = \mathfrak{a}\mathfrak{p}_i^{-e_i} - \mathfrak{a}$ and $\mathfrak{a} = \mathfrak{p}_1^{e_1} \cdots \mathfrak{p}_s^{e_s}$ is the decomposition of $\mathfrak{a}$ into powers of distinct prime ideals. Evidently if $\mathfrak{M}$ and $\mathfrak{N}$ are bounded and $\mathfrak{o}$-isomorphic, they have the same bound. On the other hand, if $\mathfrak{M}$ and $\mathfrak{N}$ are indecomposable and have the same bound $\mathfrak{p}^e$, then both of these modules may be regarded as $(\mathfrak{o} - \mathfrak{p}^e)$-modules. Hence by **16** Chapter 4, $\mathfrak{M}$ and $\mathfrak{N}$ are $(\mathfrak{o} - \mathfrak{p}^e)$-isomorphic. It follows that $\mathfrak{M}$ and $\mathfrak{N}$ are $\mathfrak{o}$-isomorphic. We recall also that an indecomposable bounded $\mathfrak{o}$-module $\mathfrak{M}$ has only one composition series and its length is the exponent e of the prime $\mathfrak{p}$ in the bound $\mathfrak{p}^e$ of $\mathfrak{M}$. Any submodule and any difference module of $\mathfrak{M}$ are indecomposable.[10]

Now let $\mathfrak{M}$ be arbitrary and suppose that $\mathfrak{M} = \mathfrak{M}_1 \oplus \cdots \oplus \mathfrak{M}_u$ where the $\mathfrak{M}_i$ are indecomposable and $\neq 0$. If the bound of $\mathfrak{M}_i$ is $\mathfrak{p}_i^{e_i}$, $\mathfrak{p}_i$ a prime, then by the Krull-Schmidt Theorem we see that the ideals $\mathfrak{p}_1^{e_1}, \cdots, \mathfrak{p}_u^{e_u}$ are invariants of $\mathfrak{M}$. If $\mathfrak{M}$ and $\mathfrak{N}$ are two $\mathfrak{o}$-isomorphic bounded modules, then they have the same invariants. On the other hand, if $\mathfrak{M}$ and $\mathfrak{N}$ have the same invariants, we may suppose that the subscripts of the indecomposable components have been chosen so that $\mathfrak{M}_i$ and $\mathfrak{N}_i$ have the same bounds Then $\mathfrak{M}_i$ and $\mathfrak{N}_i$, and consequently $\mathfrak{M}$ and $\mathfrak{N}$, are $\mathfrak{o}$-isomorphic.

For the applications to ideal theory it is more convenient to deal with the dual decomposition of 0 as a direct intersection. Here we consider submodules $\mathfrak{M}_i' \neq \mathfrak{M}$ such that $0 = \mathfrak{M}_1' \wedge \cdots \wedge \mathfrak{M}_u'$, $\mathfrak{M}_i' + (\mathfrak{M}_1' \wedge \cdots \wedge \mathfrak{M}_{i-1}' \wedge \mathfrak{M}_{i+1}' \wedge \cdots \wedge \mathfrak{M}_u') = \mathfrak{M}$ and $\mathfrak{M} - \mathfrak{M}_i'$ is indecomposable. We recall that if $\mathfrak{M} = \mathfrak{M}_1 \oplus \cdots \oplus \mathfrak{M}_u$ where the $\mathfrak{M}_i$ are indecomposable, then we obtain a dual

[10] In general, any submodule (difference module) of a bounded $\mathfrak{o}$-module $\mathfrak{M}$ is bounded. Its bound is a divisor of that of $\mathfrak{M}$.

decomposition of 0 by taking $\mathfrak{M}'_i = \mathfrak{M}_1 + \cdots + \mathfrak{M}_{i-1} + \mathfrak{M}_{i+1} + \cdots + \mathfrak{M}_u$. Conversely, any set of $\mathfrak{M}'_i$ lead to a set of $\mathfrak{M}_i$ by means of the definition $\mathfrak{M}_i = \mathfrak{M}'_1 \wedge \cdots \wedge \mathfrak{M}'_{i-1} \wedge \mathfrak{M}'_{i+1} \wedge \cdots \wedge \mathfrak{M}'_u$. It follows from what we have shown that $\mathfrak{M}$ is completely determined in the sense of $\mathfrak{o}$-isomorphism by the bounds of the modules $\mathfrak{M} - \mathfrak{M}'_i$ ($\mathfrak{o}$-isomorphic to $\mathfrak{M}_i$) where the $\mathfrak{M}'_i$ are the components of the dual decomposition.

9. Decomposition of integral $\mathfrak{o}$-ideals. The significance of the assumption that the integral right (left) $\mathfrak{o}$-ideals are bounded may now be seen. If $\mathfrak{b}$ is an integral right $\mathfrak{o}$-ideal, the boundedness of $\mathfrak{b}$ implies that the $\mathfrak{o}$-module $\mathfrak{M} = \mathfrak{o} - \mathfrak{b}$ is bounded. For if $\mathfrak{a}$ is a two-sided $\mathfrak{o}$-ideal contained in $\mathfrak{b}$, $\mathfrak{a}$ is contained in the annihilator of $\mathfrak{M}$. The bound of $\mathfrak{M}$ is the join of all two-sided $\mathfrak{o}$-ideals contained in $\mathfrak{b}$. We shall refer to this $\mathfrak{o}$-ideal also as the (right) *bound* of $\mathfrak{b}$. A similar definition holds for the left bound of an integral left $\mathfrak{o}$-ideal.

Corresponding to the dual decomposition of $\mathfrak{M} = \mathfrak{o} - \mathfrak{b}$ we obtain right $\mathfrak{o}$-ideals $\mathfrak{q}_i$ ($i = 1, \cdots, u$) such that

$$\mathfrak{q}_1 \wedge \cdots \wedge \mathfrak{q}_u = \mathfrak{b}, \qquad \mathfrak{q}_i + (\mathfrak{q}_1 \wedge \cdots \wedge \mathfrak{q}_{i-1} \wedge \mathfrak{q}_{i+1} \wedge \cdots \wedge \mathfrak{q}_u) = \mathfrak{o}$$

or, if we use the customary notation [,] for the intersection and (,) for the join, then

$$(1) \qquad [\mathfrak{q}_1, \cdots, \mathfrak{q}_u] = \mathfrak{b}, \qquad (\mathfrak{q}_i, [\mathfrak{q}_1, \cdots, \mathfrak{q}_{i-1}, \mathfrak{q}_{i+1}, \cdots, \mathfrak{q}_u]) = \mathfrak{o}.$$

The dual components of $\mathfrak{M}$ and $\mathfrak{M}'_i = \mathfrak{q}_i - \mathfrak{b}$. Since $\mathfrak{M} - \mathfrak{M}'_i$ is indecomposable, $\mathfrak{o} - \mathfrak{q}_i$, which is $\mathfrak{o}$-isomorphic to $(\mathfrak{o} - \mathfrak{b}) - (\mathfrak{b} - \mathfrak{q}_i) = \mathfrak{M} - \mathfrak{M}'_i$, is indecomposable. The bound of $\mathfrak{M} - \mathfrak{M}'_i$ is the bound of the ideal $\mathfrak{q}_i$. Evidently the converse also holds: Any decomposition of $\mathfrak{b}$ as a direct intersection of ideals (in the sense of equation (1)) such that $\mathfrak{o} - \mathfrak{q}_i$ are indecomposable $\mathfrak{o}$-modules yields a decomposition of 0 in $\mathfrak{M} = \mathfrak{o} - \mathfrak{b}$ as a direct intersection of submodules $\mathfrak{M}'_i$ such that $\mathfrak{M} - \mathfrak{M}'_i$ are indecomposable. It follows from the general theory that if $\bar{\mathfrak{b}}$ is a second integral right $\mathfrak{o}$-ideal and $\bar{\mathfrak{b}} = [\bar{\mathfrak{q}}_1, \bar{\mathfrak{q}}_2, \cdots]$ is a direct intersection such that the $\mathfrak{o} - \bar{\mathfrak{q}}_i$ are indecomposable, then a necessary and sufficient condition that $\mathfrak{o} - \mathfrak{b}$ and $\mathfrak{o} - \bar{\mathfrak{b}}$ be $\mathfrak{o}$-isomorphic is that the bounds $\bar{\mathfrak{p}}_1^{f_1}, \bar{\mathfrak{p}}_2^{f_1}, \cdots$ of $\bar{\mathfrak{q}}_1, \bar{\mathfrak{q}}_2, \cdots$ be the same (except for order) as those of $\mathfrak{q}_1, \mathfrak{q}_2, \cdots$. As in the case of principal ideal domains, we call $\mathfrak{b}$ and $\bar{\mathfrak{b}}$ (right) *similar* if $\mathfrak{o} - \mathfrak{b}$ and $\mathfrak{o} - \bar{\mathfrak{b}}$ are $\mathfrak{o}$-isomorphic. Then we have the following

THEOREM 25. *If $\mathfrak{b} = [\mathfrak{q}_1, \mathfrak{q}_2, \cdots]$ and $\bar{\mathfrak{b}} = [\bar{\mathfrak{q}}_1, \bar{\mathfrak{q}}_2, \cdots]$ are decompositions of the integral right $\mathfrak{o}$-ideals $\mathfrak{b}$ and $\bar{\mathfrak{b}}$ as direct intersections of ideals $\mathfrak{q}_i$ and $\bar{\mathfrak{q}}_i$ such that $\mathfrak{o} - \mathfrak{q}_i$ and $\mathfrak{o} - \bar{\mathfrak{q}}_i$ are indecomposable then a necessary and sufficient condition that $\mathfrak{b}$ and $\bar{\mathfrak{b}}$ be similar is that the aggregate of bounds of the $\mathfrak{q}_i$ be the same as that of the $\bar{\mathfrak{q}}_i$.*

If $\mathfrak{a} = \mathfrak{p}_1^{e_1} \cdots \mathfrak{p}_s^{e_s}$ is an integral two-sided $\mathfrak{o}$-ideal and the $\mathfrak{p}_i$ are distinct primes, $\mathfrak{a} = [\mathfrak{p}_1^{e_1}, \cdots, \mathfrak{p}_s^{e_s}]$ and $(\mathfrak{p}_i^{e_i}, [\mathfrak{p}_1^{e_1}, \cdots, \mathfrak{p}_{i-1}^{e_{i-1}}, \mathfrak{p}_{i+1}^{e_{i+1}}, \cdots]) = \mathfrak{o}$. Moreover if $\mathfrak{p}$ is a prime ideal, $\bar{\mathfrak{o}} = \mathfrak{o} - \mathfrak{p}^e$ is a primary ring whose radical is $\bar{\mathfrak{p}} = \mathfrak{p} - \mathfrak{p}^e$. Since $\bar{\mathfrak{o}} - \bar{\mathfrak{p}} = \mathfrak{d}_k$ where $\mathfrak{d}$ is a division ring and k is the capacity of $\mathfrak{p}$, $\bar{\mathfrak{o}}$ is a direct sum of k isomorphic indecomposable right ideals. It follows that there are exactly k ideals in any decomposition of $\mathfrak{p}^e$ as a direct intersection $[\mathfrak{q}_1, \mathfrak{q}_2, \cdots, \mathfrak{q}_k]$ where the $\mathfrak{o} - \mathfrak{q}_i$ are indecomposable, and all of the $\mathfrak{q}_i$ are similar. Thus the $\mathfrak{q}_i$ all have the same bound which is therefore $\mathfrak{p}^e$. Now in the general case of an arbitrary integral two-sided $\mathfrak{o}$-ideal $\mathfrak{a} = \mathfrak{p}_1^{e_1} \cdots \mathfrak{p}_s^{e_s}$, we obtain a decomposition of $\mathfrak{a} = [\mathfrak{p}_1^{e_1}, \cdots, \mathfrak{p}_s^{e_s}]$ as a direct intersection by decomposing the $\mathfrak{p}_i^{e_i}$ in this way. Hence we have

THEOREM 26. *Let $\mathfrak{a} = \mathfrak{p}_1^{e_1} \cdots \mathfrak{p}_s^{e_s}$ be an integral two sided $\mathfrak{o}$-ideal and $\mathfrak{p}_i$ a prime of capacity k_i, $\mathfrak{p}_i \neq \mathfrak{p}_j$ if $i \neq j$. Then $\mathfrak{a}$ is a direct intersection $[\mathfrak{q}_{11}, \cdots, \mathfrak{q}_{k_1 1}; \cdots; \mathfrak{q}_{1s}, \cdots, \mathfrak{q}_{k_s s}]$ where the $\mathfrak{o} - \mathfrak{q}_{ij}$ are indecomposable, and for a fixed j any pair $\mathfrak{q}_{ij}, \mathfrak{q}_{i'j}$ are similar and have the bound $\mathfrak{p}_j^{e_j}$.*

As an immediate consequence of the theory of modules we have also

THEOREM 27. *If $\mathfrak{q}$ is an integral right $\mathfrak{o}$-ideal such that $\mathfrak{o} - \mathfrak{q}$ is indecomposable, then $\mathfrak{o} - \mathfrak{q}$ has only one composition series and its length is e if the bound of $\mathfrak{q}$ is $\mathfrak{p}^e$, $\mathfrak{p}$ a prime. All the composition factors of $\mathfrak{o} - \mathfrak{q}$ are $\mathfrak{o}$-isomorphic. If $\mathfrak{q}'$ is an integral right $\mathfrak{o}$-ideal containing $\mathfrak{q}$, $\mathfrak{o} - \mathfrak{q}'$ is indecomposable.*

From Theorems 26 and 27, we obtain the

COROLLARY. *If $\mathfrak{p}$ is a prime ideal with capacity k, $\mathfrak{p}$ is a direct intersection $[\mathfrak{q}_1, \cdots, \mathfrak{q}_k]$ where the $\mathfrak{q}_i$ are maximal right $\mathfrak{o}$-ideals, are all similar and have $\mathfrak{p}$ for bound.*

10. Normal ideals. In order to obtain a satisfactory factorization theory for one-sided ideals it is necessary to consider simultaneously all of the maximal orders $\mathfrak{o}$ equivalent to a fixed order. This important remark was made first by Brandt for orders in an algebra. From now on we assume that all the maximal orders satisfy conditions II, III and IV. It may be recalled that IV holds for all maximal orders if it holds for one of them.

Definition 4. An ideal $\mathfrak{a}_{ik}$ is called *normal* if both its left order $\mathfrak{o}_i$ and its right order $\mathfrak{o}_k$ are maximal.

In the next two sections we shall develop a factorization theory for normal ideals. Here we establish the fact that this theory is valid for arbitrary right (left) $\mathfrak{o}$-ideals of a maximal order $\mathfrak{o}$ by proving that any such ideal is normal.

LEMMA 1. *Let $\mathfrak{b}$ be an integral right $\mathfrak{o}$-ideal $\neq \mathfrak{o}$ having the bound $\mathfrak{p}$, a prime ideal. Then there exists an integral left $\mathfrak{o}$-ideal $\mathfrak{c}$ with the (left) bound $\mathfrak{p}$ such that $\mathfrak{p} = \mathfrak{c}\mathfrak{b}$.*

Consider the right ideal $\mathfrak{b} - \mathfrak{p} = \bar{\mathfrak{b}}$ in the simple ring $\bar{\mathfrak{o}} = \mathfrak{o} - \mathfrak{p}$. Since $\bar{\mathfrak{b}} \neq \bar{\mathfrak{o}}$, the left ideal $\bar{\mathfrak{c}}$ of left annihilators of the elements of $\bar{\mathfrak{b}}$ is $\neq 0$. If $\mathfrak{c}$ is the integral left $\mathfrak{o}$-ideal corresponding to $\bar{\mathfrak{c}}$, then $\mathfrak{c} \neq \mathfrak{p}$ and $\mathfrak{p}^2 \leqq \mathfrak{c}\mathfrak{b} \leqq \mathfrak{p}$. Since $\mathfrak{c}\mathfrak{b}$ is a two sided $\mathfrak{o}$-ideal, either $\mathfrak{c}\mathfrak{b} = \mathfrak{p}^2$ or $\mathfrak{c}\mathfrak{b} = \mathfrak{p}$. If the former equation holds, for each y in $\mathfrak{c}$ we have $y\mathfrak{p} \leqq \mathfrak{p}^2$ and on multiplying by $\mathfrak{p}^{-1}$, $y\mathfrak{o} \leqq \mathfrak{p}$. This contradicts the fact that there exist y's in $\mathfrak{c}$ that are not in $\mathfrak{p}$. Hence $\mathfrak{c}\mathfrak{b} = \mathfrak{p}$.

LEMMA 2. *If $\mathfrak{b}$ is an integral right $\mathfrak{o}$-ideal properly contained in $\mathfrak{o}$, then $\mathfrak{b}^{-1} > \mathfrak{o}$.*

Let $\mathfrak{q}$ be a maximal right $\mathfrak{o}$-ideal so that $\mathfrak{o} > \mathfrak{q} > \mathfrak{b}$. Then the bound $\mathfrak{p}$ of $\mathfrak{q}$ is prime. If a is a regular element of $\mathfrak{p}$, $a\mathfrak{o} = \mathfrak{c}\mathfrak{p}$ where $\mathfrak{c}$ is a right $\mathfrak{o}$-ideal. By the preceding lemma, $\mathfrak{p} = \mathfrak{d}\mathfrak{q}$ for a suitable left $\mathfrak{o}$-ideal $\mathfrak{d}$ containing $\mathfrak{p}$ but $\neq \mathfrak{p}$. Hence $a\mathfrak{o} = \mathfrak{c}\mathfrak{d}\mathfrak{q}$, $\mathfrak{o} = a^{-1}\mathfrak{c}\mathfrak{d}\mathfrak{q}$ and so $a^{-1}\mathfrak{c}\mathfrak{d} \leqq \mathfrak{q}^{-1}$. If $\mathfrak{q}^{-1} = \mathfrak{o}$, $a^{-1}\mathfrak{c}\mathfrak{d} \leqq \mathfrak{o}$, $\mathfrak{c}\mathfrak{d} \leqq a\mathfrak{o} = \mathfrak{c}\mathfrak{p}$ and $\mathfrak{c}\mathfrak{d}\mathfrak{p}^{-1} \leqq \mathfrak{c}$. Thus $\mathfrak{d}\mathfrak{p}^{-1}$ is contained in the maximal order $\mathfrak{o}$. It follows that $\mathfrak{d} \leqq (\mathfrak{p}^{-1})^{-1} = \mathfrak{p}$. This contradiction shows that $\mathfrak{q}^{-1} > \mathfrak{o}$ and hence $\mathfrak{b}^{-1} > \mathfrak{o}$.

LEMMA 3. *If $\mathfrak{b}$ is a right $\mathfrak{o}$-ideal, $\mathfrak{o}$ maximal, then $\mathfrak{b}^{-1}$ is normal.*

Let $\mathfrak{o}'$ be the left order of $\mathfrak{b}$, $\mathfrak{o}''$ the right order of $\mathfrak{b}^{-1}$ and $\mathfrak{o}^*$ any order containing $\mathfrak{o}''$. Clearly $\mathfrak{o}^* \geqq \mathfrak{o}'' \geqq \mathfrak{o}'$. Consider the set $\mathfrak{b}^{-1}\mathfrak{o}^*\mathfrak{b}$. Since $(\mathfrak{b}^{-1}\mathfrak{o}^*\mathfrak{b})(\mathfrak{b}^{-1}\mathfrak{o}^*\mathfrak{b}) \leqq \mathfrak{b}^{-1}\mathfrak{o}^*\mathfrak{o}'\mathfrak{o}^*\mathfrak{b} = \mathfrak{b}^{-1}\mathfrak{o}^*\mathfrak{b}$, $\mathfrak{b}^{-1}\mathfrak{o}^*\mathfrak{b}$ is a subring of $\mathfrak{A}$. It contains $\mathfrak{o}$ since $\mathfrak{b}^{-1}\mathfrak{o}^*\mathfrak{b} \geqq \mathfrak{b}^{-1}\mathfrak{b} = \mathfrak{o}$. Now if a and b are regular elements of $\mathfrak{b}^{-1}$ and $\mathfrak{b}$ respectively, $a\mathfrak{o}^*b \leqq \mathfrak{b}^{-1}\mathfrak{o}^*\mathfrak{b}$ and $b(\mathfrak{b}^{-1}\mathfrak{o}^*\mathfrak{b})a \leqq \mathfrak{b}\mathfrak{b}^{-1}\mathfrak{o}^*\mathfrak{b}\mathfrak{b}^{-1} \leqq \mathfrak{o}'\mathfrak{o}^*\mathfrak{o}' = \mathfrak{o}^*$. This shows that $\mathfrak{b}^{-1}\mathfrak{o}^*\mathfrak{b}$ is an order and so because of the maximality of $\mathfrak{o}$, $\mathfrak{b}^{-1}\mathfrak{o}^*\mathfrak{b} = \mathfrak{o}$. It follows that $\mathfrak{b}^{-1}\mathfrak{o}^* \leqq \mathfrak{b}^{-1}$ and since $\mathfrak{o}''$ is the right order of $\mathfrak{b}^{-1}$, $\mathfrak{o}^* = \mathfrak{o}''$. This proves that $\mathfrak{o}''$ is maximal and $\mathfrak{b}^{-1}$ is normal.

THEOREM 28. *If $\mathfrak{b}$ is a right (left) $\mathfrak{o}$-ideal and $\mathfrak{o}$ is maximal, then $\mathfrak{b}$ is normal.*

First let $\mathfrak{b}$ be integral and let $\mathfrak{o} > \mathfrak{b}_1 > \mathfrak{b}_2 > \cdots > \mathfrak{b}_m = \mathfrak{b}$ be a chain of right $\mathfrak{o}$-ideals corresponding to a composition series for $\mathfrak{o} - \mathfrak{b}$. Now $\mathfrak{b}^{-1} > \mathfrak{o}$ and hence $(\mathfrak{b}^{-1})^{-1} < \mathfrak{o}$. Since $\mathfrak{b}^{-1}\mathfrak{b}\mathfrak{b}^{-1} = \mathfrak{o}\mathfrak{b}^{-1} = \mathfrak{b}^{-1}$, $\mathfrak{b} \leqq (\mathfrak{b}^{-1})^{-1}$. Hence if $m = 1$, $\mathfrak{b} = (\mathfrak{b}^{-1})^{-1}$ by the maximality of $\mathfrak{b}$. Then the theorem follows from Lemma 3. We assume now that the theorem holds for integral ideals $\mathfrak{b}'$ for which the length of $\mathfrak{o} - \mathfrak{b}'$ is less than m. Then $\mathfrak{b}_{m-1}$ is normal and if $\mathfrak{o}'$ is the left order of $\mathfrak{b}_{m-1}$, $\mathfrak{o}'$ is maximal and $\mathfrak{b}_{m-1}\mathfrak{b}_{m-1}^{-1} = \mathfrak{o}'$. We wish to show that $\mathfrak{o}' > \mathfrak{b}_m\mathfrak{b}_{m-1}^{-1}$ and that $\mathfrak{b}_m\mathfrak{b}_{m-1}^{-1}$ is maximal in $\mathfrak{o}'$. Evidently $\mathfrak{o}' = \mathfrak{b}_{m-1}\mathfrak{b}_{m-1}^{-1} \geqq \mathfrak{b}_m\mathfrak{b}_{m-1}^{-1}$ and if $\mathfrak{o}' = \mathfrak{b}_m\mathfrak{b}_{m-1}^{-1}$, $\mathfrak{o}'\mathfrak{b}_{m-1} = \mathfrak{b}_m\mathfrak{b}_{m-1}^{-1}\mathfrak{b}_{m-1} = \mathfrak{b}_m\mathfrak{o} = \mathfrak{b}_m$ contrary to the inequality $\mathfrak{b}_{m-1} > \mathfrak{b}_m$. Next let $\mathfrak{c}$ be a right $\mathfrak{o}'$-ideal such that $\mathfrak{o}' > \mathfrak{c} \geqq \mathfrak{b}_m\mathfrak{b}_{m-1}^{-1}$. Then $\mathfrak{o}'\mathfrak{b}_{m-1} = \mathfrak{b}_{m-1} \geqq \mathfrak{c}\mathfrak{b}_{m-1} \geqq \mathfrak{b}_m$ and either $\mathfrak{b}_{m-1} = \mathfrak{c}\mathfrak{b}_{m-1}$ or $\mathfrak{c}\mathfrak{b}_{m-1} = \mathfrak{b}_m$. If $\mathfrak{b}_{m-1} = \mathfrak{c}\mathfrak{b}_{m-1}$, $\mathfrak{o}' = \mathfrak{c}\mathfrak{o}' = \mathfrak{c}$. Hence $\mathfrak{c}\mathfrak{b}_{m-1} = \mathfrak{b}_m$ and $\mathfrak{c} = \mathfrak{b}_m\mathfrak{b}_{m-1}^{-1}$. This proves that $\mathfrak{b}_m\mathfrak{b}_{m-1}^{-1}$ is a maximal right $\mathfrak{o}'$-ideal contained in $\mathfrak{o}'$ and by what we have shown, $\mathfrak{b}_m\mathfrak{b}_{m-1}^{-1}$ is normal. Since, as is readily seen, the left order of $\mathfrak{b}_m\mathfrak{b}_{m-1}^{-1}$ coincides with that of $\mathfrak{b}_m$, $\mathfrak{b}_m = \mathfrak{b}$ is normal. If $\mathfrak{b}$ is not integral, there exists a regular element a such that $a\mathfrak{b} \leqq \mathfrak{o}$. Since $a\mathfrak{b}$ is an integral right $\mathfrak{o}$-ideal, its left order is maximal. Now if $\mathfrak{o}^*$ is the left order of $\mathfrak{b}$, $a\mathfrak{o}^*a^{-1}$ is the left order of $a\mathfrak{b}$. Since $a\mathfrak{o}^*a^{-1}$ is maximal, $\mathfrak{o}^*$ is maximal.

11. Brandt's groupoid. In order to obtain an extension of Theorem 22 that is applicable to one-sided ideals we require the concept of a groupoid which we now define. A system G is a *groupoid* if a product in G is defined for certain pairs of its elements subject to the following conditions:

1. For each element a_{ij}, there exist uniquely determined elements e_i and e_j in G such that the products e_ia_{ij} and $a_{ij}e_j$ are defined and $e_ia_{ij} = a_{ij}e_j = a_{ij}$. These elements are respectively the *left* and the *right unit* of a_{ij}.

2. If e is a unit for any element of G, then e is its own left unit and hence its own right unit.

3. The product ab is defined if and only if the right unit of a is the left unit of b.

4. If ab and bc are defined, $(ab)c$ and $a(bc)$ are defined and $(ab)c = a(bc)$.

5. For any element a_{ij} with the left unit e_i and the right unit e_j there exists an element a_{ij}^{-1} with left unit e_j and with right unit e_i such that $a_{ij}a_{ij}^{-1} = e_i$ and $a_{ij}^{-1}a_{ij} = e_j$. We call a_{ij}^{-1} the *inverse* of a_{ij}.

6. For any pair of units e_i and e_j there exists an element a_{ij} having e_i as left unit and e_j as right unit.[11]

Example. Let G_0 be an arbitrary group and let G be the set of $n \times n$ matrices (n finite or infinite) having one element in G_0 and the remaining elements 0. We denote the matrix having the element a of G_0 in its i-th row and j-th column by a_{ij} and we define $a_{ij}b_{jk} = (ab)_{ik}$. It may be verified that G is a groupoid. The units of G are the elements $e_i = 1_{ii}$ and the inverse of a is $(a^{-1})_{ji}$.

In an arbitrary groupoid G it is readily seen that the inverse a^{-1} of an element a is unique. We note also that $(a^{-1})^{-1} = a$ and if ab is defined, $b^{-1}a^{-1}$ is defined and $(ab)^{-1} = b^{-1}a^{-1}$. If ab is defined and e is the left unit of a, then e is the left unit of ab. For $e(ab) = (ea)b = ab$.

Let e be a unit and let $G(e)$ denote the set of elements a of G that have e as left unit and as right unit. It is readily verified that $G(e)$ is a group relative to the composition of G. If e and e' are units and c is an element having these respectively for left unit and right unit, then the mapping $x \to c^{-1}xc$ is an isomorphism between $G(e)$ and $G(e')$. If $G(e)$ is commutative, this isomorphism is independent of the choice of the element c. For if d is a second element with left unit e and right unit e', then cd^{-1} is in $G(e)$. Hence $(cd^{-1})x = x(cd^{-1})$ for any x in $G(e)$ and so $c^{-1}xc = d^{-1}xd$. In this case we call x in $G(e)$ and $x' = c^{-1}xc$ in $G(e')$ *conjunctive*.

We consider now the set G of normal ideals. Let $\mathfrak{a}$ and $\mathfrak{b}$ be normal and suppose that the left order of $\mathfrak{a}$ is $\mathfrak{o}'$ and the right order of $\mathfrak{b}$ is $\mathfrak{o}$. Then there exist regular elements a and b such that $\mathfrak{a} \leqq \mathfrak{o}'a$ and $\mathfrak{b} \leqq b\mathfrak{o}$. Hence $\mathfrak{ab} \leqq \mathfrak{o}'ab\mathfrak{o}$ and since (Theorem 5) there exists a regular element c such that $\mathfrak{o}' \leqq \mathfrak{o}c$, $\mathfrak{ab} \leqq \mathfrak{o}cab\mathfrak{o}$. We have seen that $\mathfrak{o}cab\mathfrak{o}$ is a two-sided $\mathfrak{o}$-ideal and so there exists a regular element d such that $d\mathfrak{o} \geqq \mathfrak{o}cab\mathfrak{o} \geqq \mathfrak{ab}$. This shows that $\mathfrak{ab}$ is a right $\mathfrak{o}$-ideal and in a similar manner we prove that $\mathfrak{ab}$ is a left $\mathfrak{o}'$-ideal. Since $\mathfrak{o}$ and $\mathfrak{o}'$ are maximal, the orders of $\mathfrak{ab}$ are $\mathfrak{o}'$ and $\mathfrak{o}$ and so $\mathfrak{ab}$ is normal.

We define the product $\mathfrak{ab}$ of normal ideals $\mathfrak{a}$ and $\mathfrak{b}$ to be *proper* if for any pair of normal ideals $\mathfrak{a}'$, $\mathfrak{b}'$ such that $\mathfrak{a}' \geqq \mathfrak{a}$ and $\mathfrak{b}' \geqq \mathfrak{b}$ and either $\mathfrak{a}' > \mathfrak{a}$ or $\mathfrak{b}' > \mathfrak{b}$, we have $\mathfrak{a}'\mathfrak{b}' > \mathfrak{ab}$. We wish to show that G is a groupoid relative to proper multiplication. Condition 1 holds for any normal $\mathfrak{a}_{ij}$ and its left and right orders $\mathfrak{o}_i$ and $\mathfrak{o}_j$. Condition 2 is evident. We consider 3 in the following

LEMMA. *If $\mathfrak{a}$ and $\mathfrak{b}$ are normal, $\mathfrak{ab}$ is a proper product if and only if the right order of $\mathfrak{a}$ is the left order of $\mathfrak{b}$.*

[11] It may be remarked that if G is any groupoid, we may adjoin an element 0 to G and define $0a = 0 = a0$ and $ab = 0$ if ab is undefined in G. The extended system is a special type of semi-group called completely simple (cf. A. H. Clifford [**3**]). For the present applications the definition given in the text seems to be the appropriate one.

Let $\mathfrak{o}$ be the right order of $\mathfrak{a}$ and $\mathfrak{o}'$ the left order of $\mathfrak{b}$. If $\mathfrak{o}' \neq \mathfrak{o}$, $\mathfrak{a}\mathfrak{o}'$ is a normal ideal properly containing $\mathfrak{a}$. Since $(\mathfrak{a}\mathfrak{o}')\mathfrak{b} = \mathfrak{a}(\mathfrak{o}'\mathfrak{b}) = \mathfrak{a}\mathfrak{b}$, $\mathfrak{a}\mathfrak{b}$ is not a proper product. Conversely suppose that $\mathfrak{o} = \mathfrak{o}'$ and let $\mathfrak{a}'$ be a normal ideal containing $\mathfrak{a}$ such that $\mathfrak{a}\mathfrak{b} = \mathfrak{a}'\mathfrak{b}$. Then $\mathfrak{a}'\mathfrak{b}\mathfrak{b}^{-1} = \mathfrak{a}\mathfrak{b}\mathfrak{b}^{-1}$ or $\mathfrak{a}'\mathfrak{o} = \mathfrak{a}\mathfrak{o} = \mathfrak{a}$. Thus $\mathfrak{a} \geqq \mathfrak{a}'$ and so $\mathfrak{a}' = \mathfrak{a}$.

Condition 4 is now evident. If $\mathfrak{a}_{ij}$ is normal, we set $\mathfrak{a}_{ji} = \mathfrak{a}_{ij}^{-1}$ and we obtain 5 by Theorem 20. If $\mathfrak{o}$ and $\mathfrak{o}'$ are arbitrary orders, $\mathfrak{o}'\mathfrak{o}$ is a normal ideal having $\mathfrak{o}'$ and $\mathfrak{o}$ as its orders. This proves 6 and hence

THEOREM 29. *The normal ideals form a groupoid G with respect to the operation of proper multiplication. The maximal orders are the units of G and the inverse ideal* $\mathfrak{a}^{-1}$ *is the inverse of* $\mathfrak{a}$ *in G.*

We prove next the following extension of 6.

THEOREM 30. *If* $\mathfrak{o}$ *and* $\mathfrak{o}'$ *are maximal orders,* $(\mathfrak{o}\mathfrak{o}')^{-1}$ *is an integral ideal with the right order* $\mathfrak{o}$ *and the left order* $\mathfrak{o}'$. *The ideal* $(\mathfrak{o}\mathfrak{o}')^{-1}$ *contains every integral ideal* $\mathfrak{a}$ *that has* $\mathfrak{o}$ *for right order and* $\mathfrak{o}'$ *for left order.*

Since $\mathfrak{o}\mathfrak{o}'$ is normal, $(\mathfrak{o}\mathfrak{o}')^{-1}$ is normal and has the left order $\mathfrak{o}'$ and the right order $\mathfrak{o}$. Since $\mathfrak{o} \geqq (\mathfrak{o}\mathfrak{o}')(\mathfrak{o}\mathfrak{o}')^{-1} \geqq (\mathfrak{o}\mathfrak{o}')^{-1}$, $(\mathfrak{o}\mathfrak{o}')^{-1}$ is integral. Now let $\mathfrak{a}$ be any integral ideal with the right order $\mathfrak{o}$ and the left order $\mathfrak{o}'$. Then $\mathfrak{o}\mathfrak{o}'\mathfrak{a} \leqq \mathfrak{o}\mathfrak{a} \leqq \mathfrak{o}$ and so $\mathfrak{a} \leqq (\mathfrak{o}\mathfrak{o}')^{-1}$.

The ideal $(\mathfrak{o}\mathfrak{o}')^{-1}$ is called the *distance ideal from* $\mathfrak{o}$ *to* $\mathfrak{o}'$.

We recall that the group $G(\mathfrak{o})$ of two-sided $\mathfrak{o}$-ideals is commutative. Hence if $\mathfrak{c}$ is an ideal with left order $\mathfrak{o}$ and right order $\mathfrak{o}'$, the mapping $\mathfrak{a} \to \mathfrak{c}^{-1}\mathfrak{a}\mathfrak{c} = \mathfrak{a}'$ is an isomorphism between $G(\mathfrak{o})$ and $G(\mathfrak{o}')$ independent of $\mathfrak{c}$. As in the case of an abstract groupoid, we call $\mathfrak{a}$ and $\mathfrak{a}'$ *conjunctive.* Evidently $\mathfrak{a}$ is a prime $\mathfrak{p}$ or a prime power $\mathfrak{p}^e$ if and only if $\mathfrak{a}' = \mathfrak{p}'$ or $\mathfrak{p}'^e$, $\mathfrak{p}'$ a prime of $\mathfrak{o}'$.

12. Necessity of conditions I-IV. Let $\mathfrak{A}$ be a ring with an identity in which every regular element has an inverse, and suppose that G is a set of additive subgroups $\mathfrak{a}, \mathfrak{b}, \cdots$ of $\mathfrak{A}$ that form a groupoid relative to a composition that coincides with ordinary multiplication[12] of additive subgroups when it is defined. We assume the following conditions:

1. Every $\mathfrak{a}$ in G contains a regular element.
2. Every unit $\mathfrak{o}$ of G is an order in $\mathfrak{A}$.
3. For each unit $\mathfrak{o}$ in G, every integral right (left) $\mathfrak{o}$-ideal is in G and has $\mathfrak{o}$ as its right (left) unit.
4. For any pair of units $\mathfrak{o}$ and $\mathfrak{o}'$ there is an $\mathfrak{a}$ contained in $\mathfrak{o} \wedge \mathfrak{o}'$ having $\mathfrak{o}$ as its right unit and $\mathfrak{o}'$ as its left unit.

We note then that if $\mathfrak{a}$ is in G and $\mathfrak{o}$ is its right (left) unit, $\mathfrak{a}$ is a right (left) $\mathfrak{o}$-ideal. For $\mathfrak{a}$ is a right $\mathfrak{o}$-module and if a is a regular element in $\mathfrak{a}$, $\mathfrak{a}$ contains $a\mathfrak{o}$. Since $\mathfrak{a}^{-1}\mathfrak{a} = \mathfrak{o}$, $b^{-1}\mathfrak{a} \leqq \mathfrak{o}$ if b^{-1} is a regular element in $\mathfrak{a}^{-1}$ and so $\mathfrak{a} \leqq b\mathfrak{o}$.

THEOREM 31 (Asano). *If the above conditions* 1–4 *hold, then the units of G form a set of equivalent orders satisfying conditions I–IV. The set of units of G*

[12] We recall that if $\mathfrak{a}$ and $\mathfrak{b}$ are additive subgroups, $\mathfrak{a}\mathfrak{b}$ is the smallest additive subgroup containing all ab, a in $\mathfrak{a}$ and b in $\mathfrak{b}$.

includes all the maximal orders equivalent to these orders. The groupoid G consists of the normal ideals relative to these orders with the groupoid composition as proper multiplication.

Equivalence. Let $\mathfrak{o}$ and $\mathfrak{o}'$ be two units of G and let $\mathfrak{a}$ be an ideal having $\mathfrak{o}$ as right unit and $\mathfrak{o}'$ as left unit. Then $\mathfrak{o}' = \mathfrak{a}\mathfrak{o}\mathfrak{a}^{-1} \geqq a\mathfrak{o}b$ if a is regular in $\mathfrak{a}$ and b is regular in $\mathfrak{a}^{-1}$. Similarly, $\mathfrak{o} \geqq c\mathfrak{o}'d$ for suitable regular c and d.

Boundedness of integral ideals. If a is in $\mathfrak{o}$, $a\mathfrak{o}$ is integral and hence belongs to G. Let $\mathfrak{o}'$ be the left unit of $a\mathfrak{o}$ in G and let $\mathfrak{a}$ be an ideal contained in $\mathfrak{o} \wedge \mathfrak{o}'$ and having $\mathfrak{o}$ as its left unit and $\mathfrak{o}'$ as its right unit. Then $a\mathfrak{o} = \mathfrak{o}'(a\mathfrak{o}) \geqq \mathfrak{a}(a\mathfrak{o})$ a two-sided $\mathfrak{o}$-ideal belonging to G.

Maximality. Suppose that $\mathfrak{o}$ is a unit in G and that $\mathfrak{o}^*$ is an order equivalent to $\mathfrak{o}$ and containing $\mathfrak{o}$. Then there exist elements a and b such that $\mathfrak{o}^* \leqq a\mathfrak{o}b$. If $b = dc^{-1}$ with d and c in $\mathfrak{o}$, $\mathfrak{o}^* \leqq a\mathfrak{o}c^{-1}$. We have seen that $\mathfrak{o}c$ contains an integral right ideal $g\mathfrak{o}$ and so $g^{-1}\mathfrak{o}c \geqq \mathfrak{o}$ and $g^{-1}\mathfrak{o} \geqq \mathfrak{o}c^{-1}$. Hence $\mathfrak{o}^* \leqq a\mathfrak{o}c^{-1} \leqq ag^{-1}\mathfrak{o}$. Thus if $h^{-1} = ag^{-1}$, $h\mathfrak{o}^*$ is contained in $\mathfrak{o}$ and is therefore an integral right $\mathfrak{o}$-ideal. By 3, $h\mathfrak{o}^*$ is in G and has $\mathfrak{o}$ as its right unit. If $\mathfrak{a}$ denotes the inverse of $h\mathfrak{o}^*$ in G, then $\mathfrak{a}h\mathfrak{o}^* = \mathfrak{o}$. Since $(\mathfrak{o}^*)^2 = \mathfrak{o}^*$, this implies that $\mathfrak{o}\mathfrak{o}^* = \mathfrak{o}$ and $\mathfrak{o}^* \leqq \mathfrak{o}$. Hence $\mathfrak{o}^* = \mathfrak{o}$.

Ascending chain condition. Let $\mathfrak{a}_1 \leqq \mathfrak{a}_2 \leqq \cdots$ be an ascending chain of integral right $\mathfrak{o}$-ideals. The join $\mathfrak{a}$ of the $\mathfrak{a}_i$ is an integral right $\mathfrak{o}$-ideal and hence belongs to G. We have $\mathfrak{a}_1\mathfrak{a}^{-1} \leqq \mathfrak{a}_2\mathfrak{a}^{-1} \leqq \cdots \leqq \mathfrak{a}\mathfrak{a}^{-1} = \mathfrak{o}'$ the left unit of $\mathfrak{a}$. The join of the $\mathfrak{a}_i\mathfrak{a}^{-1}$ is $\mathfrak{o}'$. Since 1 is in $\mathfrak{o}'$, it is contained in one of the $\mathfrak{a}_i\mathfrak{a}^{-1}$, say $\mathfrak{a}_m\mathfrak{a}^{-1}$. Then $\mathfrak{o}' = \mathfrak{a}_m\mathfrak{a}^{-1} = \mathfrak{a}_{m+1}\mathfrak{a}^{-1} = \cdots$. By multiplying by $\mathfrak{a}$ we obtain $\mathfrak{a}_m = \mathfrak{a}_{m+1} = \cdots$.

Restricted descending chain condition. Let $\mathfrak{b}_1 \geqq \mathfrak{b}_2 \geqq \cdots$ be a descending chain of integral right $\mathfrak{o}$-ideals all containing the two-sided $\mathfrak{o}$-ideal $\mathfrak{a}$. The $\mathfrak{b}_i$ and $\mathfrak{a}$ are in G and we have the relation $\mathfrak{o} \geqq \mathfrak{b}_1^{-1}\mathfrak{b}_2$. Hence $\mathfrak{b}_2^{-1} = \mathfrak{o}\mathfrak{b}_2^{-1} \geqq \mathfrak{b}_1^{-1}\mathfrak{o}' \geqq \mathfrak{b}_1^{-1}$ if $\mathfrak{o}'$ is the left unit of $\mathfrak{b}_2$. Thus $\mathfrak{b}_1^{-1} \leqq \mathfrak{b}_2^{-1} \leqq \cdots \leqq \mathfrak{a}^{-1}$ and $\mathfrak{a}\mathfrak{b}_1^{-1} \leqq \mathfrak{a}\mathfrak{b}_2^{-1} \leqq \cdots$ is an ascending chain of integral left $\mathfrak{o}$-ideals. It follows that $\mathfrak{a}\mathfrak{b}_m^{-1} = \mathfrak{a}\mathfrak{b}_{m+1}^{-1} = \cdots$ for a suitable index m and hence $\mathfrak{b}_m^{-1} = \mathfrak{b}_{m+1}^{-1} = \cdots$ and $\mathfrak{b}_m = \mathfrak{b}_{m+1} = \cdots$.

Now since any integral $\mathfrak{o}$-ideal is bounded and $\mathfrak{o}$ is maximal, any $\mathfrak{o}$-ideal is bounded. Hence each unit $\mathfrak{o}$ satisfies the conditions I–IV. If $\mathfrak{a}$ is any element in G, $\mathfrak{a}$ is an ideal relative to its units and since the latter are maximal, they are the orders of $\mathfrak{a}$. It follows that the inverse of $\mathfrak{a}$ in G is the ordinary inverse ideal. Hence the operations in G are the ones previously defined. It remains to show that every maximal order $\mathfrak{o}'$ equivalent to an $\mathfrak{o}$ in G is in G and every normal ideal having orders in G is in G. Let $\mathfrak{o}'$ be maximal and suppose that $\mathfrak{o}' \leqq a\mathfrak{o}b$. Then $\mathfrak{o}\mathfrak{o}' \leqq \mathfrak{o}a\mathfrak{o}b \leqq \mathfrak{o}c$ for a suitable c and so $\mathfrak{o}\mathfrak{o}'$ is a left $\mathfrak{o}$-ideal. Its orders are evidently the maximal orders $\mathfrak{o}$ and $\mathfrak{o}'$. Since the inverse ideal $(\mathfrak{o}\mathfrak{o}')^{-1}$ is the set of x's such that $(\mathfrak{o}\mathfrak{o}')x \leqq \mathfrak{o}$, $(\mathfrak{o}\mathfrak{o}')^{-1}$ is contained in $\mathfrak{o}$. It follows that $(\mathfrak{o}\mathfrak{o}')^{-1}$ belongs to G and since its left unit in G is its left order, $\mathfrak{o}'$ is in G. Finally, let $\mathfrak{b}$ be any right $\mathfrak{o}$-ideal, $\mathfrak{o}$ in G and let $\mathfrak{a}$ be a two-sided $\mathfrak{o}$-ideal contained in $\mathfrak{b}$. Since $\mathfrak{a} = \mathfrak{a}_1\mathfrak{a}_2^{-1}$ where $\mathfrak{a}_1$ and $\mathfrak{a}_2$ are integral two-sided $\mathfrak{o}$-ideals, $\mathfrak{a}$ is in G. Now $\mathfrak{c} = \mathfrak{b}^{-1}\mathfrak{a}$ is contained in $\mathfrak{o}$ and hence belongs to G. Hence $\mathfrak{b}^{-1} = \mathfrak{c}\mathfrak{a}^{-1}$ is in G and $\mathfrak{b} = (\mathfrak{b}^{-1})^{-1}$ is in G.

Example. We let $\mathfrak{o}$ be a principal ideal domain and $\mathfrak{A}$, its quotient ring. Consider the set G of additive subgroups of the form $a\mathfrak{o}b$, a and $b \neq 0$. If x is an element of $\mathfrak{A}$ such that $(a\mathfrak{o}b)x = a\mathfrak{o}b$, $(\mathfrak{o}b)x \leqq \mathfrak{o}b$ and $bx = yb$, y in $\mathfrak{o}$. Hence x is in $b^{-1}\mathfrak{o}b$ and, conversely, if x is any element in $b^{-1}\mathfrak{o}b$, $(a\mathfrak{o}b)x \leqq a\mathfrak{o}b$. It follows that if $a\mathfrak{o}b = a'\mathfrak{o}b'$, $b^{-1}\mathfrak{o}b = (b')^{-1}\mathfrak{o}b'$. Hence if we specify that $b^{-1}\mathfrak{o}b$ is the "right unit" of $a\mathfrak{o}b$, $b^{-1}\mathfrak{o}b$ is uniquely determined in G. Similarly if we define $a\mathfrak{o}a^{-1}$ to be the "left unit" of $\mathfrak{a} = a\mathfrak{o}b$, this element does not depend on the choice of a in the representation of $\mathfrak{a}$. We consider only those products $(a\mathfrak{o}b)(c\mathfrak{o}d)$ where the right order $b^{-1}\mathfrak{o}b$ of $a\mathfrak{o}b$ is the same as the left order $c\mathfrak{o}c^{-1}$ of $c\mathfrak{o}d$. Then $bc\mathfrak{o} = \mathfrak{o}bc$ and $(a\mathfrak{o}b)(c\mathfrak{o}d) = abc\mathfrak{o}d$ is in G. The set $b^{-1}\mathfrak{o}a^{-1}$ may be characterized as the totality of elements x such that $(a\mathfrak{o}b)x$ is in the left order $a\mathfrak{o}a^{-1}$. Hence if we define $(a\mathfrak{o}b)^{-1}$ as $b^{-1}\mathfrak{o}a^{-1}$, this element is uniquely determined by $a\mathfrak{o}b$ and satisfies $(a\mathfrak{o}b)(a\mathfrak{o}b)^{-1} = a\mathfrak{o}a^{-1}$, $(a\mathfrak{o}b)^{-1}(a\mathfrak{o}b) = b^{-1}\mathfrak{o}b$. Every right or left $a^{-1}\mathfrak{o}a$-ideal (integral or not) is principal and hence belongs to G. Finally for any pair of units $a^{-1}\mathfrak{o}a$ and $b^{-1}\mathfrak{o}b$ in G there is an element $b^{-1}\mathfrak{o}a$ having these respectively as right and left units. Thus G is a groupoid that satisfies conditions 1, 2 and 3. We show now that if every integral $\mathfrak{o}$-ideal is bounded, the condition 4 holds. In this case if $a = bc^{-1}$, b, c in $\mathfrak{o}$, is any element in $\mathfrak{A}$, there is an element c^* in $\mathfrak{o}$ such that $c^*\mathfrak{o} = \mathfrak{o}c^*$ and $c^{-1}c^*$ is in $\mathfrak{o}$. This is clear since $c\mathfrak{o}$ contains a two-sided $\mathfrak{o}$-ideal $c^*\mathfrak{o} = \mathfrak{o}c^*$ and so $c^* = cc'$, c' in $\mathfrak{o}$ and $c^{-1}c^* = c'$. It follows that $a\mathfrak{o}c^* = ac^*\mathfrak{o}$ is integral and has $a\mathfrak{o}a^{-1}$ as its left order and $\mathfrak{o}$ as its right order. Since the order $b^{-1}\mathfrak{o}b$ is isomorphic to $\mathfrak{o}$, it satisfies the same conditions as $\mathfrak{o}$ and so by a similar argument we may show that for any pair of orders $a^{-1}\mathfrak{o}a$ and $b^{-1}\mathfrak{o}b$ there is an $a^{-1}\mathfrak{o}a$-left, $b^{-1}\mathfrak{o}b$-right ideal contained in these orders. This shows that the present discussion is applicable directly to the principal ideal domains in which every integral ideal is bounded.

13. Factorization of normal ideals. We consider now the question of factorization of the integral elements of the groupoid G. If $\mathfrak{o}_i$ and $\mathfrak{o}_j$ are maximal orders in G, we denote the normal ideals having $\mathfrak{o}_i$ as left order and $\mathfrak{o}_j$ as right order by $\mathfrak{a}_{ij}$, $\mathfrak{b}_{ij}$, $\cdots$. The following is the fundamental lemma.

LEMMA. *A necessary and sufficient condition that* $\mathfrak{b}_{kj} \geqq \mathfrak{a}_{ij}$ $(\mathfrak{b}_{jk} \geqq \mathfrak{a}_{ji})$ *is that* $\mathfrak{a}_{ij} = \mathfrak{c}_{ik}\mathfrak{b}_{kj}$ $(\mathfrak{a}_{ji} = \mathfrak{b}_{jk}\mathfrak{c}_{ki})$ *where* $\mathfrak{c}_{ik}$ $(\mathfrak{c}_{ki})$ *is integral. Equality holds if and only if* $\mathfrak{c}_{ik} = \mathfrak{o}_i$ $(\mathfrak{c}_{ki} = \mathfrak{o}_k)$.

If $\mathfrak{a}_{ij} = \mathfrak{c}_{ik}\mathfrak{b}_{kj}$ with $\mathfrak{c}_{ik}$ in $\mathfrak{o}_k$, then $\mathfrak{a}_{ij} \leqq \mathfrak{b}_{kj}$. By the preceding lemma, $\mathfrak{a}_{ij} = \mathfrak{b}_{kj}$ only if $k = i$ and $\mathfrak{c}_{ik} = \mathfrak{o}_i$. Conversely if $\mathfrak{b}_{kj} \geqq \mathfrak{a}_{ij}$, $\mathfrak{a}_{ij} = \mathfrak{a}_{ij}\mathfrak{b}_{kj}^{-1}\mathfrak{b}_{kj} = \mathfrak{c}_{ik}\mathfrak{b}_{kj}$ where $\mathfrak{c}_{ik} = \mathfrak{a}_{ij}\mathfrak{b}_{kj}^{-1}$. Then $\mathfrak{c}_{ik} \leqq \mathfrak{a}_{ij}\mathfrak{a}_{kj}^{-1} = \mathfrak{o}_i$ and so $\mathfrak{c}_{ik}$ is integral.[13]

Suppose that $\mathfrak{a}_{ij}$ is integral and properly contained in $\mathfrak{o}_j$. Let $\mathfrak{o}_j > \mathfrak{a}_{i_1j} > \mathfrak{a}_{i_2j} > \cdots > \mathfrak{a}_{i_mj} = \mathfrak{a}_{ij}$ be a chain of integral right $\mathfrak{o}_j$-ideals corresponding to a composition series for $\mathfrak{o}_j - \mathfrak{a}_{ij}$. The composition factors of the series are then $\mathfrak{o}_j$-isomorphic to the modules $\mathfrak{a}_{i_{k-1}j} - \mathfrak{a}_{i_kj}$. By the lemma, we have $\mathfrak{a}_{i_kj} = \mathfrak{p}_{i_ki_{k-1}}\mathfrak{a}_{i_{k-1}j}$ and hence $\mathfrak{a}_{ij} = \mathfrak{p}_{ii_{m-1}}\mathfrak{p}_{i_{m-1}i_{m-2}} \cdots \mathfrak{p}_{i_1j}$ $(\mathfrak{p}_{i_1j} = \mathfrak{a}_{i_1j})$. The integral ideals $\mathfrak{p}_{i_ki_{k-1}}$ are maximal in their orders. For otherwise $\mathfrak{p}_{i_ki_{k-1}} = \mathfrak{r}_{i_kl}\mathfrak{s}_{li_{k-1}}$

[13] More generally, if $\mathfrak{b}_{kl} \geqq \mathfrak{a}_{ij}$, we have $\mathfrak{a}_{ij} = \mathfrak{c}_{ik}\mathfrak{b}_{kl}\mathfrak{d}_{lj}$ where $\mathfrak{c}_{ik} = \mathfrak{a}_{ij}(\mathfrak{o}_k\mathfrak{a}_{ij})^{-1}$ and $\mathfrak{d}_{lj} = \mathfrak{b}_{kl}^{-1}\mathfrak{a}_{ij}$ are integral.

where $\mathfrak{r}_{i_k l}$ is integral and $\neq \mathfrak{o}_{i_k}$ and $\mathfrak{s}_{l i_{k-1}}$ is integral and $\neq \mathfrak{o}_{i_{k-1}}$. Then $\mathfrak{a}_{i_{k-1}j} > \mathfrak{s}_{l i_{k-1}} \mathfrak{a}_{i_{k-1}j} > \mathfrak{a}_{i_k j}$ contrary to the irreducibility of $\mathfrak{a}_{i_{k-1}j} - \mathfrak{a}_{i_k j}$. It follows that the bound of $\mathfrak{p}_{i_k i_{k-1}}$ in $\mathfrak{o}_{i_{k-1}}$ (left bound in $\mathfrak{o}_{i_k}$) is a prime ideal. It is also clear that we may retrace our steps in the above argument: If $\mathfrak{a}_{ij} = \mathfrak{p}_{i i_{m-1}} \mathfrak{p}_{i_{m-1} i_{m-2}} \cdots \mathfrak{p}_{i_1 j}$ is any factorization of $\mathfrak{a}_{ij}$ into maximal integral ideals $\mathfrak{p}_{i_k i_{k-1}}$, then $\mathfrak{o}_j > \mathfrak{p}_{i_1 j} > \mathfrak{p}_{i_2 i_1} \mathfrak{p}_{i_1 j} > \cdots > \mathfrak{a}_{ij}$ corresponds to a composition series for $\mathfrak{o}_j - \mathfrak{a}_{ij}$.

In order to discuss the relation between different factorizations of $\mathfrak{a}_{ij}$ we require an extension of the concept of isomorphism that is applicable to modules relative to different orders of $\mathfrak{A}$. Let $\mathfrak{M}_j$ and $\mathfrak{M}_k$ be respectively $\mathfrak{o}_j$- and $\mathfrak{o}_k$-modules that are finitely generated and bounded. Then we shall say that $\mathfrak{M}_j$ and $\mathfrak{M}_k$ are *conjunctive* if the invariants of $\mathfrak{M}_j$ and $\mathfrak{M}_k$ may be paired into conjunctive pairs. If $j = k$, this is equivalent by **8** to ordinary isomorphism.

THEOREM 32. *If* $\mathfrak{c}_{ij}$ *is an integral ideal and* $\mathfrak{b}_{jk}$ *is any ideal, then* $\mathfrak{M}_j = \mathfrak{o}_j - \mathfrak{c}_{ij}$ *and* $\mathfrak{M}_k = \mathfrak{b}_{jk} - \mathfrak{c}_{ij}\mathfrak{b}_{jk}$ *are conjunctive.*

We note first that these modules are lattice isomorphic. For any submodule of $\mathfrak{M}_j$ corresponds to an ideal $\mathfrak{d}_{lj}$ such that $\mathfrak{o}_j \geqq \mathfrak{d}_{lj} \geqq \mathfrak{c}_{ij}$. Then $\mathfrak{b}_{jk} \geqq \mathfrak{d}_{lj}\mathfrak{b}_{jk} \geqq \mathfrak{c}_{ij}\mathfrak{b}_{jk}$ and by multiplying by $\mathfrak{b}_{jk}^{-1}$ we see that equality holds in the second set of equations only if it holds in the first set. Moreover, any submodule of $\mathfrak{M}_k$ has the form $\mathfrak{u}_{lk} - \mathfrak{c}_{ij}\mathfrak{b}_{jk}$ where $\mathfrak{u}_{lk}$ is an $\mathfrak{o}_k$-module contained in $\mathfrak{b}_{jk}$. It follows that $\mathfrak{u}_{lk}$ is a right $\mathfrak{o}_k$-ideal and hence is normal. Then by the lemma, $\mathfrak{u}_{lk} = \mathfrak{d}_{lj}\mathfrak{b}_{jk}$ where $\mathfrak{d}_{lj}$ is integral. Thus our correspondence between the submodules of $\mathfrak{M}_j$ and those of $\mathfrak{M}_k$ is $(1 - 1)$ and, since it preserves order, it is a lattice isomorphism. If $\mathfrak{c}_{jj}$ is the bound of $\mathfrak{M}_j$, it is readily seen that the bound of $\mathfrak{M}_k$ is the conjunctive ideal $\mathfrak{c}_{kk} = \mathfrak{b}_{jk}^{-1}\mathfrak{c}_{jj}\mathfrak{b}_{jk}$. Hence if we decompose $\mathfrak{c}_{ij}$ as a direct intersection $[\mathfrak{c}_{i_1 j}, \cdots, \mathfrak{c}_{i_s j}]$ of $\mathfrak{o}_j$-right ideals such that $\mathfrak{o}_j - \mathfrak{c}_{i_r j}$ is indecomposable, then the bound of $\mathfrak{o}_j - \mathfrak{c}_{i_r j}$ is conjunctive to that of $\mathfrak{b}_{jk} - \mathfrak{c}_{i_r j}\mathfrak{b}_{jk}$. We recall that the bounds of the $\mathfrak{o}_j - \mathfrak{c}_{i_r j}$ are the invariants of $\mathfrak{M}_j$. On the other hand by the lattice isomorphism, 0 in $\mathfrak{M}_k$ is a direct intersection of the modules $\mathfrak{M}_k^{(r)} = \mathfrak{c}_{i_r j}\mathfrak{b}_{jk} - \mathfrak{c}_{ij}\mathfrak{b}_{jk}$ and the difference modules $\mathfrak{M}_k - \mathfrak{M}_k^{(r)}$ are indecomposable. Since $\mathfrak{M}_k - \mathfrak{M}_k^{(r)}$ is isomorphic to $\mathfrak{b}_{jk} - \mathfrak{c}_{i_r j}\mathfrak{b}_{jk}$, its bound is conjunctive to that of $\mathfrak{o}_j - \mathfrak{c}_{i_r j}$ and so the invariants of $\mathfrak{M}_j$ are conjunctive to those of $\mathfrak{M}_k$.

Of course a like discussion may be made for left modules. Now we shall call the integral ideals $\mathfrak{b}_{ij}$ and $\mathfrak{c}_{kl}$ *right similar* (*left similar*) if the module $\mathfrak{o}_j - \mathfrak{b}_{ij}$ (left module $\mathfrak{o}_i - \mathfrak{b}_{ij}$) is conjunctive to $\mathfrak{o}_l - \mathfrak{c}_{kl}$ ($\mathfrak{o}_k - \mathfrak{c}_{kl}$). In the next section we shall show that two ideals are right similar if and only if they are left similar. We may therefore drop the modifiers "right" and "left" in these terms. We now state the fundamental factorization theorem.

THEOREM 33. *Any integral ideal* $\mathfrak{a}_{ij}$ *may be factored as a product* $\mathfrak{p}_{i i_{m-1}} \mathfrak{p}_{i_{m-1} i_{m-2}} \cdots \mathfrak{p}_{i_1 j}$ *of maximal integral ideals. If* $\mathfrak{a}_{ij} = \mathfrak{p}'_{i k_{n-1}} \mathfrak{p}'_{k_{n-1} k_{n-2}} \cdots \mathfrak{p}'_{k_1 j}$ *is a second factorization of this type, then the number of factors* $n = m$ *and the* $\mathfrak{p}$*'s and* $\mathfrak{p}'$*'s may be paired into similar pairs.*

We have seen that a factorization of $\mathfrak{a}_{ij}$ as $\mathfrak{p}_{i i_{m-1}} \cdots \mathfrak{p}_{i_1 j}$ corresponds to a composition series for $\mathfrak{o}_j - \mathfrak{a}_{ij}$ whose composition factors are isomorphic to certain modules $\mathfrak{a}_{i_{k-1}j} - \mathfrak{p}_{i_k i_{k-1}}\mathfrak{a}_{i_{k-1}j}$. By the preceding theorem these are conjunctive to the modules $\mathfrak{o}_{i_{k-1}} - \mathfrak{p}_{i_k i_{k-1}}$. Our theorem is therefore an immediate consequence of the Jordan-Hölder theorem.

The "necessity" part of the following theorem is an immediate consequence of Theorem 27.

THEOREM 34. *A necessary and sufficient condition that $\mathfrak{o}_j - \mathfrak{q}_{ij}$ be indecomposable is that $\mathfrak{q}_{ij}$ have only one factorization as a product of maximal integral ideals. If the condition holds, then the maximal factors of $\mathfrak{q}_{ij}$ are all similar.*

To prove the sufficiency, suppose that $\mathfrak{q}_{ij} = [\mathfrak{q}_{i_1 j}, \mathfrak{q}_{i_2 j}]$ and $(\mathfrak{q}_{i_1 j}, \mathfrak{q}_{i_2 j}) = \mathfrak{o}_j$. Then $\mathfrak{q}_{ij} = \mathfrak{r}_{ii_1}\mathfrak{q}_{i_1 j} = \mathfrak{r}_{ii_2}\mathfrak{q}_{i_2 j}$ where the $\mathfrak{r}$'s are integral ideals. Hence if $\mathfrak{q}_{i_1 j} \neq \mathfrak{q}_{ij}$ and $\mathfrak{q}_{i_2 j} \neq \mathfrak{q}_{ij}$, we have two distinct factorizations of $\mathfrak{q}_{ij}$.

The following corollaries are evident.

COROLLARY 1. *If $\mathfrak{o}_j - \mathfrak{q}_{ij}$ is indecomposable and $\mathfrak{q}_{kl}$ is a factor of $\mathfrak{q}_{ij}$, $\mathfrak{o}_l - \mathfrak{q}_{kl}$ is indecomposable.*

COROLLARY 2. *The module $\mathfrak{o}_j - \mathfrak{q}_{ij}$ is indecomposable if and only if the left module $\mathfrak{o}_i - \mathfrak{q}_{ij}$ is indecomposable.*

We recall that if $\mathfrak{o}_i - \mathfrak{q}_{ij}$ is indecomposable then its (right) bound has the form $\mathfrak{p}_{jj}^e$ where $\mathfrak{p}_{jj}$ is prime and e is the length of a composition series of $\mathfrak{o}_i - \mathfrak{q}_{ij}$. Evidently e may be characterized as the number of maximal factors in a factorization of $\mathfrak{q}_{ij}$. This together with the corresponding result for left modules yields

COROLLARY 3. *If $\mathfrak{o}_j - \mathfrak{q}_{ij}$ is indecomposable and the bound of $\mathfrak{q}_{ij}$ is $\mathfrak{p}_{jj}^e$, $\mathfrak{p}_j$ a prime, then the left bound of $\mathfrak{q}_{ij}$ has the form $\mathfrak{p}_{ii}^e$, $\mathfrak{p}_{ii}$ a prime.*

If $\mathfrak{p}_{jj}$ is a prime ideal, $\mathfrak{o}_j - \mathfrak{p}_{jj}$ is a simple ring and hence the composition factors of the module $\mathfrak{o}_j - \mathfrak{p}_{jj}$ are all isomorphic. This implies the following

THEOREM 35. *If $\mathfrak{p}_{jj}$ is prime, all the maximal factors of $\mathfrak{p}_{jj}$ are similar.*

We show finally that the order of the similarity classes of maximal ideals appearing in a factorization of any integral ideal is arbitrary: Thus if $\mathfrak{a}_{ij} = \mathfrak{p}_{ii_{m-1}}\mathfrak{p}_{i_{m-1}i_{m-2}} \cdots \mathfrak{p}_{i_1 j}$ where the $\mathfrak{p}$'s are maximal and if the class of $\mathfrak{p}_{i_k i_{k-1}}$ is C_k, there exists a factorization of $\mathfrak{a}_{ij}$ as $\mathfrak{p}'_{ik_{m-1}} \cdots \mathfrak{p}'_{k_1 j}$ where the corresponding similarity classes $C'_1, \cdots, C'_m$ is any prescribed permutation of $C_1, \cdots, C_m$. Evidently it suffices to prove the following

THEOREM 36. *If $\mathfrak{p}_{ij}$, $\mathfrak{p}_{jk}$ are maximal integral ideals, then $\mathfrak{p}_{ij}\mathfrak{p}_{jk} = \mathfrak{p}'_{il}\mathfrak{p}'_{lk}$ where $\mathfrak{p}_{ij}$ and $\mathfrak{p}'_{lk}$ are similar and $\mathfrak{p}_{jk}$ and $\mathfrak{p}'_{il}$ are right (left) similar.*

If $\mathfrak{o}_k - \mathfrak{p}_{ij}\mathfrak{p}_{jk}$ is indecomposable, $\mathfrak{p}_{ij}$ and $\mathfrak{p}_{jk}$ are right similar. Hence we may take $\mathfrak{p}'_{il} = \mathfrak{p}_{ij}$, $\mathfrak{p}'_{lk} = \mathfrak{p}'_{jk}$. Otherwise we have $\mathfrak{p}_{ij}\mathfrak{p}_{jk} = [\mathfrak{p}'_{i_1 k}, \mathfrak{p}'_{i_2 k}]$, $(\mathfrak{p}'_{i_1 k}, \mathfrak{p}'_{i_2 k}) = \mathfrak{o}_k$. Then we may suppose that $\mathfrak{p}'_{i_1 k}$ is similar to $\mathfrak{p}_{ij}$ and $\mathfrak{p}'_{i_2 k}$ is similar to $\mathfrak{p}_{jk}$. Since $\mathfrak{p}_{ij}\mathfrak{p}_{jk} = \mathfrak{r}_{ii_1}\mathfrak{p}'_{i_1 k} = \mathfrak{r}_{ii_2}\mathfrak{p}'_{i_2 k}$, we may take $\mathfrak{p}'_{lk} = \mathfrak{p}'_{i_1 k}$ and $\mathfrak{p}'_{il} = \mathfrak{r}_{ii_1}$.

14. Similarity of normal ideals. If $\mathfrak{a}_{ij}$ is any integral ideal, its bound $\mathfrak{a}_{jj}$ is a two-sided $\mathfrak{o}_j$-ideal of the form $\mathfrak{b}_{ji}\mathfrak{a}_{ij}$, $\mathfrak{b}_{ji}$ integral, and having the property that it is a divisor of any ideal of this form. A similar characterization holds for the left bound $\mathfrak{a}_{ii}$. Thus $\mathfrak{a}_{ii}$ is a divisor of $\mathfrak{a}_{ij}\mathfrak{b}_{ji}$ and the number of maximal factors of $\mathfrak{a}_{ii}$ does not exceed that of $\mathfrak{a}_{ij}\mathfrak{b}_{ji}$, or that of $\mathfrak{a}_{jj}$. By symmetry the numbers of maximal factors of $\mathfrak{a}_{ii}$ and of $\mathfrak{a}_{jj}$ are equal and so $\mathfrak{a}_{ii} = \mathfrak{a}_{ij}\mathfrak{b}_{ji}$. Then $\mathfrak{a}_{ii} = \mathfrak{a}_{ij}\mathfrak{a}_{jj}\mathfrak{a}_{ij}^{-1}$ is conjunctive with $\mathfrak{a}_{jj}$. We may use this fact to prove the following

LEMMA. *If $\mathfrak{a}_{ij}$ and $\mathfrak{b}_{kl}$ are right similar and $\mathfrak{o}_j - \mathfrak{a}_{ij}$ is indecomposable, then $\mathfrak{a}_{ij}$ and $\mathfrak{b}_{kl}$ are left similar.*

Since $\mathfrak{o}_i - \mathfrak{a}_{ij}$ and $\mathfrak{o}_k - \mathfrak{b}_{kl}$ are indecomposable, it suffices to show that these left modules have conjunctive bounds, i.e. the left bound of $\mathfrak{a}_{ij}$ is conjunctive to that of $\mathfrak{b}_{kl}$. By assumption the right bound of $\mathfrak{a}_{ij}$ is conjunctive to that of $\mathfrak{b}_{kl}$. Since the two bounds of an ideal are conjunctive this result is clear.

Now suppose that $\mathfrak{o}_j - \mathfrak{a}_{ij}$ is decomposable and let $\mathfrak{a}_{ij} = [\mathfrak{a}_{i_1j}, \mathfrak{a}_{i_2j}]$, $(\mathfrak{a}_{i_1j}, \mathfrak{a}_{i_2j}) = \mathfrak{o}_j$. Then $\mathfrak{a}_{ij} = \mathfrak{a}'_{ii_1}\mathfrak{a}_{i_1j} = \mathfrak{a}'_{ii_2}\mathfrak{a}_{i_2j}$. The intersection $\bar{\mathfrak{a}}_{ik} = [\mathfrak{a}'_{ii_1}, \mathfrak{a}'_{ii_2}]$ contains $\mathfrak{a}_{ij}$ and so $\mathfrak{a}_{ij} = \bar{\mathfrak{a}}_{ik}\mathfrak{c}_{kj}$. Since $\mathfrak{c}_{kj} = \bar{\mathfrak{a}}_{ik}^{-1}\mathfrak{a}_{ij} \geqq \mathfrak{a}'^{-1}_{ii_1}\mathfrak{a}_{ij} = \mathfrak{a}_{i_1j}$ and similarly $\mathfrak{c}_{kj} \geqq \mathfrak{a}_{i_2j}$, we have $\mathfrak{c}_{kj} = \mathfrak{o}_j$, or $\mathfrak{a}_{ij} = [\mathfrak{a}'_{ii_1}, \mathfrak{a}'_{ii_2}]$. Next let $(\mathfrak{a}'_{ii_1}, \mathfrak{a}'_{ii_2}) = \mathfrak{c}_{il}$. Then $\mathfrak{c}_{il}^{-1}\mathfrak{a}_{ij} \leqq \mathfrak{a}'^{-1}_{ii_1}\mathfrak{a}_{ij} = \mathfrak{a}_{i_1j}$ and $\mathfrak{c}_{il}^{-1}\mathfrak{a}_{ij} \leqq \mathfrak{a}_{i_2j}$ which implies that $\mathfrak{c}_{il} = \mathfrak{o}_i$. Thus $\mathfrak{a}_{ij}$ is a direct intersection of the $\mathfrak{o}_i$-left ideals $\mathfrak{a}'_{ii_1}$ and $\mathfrak{a}'_{ii_2}$.

We have seen that $\mathfrak{o}_{i_2} - \mathfrak{a}'_{ii_2}$ and $\mathfrak{a}_{i_2j} - \mathfrak{a}'_{ii_2}\mathfrak{a}_{i_2j} = \mathfrak{a}_{i_2j} - \mathfrak{a}_{ij}$ are conjunctive. Since $\mathfrak{o}_j - \mathfrak{a}_{i_1j} = (\mathfrak{a}_{i_1j}, \mathfrak{a}_{i_2j}) - \mathfrak{a}_{i_1j}$ is $\mathfrak{o}_j$-isomorphic to $\mathfrak{a}_{i_2j} - [\mathfrak{a}_{i_1j}, \mathfrak{a}_{i_2j}] = \mathfrak{a}_{i_2j} - \mathfrak{a}_{ij}$, it follows that $\mathfrak{o}_{i_2} - \mathfrak{a}'_{ii_2}$ and $\mathfrak{o}_j - \mathfrak{a}_{i_1j}$ are conjunctive. Hence $\mathfrak{a}_{i_1j}$ and $\mathfrak{a}'_{ii_2} = \mathfrak{a}_{ij}\mathfrak{a}_{i_2j}^{-1}$ are right similar and by symmetry, these ideals are also left similar. Likewise $\mathfrak{a}_{i_2j}$ and $\mathfrak{a}'_{ii_1}$ are similar.

We consider now the general case where $\mathfrak{a}_{ij} = [\mathfrak{a}_{i_1j}, \cdots, \mathfrak{a}_{i_sj}]$ and where, if $\mathfrak{b}_{k_rj}$ denotes $[\mathfrak{a}_{i_1j}, \cdots, \mathfrak{a}_{i_{r-1}j}, \mathfrak{a}_{i_{r+1}j}, \cdots, \mathfrak{a}_{i_sj}]$, then $(\mathfrak{a}_{i_rj}, \mathfrak{b}_{k_rj}) = \mathfrak{o}_j$. We write $\mathfrak{a}_{ij} = \mathfrak{a}'_{ik_r}\mathfrak{b}_{k_rj} = \mathfrak{b}'_{ii_r}\mathfrak{a}_{i_rj}$ and shall show that $\mathfrak{a}_{ij}$ is a direct intersection of the $\mathfrak{o}_i$-left ideals $\mathfrak{a}'_{ik_r}$. This has been proved above if $s = 2$. Hence we may suppose that it holds for a decomposition into $(s - 1)$ components. It is readily seen that $\mathfrak{b}'_{ii_r} = \mathfrak{a}_{ij}\mathfrak{a}_{i_rj}^{-1} = [\mathfrak{a}_{i_1j}, \cdots, \mathfrak{a}_{i_sj}]\mathfrak{a}_{i_rj}^{-1} = [[\mathfrak{a}_{i_1j}, \mathfrak{a}_{i_rj}]\mathfrak{a}_{i_rj}^{-1}, \cdots, [\mathfrak{a}_{i_sj}, \mathfrak{a}_{i_rj}]\mathfrak{a}_{i_rj}^{-1}]$ and if we delete the term $(\mathfrak{a}_{i_qj} \wedge \mathfrak{a}_{i_rj})\mathfrak{a}_{i_rj}^{-1}$, $q \neq r$ in the last expression we obtain $\mathfrak{b}_{k_qj}\mathfrak{a}_{i_rj}^{-1}$. Since $(\mathfrak{b}_{k_qj}, [\mathfrak{a}_{i_qj}, \mathfrak{a}_{i_rj}]) = [(\mathfrak{b}_{k_qj}, \mathfrak{a}_{i_qj}), \mathfrak{a}_{i_rj}]$ by Dedekind's law, and $(\mathfrak{b}_{k_qj}, \mathfrak{a}_{i_qj}) = \mathfrak{o}_j$, we have $(\mathfrak{b}_{k_qj}, [\mathfrak{a}_{i_qj}, \mathfrak{a}_{i_rj}]) = \mathfrak{a}_{i_rj}$. Hence $(\mathfrak{b}_{k_qj}\mathfrak{a}_{i_rj}^{-1}, [\mathfrak{a}_{i_qj}, \mathfrak{a}_{i_rj}]\mathfrak{a}_{i_rj}^{-1}) = \mathfrak{o}_j$, and so the decomposition of $\mathfrak{b}'_{ii_r}$ into the $[\mathfrak{a}_{i_qj}, \mathfrak{a}_{i_rj}]\mathfrak{a}_{i_rj}^{-1}$, $q \neq r$, is direct. Since $\mathfrak{b}'_{ii_r} = \mathfrak{a}'_{ik_r}(\mathfrak{b}_{k_rj}\mathfrak{a}_{i_rj}^{-1})$, we conclude from the induction hypothesis that $\mathfrak{b}'_{ii_r}$ is a direct intersection of the $\mathfrak{o}_i$-left ideals $\mathfrak{a}'_{ik_q}$. Since $\mathfrak{a}_{ij}$ is a direct intersection of $\mathfrak{a}'_{ik_r}$ and $\mathfrak{b}'_{ii_r}$, $\mathfrak{a}_{ij}$ is a direct intersection of all the $\mathfrak{a}'_{ik_r}$'s. We state this result as

THEOREM 37. *If $\mathfrak{a}_{ij}$ is a direct intersection of the right $\mathfrak{o}_j$-ideals $\mathfrak{a}_{i_rj}$ and $\mathfrak{b}_{k_rj}$ denotes $[\mathfrak{a}_{i_1j}, \cdots, \mathfrak{a}_{i_{r-1}j}, \mathfrak{a}_{i_{r+1}j}, \cdots, \mathfrak{a}_{i_sj}]$, then $\mathfrak{a}_{ij}$ is a direct intersection of the $\mathfrak{o}_i$-left ideals $\mathfrak{a}_{ij}\mathfrak{b}_{k_rj}^{-1} = \mathfrak{a}'_{ik_r}$.*

Since $\mathfrak{a}_{ij} = [\mathfrak{a}_{i_rj}, \mathfrak{b}_{k_rj}]$, $(\mathfrak{a}_{i_rj}, \mathfrak{b}_{k_rj}) = \mathfrak{o}_j$, the ideal $\mathfrak{a}_{i_rj}$ is similar to $\mathfrak{a}'_{ik_r} = \mathfrak{a}_{ij}\mathfrak{b}_{k_rj}^{-1}$. In conjunction with the lemma this implies

THEOREM 38. *If $\mathfrak{a}_{ij} = [\mathfrak{a}_{i_1j}, \cdots, \mathfrak{a}_{i_rj}] = [\mathfrak{a}_{ij_1}, \cdots, \mathfrak{a}_{ij_r}]$ are direct decompositions of $\mathfrak{a}_{ij}$ into $\mathfrak{o}_j$-right ideals and $\mathfrak{o}_j$-left ideals such that $\mathfrak{o}_j - \mathfrak{a}_{i_kj}$ and $\mathfrak{o}_i - \mathfrak{a}_{ij_k}$ are indecomposable, then the divisors $\mathfrak{a}_{i_kj}$ and $\mathfrak{a}_{ij_k}$ may be paired into pairs that are right and left similar.*

Evidently this implies

THEOREM 39. *Any two ideals which are right (left) similar are also left (right) similar.*

We shall prove finally that two two-sided ideals are conjunctive if and only if they are similar. For this purpose we require the

LEMMA. *If $\mathfrak{p}_{ii}$ and $\mathfrak{p}_{jj}$ are conjunctive prime ideals then their capacities are equal.*

We factor $\mathfrak{p}_{ii}$ as $\mathfrak{p}_{ii_{k-1}}\mathfrak{p}_{i_{k-1}i_{k-2}} \cdots \mathfrak{p}_{i_1 i}$ where the $\mathfrak{p}_{i_r i_{r-1}}$ are maximal. Then since k is the length of a composition series for $\mathfrak{o}_i - \mathfrak{p}_{ii}$, k is the capacity of $\mathfrak{p}_{ii}$. We may suppose that $\mathfrak{p}_{jj} = \mathfrak{q}_{ji}\mathfrak{p}_{ii}\mathfrak{q}_{ji}^{-1}$ where $\mathfrak{q}_{ji}$ is a maximal integral ideal. Now consider $\mathfrak{p}_{i_1 i}\mathfrak{q}_{ji}^{-1}$. If $\mathfrak{p}_{i_1 i} = \mathfrak{q}_{ji}$, we evidently have $\mathfrak{p}_{i_1 i}\mathfrak{q}_{ji}^{-1} = \mathfrak{o}_j = \mathfrak{q}_{j_1 j}'^{-1}\mathfrak{p}_{j_1 j}'$ where $\mathfrak{p}_{j_1 j}' = \mathfrak{q}_{j_1 j}'$ is a maximal integral ideal. On the other hand if $\mathfrak{p}_{i_1 i} \neq \mathfrak{q}_{j_1 j}'$, $(\mathfrak{p}_{i_1 i}, \mathfrak{q}_{ji}) = \mathfrak{o}_i$ and so $[\mathfrak{p}_{i_1 i}, \mathfrak{q}_{ji}]$ is a direct intersection of $\mathfrak{p}_{i_1 i}$ and $\mathfrak{q}_{ji}$. Then $[\mathfrak{p}_{i_1 i}, \mathfrak{q}_{ji}] = \mathfrak{q}_{j_1 i_1}'\mathfrak{p}_{i_1 i} = \mathfrak{p}_{j_1 j}'\mathfrak{q}_{ji}$ and so again $\mathfrak{p}_{i_1 i}\mathfrak{q}_{ji}^{-1} = \mathfrak{q}_{j_1 i_1}'^{-1}\mathfrak{p}_{j_1 j}'$ where $\mathfrak{p}'$ and $\mathfrak{q}'$ are maximal. Thus $\mathfrak{p}_{jj} = \mathfrak{q}_{ji}\mathfrak{p}_{ii_{k-1}} \cdots \mathfrak{p}_{i_2 i_1}\mathfrak{q}_{j_1 i_1}'^{-1}\mathfrak{p}_{j_1 j}' = \mathfrak{q}_{ji}\mathfrak{p}_{ii_{k-1}} \cdots \mathfrak{p}_{i_3 i_2}\mathfrak{q}_{j_2 i_2}'^{-1}\mathfrak{p}_{j_2 j_1}'\mathfrak{p}_{j_1 j}' = \cdots = \mathfrak{q}_{ji}\mathfrak{q}_{j_k i}'^{-1}\mathfrak{p}_{j_k j_{k-1}}' \cdots \mathfrak{p}_{j_1 j}' = \mathfrak{q}_{lj}'^{-1}\mathfrak{q}_{l j_k}''\mathfrak{p}_{j_k j_{k-1}}' \cdots \mathfrak{p}_{j_1 j}' \equiv \mathfrak{q}_{lj}'^{-1}\mathfrak{r}_{lj}$. Since $\mathfrak{q}_{lj}'$ is maximal, $\mathfrak{r}_{lj} = \mathfrak{q}_{lj}'\mathfrak{p}_{jj}$ has $k' + 1$ maximal factors if k' is the capacity of $\mathfrak{p}_{jj}$. On the other hand, the factorization $\mathfrak{r}_{lj} = \mathfrak{q}_{l j_k}''\mathfrak{p}_{j_k j_{k-1}}' \cdots \mathfrak{p}_{j_1 j}'$ shows that $\mathfrak{r}_{lj}$ has $k + 1$ maximal factors and so we have proved that $k' = k$.

THEOREM 40. *A necessary and sufficient condition that $\mathfrak{a}_{ii}$ and $\mathfrak{a}_{jj}$ be similar is that they be conjunctive.*

If $\mathfrak{a}_{ii}$ is an intersection of right $\mathfrak{o}_i$-ideals, it is clear that $\mathfrak{a}_{ii}$ is also the intersection of the bounds of these ideals. Hence it follows directly from the definition of (right) similarity that if $\mathfrak{a}_{ii}$ and $\mathfrak{a}_{jj}$ are similar, they are conjunctive. Suppose, conversely, that $\mathfrak{a}_{ii}$ and $\mathfrak{a}_{jj}$ are conjunctive. Then the prime powers $\mathfrak{p}_{ii}^e$, $\mathfrak{p}_{jj}^e$ of these ideals may be paired into conjunctive pairs. Now $\mathfrak{o}_i - \mathfrak{p}_{ii}^e$ is decomposable as a direct sum of k isomorphic indecomposable modules, k the capacity of $\mathfrak{p}_{ii}$, and each of these submodules has the bound $\mathfrak{p}_{ii}^e$. Hence by the preceding theorem $\mathfrak{o}_j - \mathfrak{p}_{jj}^e$ is a direct sum of k indecomposable modules, each having the bound $\mathfrak{p}_{jj}^e$. Thus $\mathfrak{p}_{ii}^e$ and $\mathfrak{p}_{jj}^e$ are similar and consequently $\mathfrak{a}_{ii}$ and $\mathfrak{a}_{jj}$ are similar.

BIBLIOGRAPHY

Albert, A. A. [1]: *On the rank equation of any normal division algebra*, Bull. Amer. Math. Soc., v. 35 (1929), 335–338; [2]: *The rank function of any simple algebra*, Proc. Nat. Acad. Sci., v. 15 (1929), 372–376; [3]: *On direct products*, Trans. Amer. Math. Soc., v. 33 (1931), 690–711; [4]: *On the construction of cyclic algebras with a given exponent*, Amer. J. Math., v. 54 (1932), 1–13; [5]: *On normal simple algebras*, Trans. Amer. Math. Soc., v. 34 (1932), 620–625; [6]: *A note on normal division algebras of order sixteen*, Bull. Amer. Math. Soc., v. 38 (1932), 703–706; [7]: *Normal division algebras over a modular field*, Trans. Amer. Math. Soc., v. 36 (1934), 388–394; [8]: *On normal Kummer fields over a non-modular field*, Trans. Amer. Math. Soc., v. 36 (1934), 885–892; [9]: *Involutorial simple algebras and real Riemann matrices*, Ann. of Math., v. 36 (1935), 886–964; [10]: *Normal division algebras of degree p^e over F of characteristic p*, Trans. Amer. Math. Soc., v. 39 (1936), 183–188; [11]: *Simple algebras of degree p^e over a centrum of characteristic p*, Trans. Amer. Math. Soc., v. 40 (1936), 112–126; [12]: *Modern Higher Algebra*, Chicago, 1937; [13]: *On cyclic algebras*, Ann. of Math., v. 39 (1938), 669–682; [14]: *Non-cyclic algebras with pure maximal subfields*, Bull. Amer. Math. Soc., v. 44 (1938), 576–579; [15]: *A note on normal division algebras of prime degree*, Bull. Amer. Math. Soc., v. 44 (1938), 649–652; [16]: *Structure of Algebras*, New York, 1939; [17]: *On p-adic fields and rational division algebras*, Ann. of Math., v. 41 (1940), 674–692; [18]: *Non-associative algebras, I. Fundamental concepts and isotopy*, Ann. of Math., v. 43 (1942), 685–707; [19]: *Non-associative algebras, II. New simple algebras*, Ann. of Math., v. 43 (1942), 708–723.

Albert, A. A., and Hasse, H. [1]: *A determination of all normal division algebras over an algebraic number field*, Trans. Amer. Math. Soc., v. 34 (1932), 722–726.

Artin, E. [1]: *Über einen Satz von Herrn J. H. Maclagan Wedderburn*, Abh. Math. Sem. Univ. Hamburg, v. 5 (1927), 245–250; [2]: *Zur Theorie der hyperkomplexen Zahlen*, Abh. Math. Sem. Univ. Hamburg, v. 5 (1927), 251–260; [3]: *Zur Arithmetik hyperkomplexer Zahlen*, Abh. Math. Sem. Univ. Hamburg, v. 5 (1927), 261–289; [4]: *Galois Theory*, Notre Dame, 1942.

Artin, E., and Whaples, G. [1]: *The theory of simple rings*, Amer. J. Math., v. 65 (1943), 87–107.

Asano, K. [1]: *Über die Darstellungen einer endlichen Gruppe durch reele Kollineationen*, Proc. Imp. Acad. Tokyo, v. 9 (1933), 574–576; [2]: *Nichtkommutative Hauptidealringe. I*, Act, Sci. Ind., No. 696, Paris 1938; [3]: *Arithmetische Idealtheorie in nichtkommutativen Ringen*, Jap. J. Math., v. 15 (1939), 1–36; [4]: *Über verallgemeinerte Abelsche Gruppe mit hyperkomplexem Operatorenring und ihre Anwendungen*, Jap. J. Math., v. 15 (1939), 231–253; [5]: *Über Ringe mit Vielfachenkettensatz*, Proc. Imp. Acad. Tokyo, v. 15 (1939), 288–291.

Asano, K., and Nakayama, T. [1]: *Über halblineare Transformationen*, Math. Ann., v. 115 (1937), 87–114.

Asano, K., and Shoda, K. [1]: *Zur Theorie der Darstellungen einer endlichen Gruppe durch Kollineationen*, Compositio Math., v. 2 (1935), 230–240.

Baer, R. [1]: *A Galois theory of linear systems over commutative fields*, Amer. J. Math., v. 62 (1940), 551–588; [2]: *Inverses and zero-divisors*, Bull. Amer. Math. Soc., v. 48 (1942), 630–638.

Birkhoff, G. [1]: *On the representability of Lie algebras and Lie groups by matrices*, Ann. of Math., v. 38 (1937), 526–532; [2]: Lattice Theory, New York, 1940.

Birkhoff, G., and MacLane, S., [1]: *A Survey of Modern Algebra*, New York, 1941.

Brandt, H. [1]: *Idealtheorie in einer Dedekindschen Algebra*, Jber. Deutsch. Math. Verein, v. 37 (1928), 5–7; [2]: *Primidealzerlegung in einer Dedekindschen Algebra*, Schweizerische Naturforschende Gesellschaft. Verhandlungen, v. 28 (1929), 288–290; [3]: *Zur Idealtheorie Dedekindscher Algebren*, Comment. Math. Helv., v. 2 (1930), 13–17; [4]: *Über die Axiome des Gruppoids*, Vierteljschr. Naturforsch. Ges. Zurich, v. 85 (1940), 95–104.

Brauer, R. [1]: *Untersuchungen über der arithmetischen Eigenschaften von Gruppen linearer Substitutionen, I*, Math. Z., v. 28 (1928), 677–696; [2]: *ibid., II*, v. 31 (1930), 733–747; [3]: *Über Systeme hyperkomplexer Zahlen*, Math. Z., v. 30 (1929), 79–107; [4]: *Über die algebraische Struktur von Schiefkörpern*, J. Reine Angew. Math., v. 166 (1932), 241–252; [5]: *Über die Konstruktion der Schiefkörper, die von endlichen Rang in bezug auf ein gegebenes Zentrum sind*, J. Reine Angew. Math., v. 168 (1932), 44–64; [6]: *Über den Index und den Exponenten von Divisionsalgebren*, Tôhoku Math. J., v. 37 (1933), 77–87; [7]: *Über die Kleinsche Theorie der algebraischen Gleichungen*, Math. Ann., v. 110 (1934), 473–500; [8]: *Eine Bedingung für vollständige Reduzibilität von Darstellungen gewöhnlicher und infinitesimaler Gruppen*, Math. Z., v. 41 (1936), 330–339; [9]: *On algebras which are connected with the semi-simple continuous groups*, Ann. of Math., v. 38 (1937), 857–872; [10]: *On normal division algebras of index five*, Proc. Nat. Acad. Sci., v. 24 (1938), 243–246; [11]: *On modular and p-adic representations of algebras*, Proc. Nat. Acad. Sci., v. 25 (1939), 252–258; [12]: *Investigations on group characters*, Ann. of Math., v. 42 (1941), 936–958; [13]: *On sets of matrices with coefficients in a division ring*, Trans. Amer. Math. Soc., v. 49 (1941), 502–548; [14]: *On the nilpotency of the radical of a ring*, Bull. Amer. Math. Soc., v. 48 (1942), 752–758.

Brauer, R., Hasse, H., and Noether, E. [1]: *Beweis eines Hauptsatzes in der Theorie der Algebren*, J. Reine Angew. Math., v. 167 (1932), 399–404.

Brauer, R., and Nesbitt, C. [1]: *On the regular representations of algebras*, Proc. Nat. Acad. Sci., v. 23 (1937), 236–240; [2]: *On the modular representations of groups of finite order. I*, Toronto Studies, 1937; [3]: *On the modular characters of groups*, Ann. of Math., v. 42 (1941), 556–590.

Brauer, R., and Noether, E. [1]: *Über minimale Zerfällungskörper irreduzibler Darstellungen*, S.-B. Preuss. Akad. Wiss., v. 32 (1927), 221–228.

Brauer, R., and Weyl, H. [1]: *Spinors in n dimensions*, Amer. J. Math., v. 57 (1935), 425–449.

Casimir, H., and van der Waerden, B. L. [1]: *Algebraischer Beweis der vollständigen Reduzibilität der Darstellungen halbeinfacher Liescher Gruppen*, Math. Ann., v. 111 (1935), 1–12.

Chevalley, C. [1]: *Sur la théorie du corps de classes dans les corps finis et les corps locaux*, J. Fac. Sci. Imp. Univ. Tokyo, v. 9 (1933), 366–476; [2]; *La théorie du symbole de restes normiques*, J. Reine Angew. Math., v. 169 (1933), 140–156; [3]: *Sur certains idéaux d'une algèbre simple*, Abh. Math. Sem. Univ. Hamburg, v. 10 (1934), 83–105; [4]: *Démonstration d'une hypothèse de M. Artin*, Abh. Math. Sem. Univ. Hamburg, v. 11 (1936), 73–75; [5]: *Généralisation de la théorie du corps de classes pour les extensions infinies*, J. Math. Pures Appl., Sér. 9, v. 15 (1936), 359–371; [6]: *L'arithmétique dans les algèbres de matrices*, Act. Sci. Ind. No. 323, Paris, 1936; [7]: *La théorie du corps de classes*, Ann. of Math., v. 41 (1940), 394–418; [8]: *An algebraic proof of a property of Lie groups*, Amer. J. Math., v. 63 (1941), 785–793; [9]: *On the composition of fields*, Bull. Amer. Math. Soc., v. 48 (1942), 482–487.

Clifford, A. H. [1]: *Representations induced in an invariant subgroup*, Ann. of Math., v. 38 (1937), 533–550; [2]: *Semigroups admitting relative inverses*, Ann. of Math., v. 42 (1941), 1037–1049; [3]: *Matrix representations of completely simple semigroups*, Amer. J. Math., v. 64 (1942), 327–342.

Deuring, M. [1]: *Galoissche Theorie und Darstellungstheorie*, Math. Ann., v. 107 (1932), 140–144; [2]: *Algebren*, Ergebnisse der Math., v. 4, Berlin, 1935.

Dickson, L. E. [1]: *Linear Algebras*, Cambridge, 1914; [2]: *Algebras and their Arithmetics*, Chicago, 1923; [3]: *Algebren und ihre Zahlentheorie*, Zurich, 1927.

DORROH, J. L. [1]: *Concerning adjunctions to algebras*, Bull. Amer. Math. Soc., v. 38 (1932), 85–88; [2]: *Concerning the direct product of algebras*, Ann. of Math., v. 36 (1935), 882–885.

EICHLER, M. [1]: *Bestimmung der Idealklassenzahl in gewissen normalen einfachen Algebren*, J. Reine Angew. Math., v. 176 (1937), 192–202; [2]: *Über die Einheiten der Divisionsalgebren*, Math. Ann., v. 114 (1937), 635–654; [3]: *Allgemeine Kongruenzklasseneinteilungen der Ideale einfacher Algebren über algebraischen Zahlkörpern und ihre L-Reihen*, J. Reine Angew. Math., v. 179 (1938), 227–251; [4]: *Über die Idealklassenzahl hyperkomplexer Systeme*, Math. Z., v. 43 (1938), 481–494; [5]: *Zur Einheitentheorie der einfachen Algebren*, Comment. Math. Helv., v. 11 (1939), 253–272.

ETHERINGTON, I. M. H. [1]: *Genetic algebras*, Proc. Roy. Soc. Edinburgh, v. 59 (1939), 242–258.

EVERETT, C. J., JR. [1]: *Rings as groups with operators*, Bull. Amer. Math. Soc., v. 45 (1939) 274–279; [2]: *Annihilator ideals and representation iteration for abstract rings*, Duke Math. J., v. 5 (1939), 623–627; [3]: *Vector spaces over rings*, Bull. Amer. Math. Soc., v. 48 (1942), 312–316; [4]: *An extension theory for rings*, Amer. J. Math., v. 64 (1942), 363–370.

FITTING, H. [1]: *Die Theorie der Automorphismenringe Abelscher Gruppen und ihr Analogon bei nicht kommutativen Gruppen*, Math. Ann., v. 107 (1932), 514–542; [2]: *Über die direkten Produktzerlegungen einer Gruppe in direkt unzerlegbare Faktoren*, Math. Z., v. 39 (1935), 16–30; [3]: *Primärkomponentenzerlegung in nichtkommutativen Ringen*, Math. Ann., v. 111 (1935), 19–41; [4]: *Über die Existenz gemeinsamer Verfeinerungen bei direkten Produktzerlegungen einer Gruppe*, Math. Z., v. 41 (1936), 380–395; [5]: *Die Determinantenideale eines Moduls*, Jber. Deutsch. Math. Verein. v. 46 (1936), 195–228; [6]: *Über der Zussamenhang zwischen dem Begriff der Gleichartigkeit zweier Ideale und dem Äquivalenzbegriff der Elementarteilertheorie*, Math. Ann., v. 112 (1936), 572–582; [7]: *Der Normenbegriff für Ideale eines Ringes beliebiger Struktur*, J. Reine Angew. Math., v. 178 (1937), 107–122.

GELFAND, I. [1]: *Normierte Ringe*, Rec. Math. [Mat. Sbornik] N.S., v. 9 (1941), 3–23; [2]: *Ideale und primäre Ideale in normierten Ringen*, Rec. Math. [Mat. Sbornik] N.S., v. 9 (1941), 41–48.

GOLOWIN, O. N. [1]: *On factors without centres in direct decompositions of groups* (*Russian*) Rec. Math. [Mat. Sbornik] N.S., v. 6 (1939), 423–426.

HAANTJES, J. [1]: *Halblineare Transformationen*, Math. Ann., v. 114 (1937), 293–304.

HALL, M. [1]: *A type of algebraic closure*, Ann. of Math., v. 40 (1939), 360–369; [2]: *The position of the radical in an algebra*, Trans. Amer. Math. Soc., v. 48 (1940), 391–404.

HARRISON, G. [1]: *The structure of algebraic moduls*, Proc. Nat. Acad. Sci., v. 28 (1942), 410–413.

HASSE, H. [1]: *Darstellbarkeit von Zahlen durch quadratische Formen in einem beliebigen algebraischen Zahlkörper*, J. Reine Angew. Math., v. 153 (1924), 113–130; [2]: *Äquivalenz quadratischer Formen in einem beliebigen algebraischen Zahlkörper*, J. Reine Angew. Math., v. 153 (1924), 158–162; [3]: *Über p-adische Schiefkörper und ihre Bedeutung für die Arithmetik hyperkomplexer Zahlsysteme*, Math. Ann., v. 104 (1931), 495–534; [4], *The theory of cyclic algebras over an algebraic number field*, Trans. Amer. Math. Soc., v. 34 (1932), 171–214; and *Additional note to the author's "Theory of cyclic algebras over an algebraic number field,"* Trans. Amer. Math. Soc., v. 34 (1932), 727–730; [5]: *Die Struktur der R. Brauerschen Algebrenklassengruppe über einem algebraischen Zahlkörper*, Math. Ann., v. 107 (1932), 731–760; [6]: *Über gewisse Ideale in einer einfachen Algebra*, Act. Sci. Ind. No. 109, Paris, 1934.

HASSE, H., AND SCHILLING, O. F. G. [1]: *Die Normen aus einer normalen Divisionsalgebra über einem algebraischen Zahlkörper*, J. Reine Angew. Math., v. 174 (1936), 248–252.

HOPKINS, C. [1]: *Nilrings with minimal condition for admissible left ideals*, Duke Math. J., v. 4 (1938), 664–667; [2]: *Rings with minimal conditions for left ideals*, Ann. of Math., v. 40 (1939), 712–730.

HENKE, K. [1]: *Zur arithmetischen Idealtheorie hyperkomplexer Zahlen*, Abh. Math. Sem. Univ. Hamburg, v. 11 (1936), 311–332.

HIGMAN, G. [**1**]: *The units of group-rings*, Proc. London Math. Soc., v. 46 (1940), 231–240.

HOCHSCHILD, G. P. [**1**]: *Semi-simple algebras and generalized derivations*, Amer. J. Math., v. 64 (1941), 677–694.

INGRAHAM, M. H., AND WOLF, M. C. [**1**]: *Relative linear sets and similarity of matrices whose elements belong to a division algebra*, Trans. Amer. Math. Soc., v. 42 (1937), 16–31.

JACOBSON, N. [**1**]: *Non-commutative polynomials and cyclic algebras*, Ann. of Math., v. 35 (1934), 197–208; [**2**]: *Totally disconnected locally compact rings*, Amer. J. Math., v. 58 (1936), 433–449; [**3**]: *Pseudo-linear transformations*, Ann. of Math., v. 38 (1937), 484–507; [**4**]: *Abstract derivations and Lie algebras*, Trans. Amer. Math. Soc., v. 42 (1937), 206–224; [**5**]: *p-algebras of exponent p.*, Bull. Amer. Math. Soc., v. 43 (1937), 667–670; [**6**]: *A note on topological fields*, Amer. J. Math., v. 59 (1937), 889–894; [**7**]: *A note on non-associative algebras*, Duke Math. J., v. 3 (1937), 544–548; [**8**]: *Simple Lie algebras over a field of characteristic zero*, Duke Math. J., v. 4 (1938), 534–551; [**9**]: *Normal semi-linear transformations*, Amer. J. Math., v. 61 (1939), 45–58; [**10**]: *The fundamental theorem of the Galois theory for quasi-fields*, Ann. of Math., v. 41 (1940), 1–7; [**11**]: *Restricted Lie algebras of characteristic p*, Trans. Amer. Math. Soc., v. 50 (1941), 15–25; [**12**]: *Classes of restricted Lie algebras of characteristic p. I*, Amer. J. Math., v. 63 (1941), 481–515; [**13**]: *ibid., II*, Duke Math. J., v. 10 (1943), 107–121.

JACOBSON, N., AND TAUSSKY, O. [**1**]: *Locally compact rings*, Proc. Nat. Acad. Sci., v. 21 (1935), 106–108.

JENNINGS, S. A. [**1**]: *The structure of the group ring of a p-group over a modular field*, Trans. Amer. Math. Soc., v. 50 (1941), 175–185; [**2**]: *Central chains of ideals in an associative ring*, Duke Math. J., v. 9 (1942), 341–355.

JORDAN, P., VON NEUMANN, J., AND WIGNER, E. [**1**]: *On an algebraic generalization of the quantum mechanical formalism*, Ann. of Math., v. 35 (1934), 29–64.

KALISCH, G. K. [**1**]: *On special Jordan algebras*, Dissertation, University of Chicago, Chicago, 1942.

KAWADA, Y., AND KONDÔ, K. [**1**]: *Idealtheorie in nichtkommutativen Halbgruppen*, Jap. J. Math., v. 16 (1939), 37–45.

KIOKEMEISTER, F. [**1**]: *The parastrophic criterion for the factorization of primes*, Trans. Amer. Math. Soc., v. 50 (1941), 140–159.

KOŘINEK, V. [**1**]: *Maximale kommutative Körper in einfachen Systemen hyperkomplexer Zahlen*, Mém. Soc. Sci. Bohème, 1932, No. 1, (1933), 1–24; [**2**]: *Une remarque concernant l'arithmétique des nombres hypercomplexes*, Mém. Soc. Roy. Sci. Bohème, 1932, No. 4 (1933), 1–8; [**3**]: *Sur la décomposition d'un groupe en produit direct des sousgroupes*, Casopis Pěst. Mat. Fys., v. 66 (1937), 261–286; also correction, v. 67 (1938), 209–210.

KOETHE, G. [**1**]: *Schiefkörper unendlichen Ranges über dem Zentrum*, Math. Ann., v. 105 (1931), 15–39; [**2**]: *Verallgemeinerte Abelsche Gruppen mit hyperkomplexen Operatorenring*, Math. Z., v. 39 (1934), 29–44.

KRULL, W. [**1**]: *Über verallgemeinerte endliche Abelsche Gruppen*, Math. Z., v. 23 (1925), 161–196.

KUROSCH, A. [**1**]: *Ringtheoretische Probleme, die mit dem Burnsideschen Problem über periodische Gruppen in Zusammenhang stehen*, Bull. Acad. Sci. URSS, Sér. Math., v. 5 (1941), 233–240.

LANDHERR, W. [**1**]: *Über einfache Liesche Ringe*, Abh. Math. Sem. Univ. Hamburg, v. 11 (1934), 41–64; [**2**]: *Liesche Ringe vom Typus A über einem algebraischen Zahlkörper (Die lineare Gruppe) und hermitesche Formen über einem Schiefkörper*, Abh. Math. Sem. Hansischen Univ., v. 12 (1938), 200–241.

LEVITZKI, J. [**1**]: *Über nilpotente Subringe*, Math. Ann., v. 105 (1931), 620–627; [**2**]: *On the equivalence of the nilpotent elements of a semi-simple ring*, Compositio Math., v. 5 (1938), 392–402; [**3**]: *On rings which satisfy the minimum condition for the right-hand ideals*, Compositio Math., v. 7 (1939), 214–222.

MACDUFFEE, C. C. [**1**]: *The Theory of Matrices*, Erg. der Math., v. 2, Berlin, 1933; [**2**]: *Matrices with elements in a principal ideal ring*, Bull. Amer. Math. Soc., v. 39 (1933),

564–584; [3]: *Modules and ideals in a Frobenius algebra*, Monatsh. Math. Phys., v. 48 (1939), 292–313; [4]: *An Introduction to Abstract Algebra*, New York, 1940.

MacLane, S., and Schilling, O. F. G., [1]: *A formula for the direct product of crossed product algebras*, Bull. Amer. Math. Soc., v. 48 (1942), 108–114; [2]: *Groups of algebras over an algebraic number field*, Amer. J. Math., v. 65 (1943), 299–308.

Maeda, F. [1]: *Ring-decomposition without chain condition*, J. Sci. Hirosima Univ., Ser. A, v. 8 (1938), 145–167.

Malcev, A. [1]: *On the immersion of an algebraic ring into a field*, Math. Ann., v. 113 (1936), 686–691.

McCoy, N. H. [1]: *On the characteristic roots of matric polynomials*, Bull. Amer. Math. Soc., v. 42 (1936), 592–600; [2]: *Quasi-commutative rings and differential ideals*, Trans. Amer. Math. Soc., v. 39 (1936), 101–116; [3]: *Subrings of direct sums*, Amer. J. Math., v. 60 (1938), 374–382; [4]: *Subrings of infinite direct sums*, Duke Math. J., v. 4 (1938), 486–494; [5]: *Generalized regular rings*, Bull. Amer. Math. Soc., v. 45 (1939), 175–178; [6]: *Algebraic properties of certain matrices over a ring*, Duke Math. J., v. 9 (1942), 322–340.

McCoy, N. H., and Montgomery, D. [1]: *A representation of generalized Boolean rings*, Duke Math. J., v. 3 (1937), 455–459.

Moriya, M. [1]: *Zur Bewertung der einfachen Algebren*, Proc. Imp. Acad. Japan, v. 13 (1937), 392–395.

Murray, F. J., and von Neumann, J. [1]: *On rings of operators*, Ann. of Math., v. 37 (1936), 116–229; [2]: *ibid., II*, Trans. Amer. Math. Soc., v. 41 (1937), 208–248.

Nakayama, T. [1]: *Über die Beziehungen zwischen den Faktorensystemen und der Normklassengruppe eines galoisschen Erweiterungskörpers*, Math. Ann., v. 112 (1935), 85–91; [2]: *Über die direkte Zerlegung einer Divisionsalgebra*, Jap. J. Math., v. 12 (1935), 65–70; [3]: *Über die Algebren über einem Körper von der Primzahlcharakteristik*, Proc. Imp. Acad. Tokyo, v. 11 (1935), 305–306; [4]: *ibid., II*, v. 12 (1936), 113–114; [5]: *Eine Bemerkung über die Summe und den Durchschnitt von zwei Idealen in einer Algebra*, Proc. Imp. Acad. Tokyo, v. 12 (1936), 179–182; [6]; *Über die Klassifikation halblinearer Transformationen*, Proc. Phys.-Math. Soc. Japan, v. 19 (1937), 99–107; [7]: *Divisionsalgebren über diskret bewerteten perfekten Körpern*, J. Reine, Angew. Math., v. 178 (1937), 11–13; [8]: *A note on the elementary divisor theory in non-commutative domains*, Bull. Amer. Math. Soc., v. 44 (1938), 719–723; [9]: *Some studies on regular representations, induced representations and modular representations*, Ann. of Math., v. 39 (1938), 361–369; [10]: *On Frobeniusean Algebras, I*, Ann. of Math., v. 40 (1939), 611–633; [11]: *A remark on the sum and the intersection of two normal ideals of an algebra*, Bull. Amer. Math. Soc., v. 46 (1940), 469–472; also correction, v. 47 (1941), 332; [12]: *Note on uni-serial and generalized uni-serial rings*, Proc. Imp. Acad. Tokyo, v. 16 (1940), 285–289; [13]: *Normal basis of a quasi-field*, Proc. Imp. Acad. Tokyo, v. 16 (1940), 532–536; [14]: *On Frobeniusean algebras, II*, Ann. of Math., v. 42 (1941), 1–21; [15]: *Algebras with anti-isomorphic left and right ideal lattices*, Proc. Imp. Acad. Tokyo, v. 17 (1941), 53–56.

Nakayama, T., and Nesbitt, C. [1]: *Note on symmetric algebras*, Ann. of Math., v. 39 (1938), 659–668.

Nakayama, T., and Shoda, K. [1]: *Über die Darstellung einer endlichen Gruppe durch halblineare Transformationen*, Jap. J. Math., v. 12 (1936), 109–122.

Nehrkorn, H. [1]: *Über absolute Idealklassengruppen und Einheiten in algebraischen Zahlkörpern*, Abh. Math. Sem. Univ. Hamburg, v. 9 (1933), 318–334.

Nesbitt, C. [1]: *On the regular representations of algebras*, Ann. of Math., v. 39 (1938), 634–658.

Neuhaus, A. [1]: *Products of normal semi-fields*, Trans. Amer. Math. Soc., v. 49 (1941), 106–121.

von Neumann, J. [1]: *On regular rings*, Proc. Nat. Acad. Sci., v. 22 (1936), 707–713; [2]: *Algebraic theory of continuous geometries*, Proc. Nat. Acad. Sci., v. 23 (1937), 16–22; [3]: *Continuous rings and their arithmetics*, Proc. Nat. Acad. Sci., v. 23 (1937), 341–349; [4]: *On rings of operators, III*, Ann. of Math., v. 41 (1940), 94–161.

NIVEN, I. [1]: *Equations in quaternions*, Amer. Math. Monthly, v. 48 (1941), 654–661; [2]: *The roots of a quaternion*, Amer. Math. Monthly, v. 49 (1942), 386–388.

NOETHER, E. [1]: *Der Diskriminantensatz für die Ordnungen eines algebraischen Zahl-oder Funktionenkörpers*, J. Reine Angew. Math., v. 157 (1927), 82–104; [2]: *Hyperkomplexe Grössen und Darstellungstheorie*, Math. Z., v. 30 (1929), 641–692; [3]: *Hyperkomplexe Systeme in ihre Beziehungen zur kommutativen Algebra und Zahlentheorie*, Zurich Congress Proceedings, 1932, 189–194; [4]: *Der Hauptgeschlechtssatz für relativ-galiossche Zahlkörper*, Math. Ann., v. 108 (1933), 411–419; [5]: *Nichtkommutativ Algebra*, Math. Z., v. 37 (1933), 514–541; [6]: *Zerfallende verschränkte Produkte und ihre Maximalordnungen*, Act. Sci. Ind., Paris, 1934.

NOETHER, E., AND SCHMEIDLER, W. [1]: *Moduln in nichtkommutativen Bereichen, insbesondere aus Differential-und Differenzenausdrücken*, Math. Z., v. 8 (1920), 1–35.

ORE, O. [1]: *Linear equations in non-commutative fields*, Ann. of Math., v. 32 (1931), 463–477; [2]: *Theory of non-commutative polynomials*, Ann. of Math., v. 34 (1933), 480–508.

OSIMA, M. [1]: *Über die Darstellung einer Gruppe durch halblineare Transformationen*, Proc. Phys.-Math. Soc. Japan, v. 20 (1938), 1–5.

PERLIS, S. [1]: *A characterization of the radical of an algebra*. Bull. Amer. Math. Soc., v. 48 (1942), 128–132.

PICKERT, G. [1]: *Neue Methoden in der Strukturtheorie der kommutativ-assoziativen Algebren*, Math. Ann., v. 116 (1938), 217–280.

REES, D. [1]: *On semi-groups*, Proc. Cambridge Philos. Soc., v. 36 (1940), 387–400; [2]: *Note on semi-groups*, Proc. Cambridge Philos. Soc., v. 37 (1941), 434–435.

RINEHART, R. F. [1]: *Some properties of the discriminant matrices of a linear associative algebra*, Bull. Amer. Math. Soc., v. 42 (1936), 570–576; [2]: *Commutative algebras which are polynomial algebras*, Duke Math. J., v. 4 (1938), 725–736; [3]: *An interpretation of the index of inertia of the discriminant matrices of a linear associative algebra*, Trans. Amer. Math. Soc., v. 46 (1939), 307–327.

SCHIFFMAN, M. [1]: *The ring of automorphisms of an Abelian group*, Duke Math. J., v. 6 (1940), 579–597.

SCHILLING, O. F. G. [1]: *Über gewisse Beziehungen zwischen der Arithmetik hyperkomplexer Zahlsysteme und algebraischer Zahlkörper*, Math. Ann., v. 111 (1935), 372–398; [2]: *Einheitentheorie in rationalen hyperkomplexen Systemen*, J. Reine Angew. Math., v. 175 (1936), 246–251; [3]: *Über die Darstellungen endlicher Gruppen*, J. Reine Angew. Math., v. 174 (1936), 188; [4]: *The structure of certain rational infinite algebras*, Duke Math. J., v. 3 (1937), 303–310; [5]: *Arithmetic in a special class of algebras*, Ann. of Math., v. 38 (1937), 116–119; [6]: *Units in p-adic algebras*, Amer. J. Math., v. 61 (1939), 883–896.

SCHREIER, O. [1]: *Über den Jordan-Hölderschen Satz*, Abh. Math. Sem. Univ. Hamburg, v. 6 (1928), 300–302.

SCHMIDT, O. [1]: *Über unendliche Gruppen mit endlicher Kette*, Math. Z., v. 29 (1928), 34–41.

SCHNEIDMÜLLER, V. I. [1]: *On rings with finite decreasing chains of subrings*, C. R. (Doklady) Acad. Sci. URSS (N.S.), v. 28 (1940), 579–581.

SCHUR, I. [1]: *Über die Darstellung der endlichen Gruppen durch gebrochene lineare Substitutionen*, J. Reine Angew. Math., v. 127 (1904), 20–50; [2]: *Einige Bemerkungen zu der vorstehenden Arbeit des Herrn A. Speiser*, Math. Z., v. 5 (1919), 7–10.

SCORZA, G. [1]: *Corpi Numerici e Algebre*, Messina, 1921.

SCOTT, W. M. [1]: *On matrix algebras over an algebraically closed field*, Ann. of Math., v. 43 (1942), 147–160.

SEGRE, C. [1]: *Un nuovo campo di ricerche geometriche*, Atti Accad. Sci. Torino, v. 25 (1889), 276–301.

SERBIN, H. [1]: *Factorization in principal ideal rings*, Duke Math. J., v. 4 (1938), 656–663.

SHODA, K. [1]: *Über die Automorphismen einer endlichen Abelschen Gruppe*, Math. Ann. v. 100 (1928), 674–686; [2]: *Über die mit einer Matrix vertauschbaren Matrizen*, Math. Z., v. 29 (1929), 696–712; [3]: *Über die Galoissche Theorie der halbeinfachen hyperkomplexen Systeme*, Math. Ann., v. 107 (1932), 252–258; [4]: *Über die Äquivalenz der Darstellungen endlicher Gruppen durch halblineare Transformationen*, Proc. Imp. Acad. Japan, v. 14 (1938), 278–280; [**5**]: *Über die Invarianten der endlichen Gruppen halblinearer Transformationen*, Proc. Imp. Acad. Japan, v. 14 (1938), 281–285.

SKOLEM, T. [**1**]: *Zur Theorie der associativen Zahlensysteme*, Oslo, 1927.

SPEISER, A. [**1**]: *Zahlentheoretische Sätze aus der Gruppentheorie*, Math. Z., v. 5 (1919), 1–6; [**2**]: *Idealtheorie in rationalen Algebren*, Dickson's Algebren, Chapter 13, Zurich, 1927; [**3**]: *Zahlentheorie in rationalen Algebren*, Comment. Math. Helv., v. 8 (1936), 391–406.

STAUFFER, R. [**1**]: *The construction of a normal basis in a separable normal extension field*, Amer. J. Math., v. 58 (1936), 585–597.

STONE, M. H., [**1**]: *The theory of representations of Boolean algebras*, Trans. Amer. Math. Soc., v. 40 (1936), 37–111.

TEICHMÜLLER, O. [**1**]: *Verschränkte Produkte mit Normalringen*, Deutsche Math., v. 1 (1936), 92–102; [**2**]: *Multiplikation zyklischer Normalringe*, Deutsche Math., v. 1 (1936) 197–238; [**3**]: *p-Algebren*, Deutsche Math., v. 1 (1936), 362–388; [**4**]: *Zerfällende zyklische p-Algebren*, J. Reine Angew. Math., v. 176 (1937), 157–160; [**5**]: *Der Elementarteilersatz für nichtkommutative Ringe*, S.-B. Preuss. Akad. Wiss., 1937; [**6**]: *Über die sogenannte nichtkommutative Galoissche Theorie und die Relationen* $\xi_{\lambda,\mu,\nu}\xi_{\lambda,\mu\nu,\pi}\xi^{\lambda}_{\mu,\nu,\pi} = \xi_{\lambda,\mu,\nu\pi}\xi_{\lambda\mu,\nu,\pi}$ Deutsche Math., v. 5 (1940), 138–149.

TSEN, C. C. [**1**]: *Divisionsalgebren über Funktionenkörpern*, Nach. Ges. Wiss. Göttingen, 1933, 335–339; [**2**]: *Algebren über Funktionenkörpern*, Göttingen dissertation, 1934; [**3**]: *Zur Stufentheorie der quasialgebraisch-Abgeschlossenheit kommutativer Körper*, J. Chinese Math. Soc., v. 1 (1936), 81–92.

UZKOW, A. I. [**1**]: *Abstract foundation of Brandt's theory of ideals* (*Russian*), Rec. Math. [Mat. Sbornik] N. S. v. 6 (1939), 263–281.

VAN DER WAERDEN, B. L. [**1**]: *Moderne Algebra*, v. 1 and 2, Berlin, 1931; 2nd edition, v. 1, 1937, 2, 1940; [**2**]: *Die Klassifikation der einfachen Lieschen Gruppen*, Math. Z., v. 37 (1933), 446–462; [**3**]: *Gruppen von linearen Transformationen*, Erg. der Math., v. 4, Berlin, 1935.

WARNING, E. [**1**]: *Bemerkung zur vorstehenden Arbeit von Herrn Chevalley*, Abh. Math. Sem. Univ. Hamburg, v. 11 (1936), 76–83.

WEDDERBURN, J. H. M. [**1**]: *A theorem on finite algebras*, Trans. Amer. Math. Soc., v. 6 (1905) 349–352; [**2**]: *On hypercomplex numbers*, Proc. London Math. Soc., Ser. 2, v. 6 (1908), 77–117; [**3**]: *A type of primitive algebra*, Trans. Amer. Math. Soc., v. 15 (1914), 162–166; [**4**]: *On division algebras*, Trans. Amer. Math. Soc., v. 22 (1921), 129–135; [**5**]: *Algebras which do not possess a finite basis*. Trans. Amer. Math. Soc., v. 26 (1924), 395–426; [**6**]: *A theorem on simple algebras*, Bull. Amer. Math. Soc., v. 31 (1925), 11–13; [**7**]: *Noncommutative domains of integrity*, J. Reine Angew. Math., v. 167 (1932), 129–141; [**8**]: *Lectures on Matrices*, New York, 1934; [**9**]: *Note on algebras*, Ann. of Math., v. 38 (1937), 854–856.

WEYL, H. [**1**]: *Note on matric algebras*, Ann. of Math., v. 38 (1937), 477–483; [**2**]: *Commutator algebra of a finite group of collineations*, Duke Math. J., v. 3 (1937), 200–212; [**3**]: *The Classical Groups*, Princeton, 1939.

WHAPLES, G. [**1**]: *Non-analytic class field theory and Grunwald's theorem*, Duke Math. J., v. 9 (1942), 455–473.

WHITEHEAD, J. H. C. [**1**]: *On the decomposition of an infinitesimal group*, Proc. Cambridge Philos. Soc., v. 32 (1936), 229–237; [**2**]: *Certain equations in the algebra of a semi-simple infinitesimal group*, Quart. J. Math., Oxford Ser., v. 8 (1937), 220–237; [**3**]: *Note on linear associative algebras*, J. London Math. Soc., v. 16 (1941), 118–125.

WHITNEY H. [**1**]: *Tensor products of abelian groups*, Duke Math. J., v. 4 (1938), 495–528.

WITT, E. [**1**]: *Über die Kommutativität endlicher Schiefkörper*, Abh. Math. Sem. Univ. Hamburg, v. 8 (1930), 413; [**2**]: *Zerlegung reeler algebraischer Funktionen in Quadrate. Schiefkörper über reelen Funktionenkörper*, J. Reine Angew. Math., v. 171 (1934), 4–11; [**3**]: *Riemann-Rochscher Satz und ζ-Funktion in Hyperkomplexen*, Math. Ann., v. 110 (1934), 12–28; [**4**]: *Zwei Regeln über verschränkte Produkte*, J. Reine Angew. Math., v. 173 (1935), 191–192; [**5**]: *Theorie der quadratischen Formen in beliebigen Körpern*, J. Reine Angew.

Math., v. 176 (1937), 31–44; [6]: *Zyklische Körper und Algebren der Charakteristik p vom Grad p^n:*, J. Reine Angew. Math., v. 176 (1937), 126–140; [7]: *Schiefkörper über diskret bewerteten Körpern*, J. Reine Angew. Math., v. 176 (1937), 153–156; [8]: *Treue Darstellung Liescher Ringe*, J. Reine Angew. Math., v. 177 (1937), 152–160.

WOLF, L. A. [1]: *Similarity of matrices in which the elements are real quaternions*, Bull. Amer. Math. Soc., v. 42 (1936), 737–743.

ZASSENHAUS, H. [1]: *Zum Satz von Jordan-Hölder-Schreier*, Abh. Math. Sem. Univ. Hamburg, v. 10 (1934), 106–108; [2]: *Lehrbuch der Gruppentheorie*, Leipzig, 1937; [3]: *Über Liesche Ringe mit Primzahlcharakteristik*, Abh. Math. Sem. Univ. Hamburg, v. 13 (1939), 1–100; [4]: *Darstellungstheorie nilpotenter Lie-Ringe bei charakteristik $p > 0$*. J. Reine Angew. Math., v. 182 (1940), 150–155.

ZORN, M. [1]: *Theorie der alternativen Ringe*, Abh. Math. Sem. Univ. Hamburg, v. 8 (1930), 123–147; [2]: *Note zur analytischen hyperkomplexen Zahlentheorie*, Abh. Math. Sem. Univ. Hamburg, v. 9 (1933), 197–201; [3]: *Alternativkörper und quadratische systeme*, Abh. Math. Sem. Univ. Hamburg, v. 9 (1933), 395–402; [4]: *On a theorem of Engel*, Bull. Amer. Math. Soc., v. 43 (1937), 401–404; [5]: *Alternative rings and related questions, I: Existence of the radical*, Ann. of Math., v. 42 (1941), 676–686.

INDEX